Applied Fuzzy Systems

APPLIED FUZZY SYSTEMS

Edited by

Toshiro Terano

Kiyoji Asai

Michio Sugeno

Translated by Charles Aschmann

AP PROFESSIONAL

Boston San Diego New York
London Sydney Tokyo Toronto

This book is printed on acid-free paper ∞

AP PROFESSIONAL
955 Massachusetts Avenue, Cambridge, MA 02139

An imprint of ACADEMIC PRESS, INC.
A Division of HARCOURT BRACE & COMPANY

United Kingdom Edition published by
ACADEMIC PRESS LIMITED
24–28 Oval Road, London NW1 7DX

Library of Congress Cataloging-in-Publication Data

[Ouyou fajii shisutemu nyūmon. English]
Applied fuzzy systems/edited by T. Terano, K. Asai, and M. Sugeno;
translated by Charles G. Aschmann III.
p. cm.
Translation of: Ouyou fajii shisutemu nyūmon.
ISBN 0-12-685242-1
1. Automatic control. 2. Fuzzy systems. I. Terano, Toshirō,
1923– . II. Asai, Kiyoji, 1923– . III. Sugeno, Michio, 1940–
. IV. Title: Applied fuzzy systems.
TJ213.09313 1994
629.8–dc20 94-3001
CIP

Printed in the United States of America
94 95 96 97 BB 9 8 7 6 5 4 3 2 1

Contents

Preface

Fuzzy Systems Theory and Its Applications, which was published two years ago, was received more favorably than expected and even now sales are continuing to grow. This is probably because the explanations of theory and applications are just what the readers desired. But if we think about it, the situation of research on fuzzy systems has undergone great change over the past two years. Fuzzy theory, which at that time could only be viewed as a special theory that only a few specialists were working with, is now attracting the public's attention as a new tool that ties together people and information systems. The rapid progress of applications research is eye-opening, and as a frontier technology in which Japan is taking the lead, it is cause for national pride.

Recently, we have seen several remarkable trends working together under these conditions. One of them is the progress of large-scale national projects. With the assistance of the Ministry of International Trade and Industry, the Laboratory for International Fuzzy Engineering Research (LIFE), which was established in April of this year, is aiming at the development of high-level artificial intelligence and information processing systems, as well as the realization of a new fuzzy computer through the cooperative organization of industry, universities and government, with a budget of 5 billion yen over six years. In addition, research into fuzzy systems is being driven by 1 billion yen of funds for the promotion of science, coming from the Science and Technology Agency over a period of five years. Plans are also going ahead for the establishment of a Fuzzy

Systems Research Institute in Iizuka in Kyushu, and the aim is toward creating a Japanese Silicon Valley. In a few years, after the results of these have blossomed, Japan will occupy an internationally solid position in fuzzy systems research.

Another trend is the use of fuzzy theory in fields outside of engineering. The main emphasis of the foregoing projects is engineering, but recently the need for the use of fuzzy theory in information processing (diagnosis, evaluation, forecasting) for medicine, management, marketing, securities, meteorology, disaster prevention, education and other fields has increased in the extreme. Since these are high-level decision-making systems centered on human beings, man/machine communication is more important than anything else, and fuzzy theory hits the nail right on the head. Research in this direction has just started, but it holds potential for large growth in the near future.

The third trend is not related to specialized fields; it is brought about by the societal issue of whether fuzzy concepts are necessary for survival in the upcoming information society. This tendency can be seen remarkably well in what has been published. In the last two years 13 separate books related to fuzzy theory and systems have been published in Japan alone, and of these, six were books for liberal arts education. This phenomenon is unique to Japan, and looking at it leads us to guess that fuzzy concepts are compatible with the Japanese style of behavior. The authors feel that this is a very important point. The reason is that information technology is not the exclusive property of a select group of specialists; it is something that everyone must get used to doing in daily life, and if the concepts of fuzzy theory spread, a great amount of human intimacy will be possible in information technology and systems engineering. When industrial robots were first produced, Japanese workers actively worked for their introduction as something that would help them in their work, but workers in other countries rejected them as something that would take away their jobs. It seems that this same sort of thing is liable to occur in the coming information society. If we compare countries in which all of the people will get used to using information systems with those in which information will remain exclusive to a few experts, the former might be like human beings with their advanced cerebrum and the latter like animals that only have a cerebellum. The comparative merits and demerits are that clear.

Why does fuzzy theory have the wide applications just mentioned? It is because fuzzy theory is the only theory that can deal with the meaning of human language mathematically. Up to now, research into information processing has been so computer-oriented that it ignored the basic issues of what information means to people and how they process it. Research into fuzzy theory must inevitably return to these questions. This means that research into fuzzy theory and systems is human research.

This book is an easily understood explanation of the various fuzzy applications that suddenly developed after the publication of *Fuzzy Systems Theory and Its Applications*, for people who are thinking about introducing fuzzy theory into their own fields. Repetition of information from the previous volume was avoided as much as possible, but the contents take into consideration the need for being able to understand fuzzy theory from this book alone. Still, it will be more effective if read in combination with the previous volume.

Toshiro Terano
for the authors

Contributors

Numbers following the authors' names indicate the section(s) to which they contributed.

Kiyoji Asai, 5.1
Hidetomo Ichihashi, 5.3
Osamu Itoh, 3.5
Mamoru Inaba, 3.2
Sosuke Iwai, 4.4
Motohide Umano, 6.2.3
Masatoshi Ohuchi, 4.2
Masatoshi Sakawa, 5.4
Michio Sugeno, 3.1, 4.1
Eiichiro Tazaki, 4.5
Hideo Tanaka, 5.2
Toshiro Terano, 1, 3.8
Teruhide Niidome, 3.3
Kaoru Hirota, 2
Jun-ichiroh Fujimoto, 3.7
Hitoshi Furuta, 4.3
Masao Mukaidono, 6.2.1, 6.2.2
Shuta Murakami, 3.4
Seiji Yasunobu, 3.6
Takeshi Yamakawa, 6.1
Junzo Watada, 5.5

Chapter 1

Introduction to Fuzzy Systems

Unlike a few years ago, the word "fuzzy" is now a household word. This is a very good thing, but on the other hand, we fear that it is just the word, without its being given any deeper consideration. This can be said of any field, but when something new comes into being, there are motives and awareness of goals and problems. If the field solidifies and enters a period of development, the motives are considered clear and these are no longer viewed as problems. People then tend to turn their attention to detailed, advanced techniques. However, we must not lose awareness of the basic problems, if we are to maintain the vitality of the field and keep it developing. In this chapter, we will take a limited look at the basic problems addressed by fuzzy systems.

1.1 BALANCED TECHNOLOGY

The word "fuzzy" has become common knowledge and comes up in everyday conversation. This comes as quite a surprise to those of us who specialize in research into fuzzy systems. However, we have some idea of why it has come about. It is related to the fact that our current society is one in which technology, especially information technology, is highly developed. In other words, when our everyday lives are rationalized to the extreme and speeded up as they have never been, we not only receive the benefits, but also unconsciously experience aversion to the unfeeling nature of technology. We human beings need a balanced life. Of course there are many worthwhile human activities such as leisure, the arts, study, and social activities, but since technology has come to have such an intimate relationship with everyday life, there is a desire to make technology itself more humane. To put it another way, it might be that we are in need of technology that offers not only rationality and economic benefits, but also things like latitude, interest, humaneness, depth of appreciation, beauty and change. This could be the balancing of human senses [1].

Setting aside the problem of technology, the more complex society becomes, the greater the increase in rules. Systemization and rationalization progress, but this gives a constricted feeling. When we think that there should be some way to make a more accommodating, gentle society, we find that the concept of fuzziness is just exactly right.

Leaving aside the question of whether this is right or wrong, let us consider one field of advanced technology, artificial intelligence. Artificial intelligence has advanced with the goal of computerizing the thinking functions of human beings. And there are many researchers who believe that the ability to do so is at hand. However, the authors are suspicious.

Science and technology have advanced to this point by placing logical correctness in a position of supremacy. In the field of cerebral physiology, it is known that logic is a left-hemisphere activity. In contrast to this, the right hemisphere administers information processing such as sight and hearing, intuition and feelings. In the West, it is felt that the more logically true something is, the more human it is, and the left hemisphere is called the superior brain, the right hemisphere the inferior one; but we wonder if this is really true. For example, imagination, invention and creation are thought to be the highest level of the intelligent activities of human beings, and when they come about, numerous images first flash intuitively in the right hemisphere. The next stage is the one in which the left hemisphere first analyzes and confirms their soundness by means of logic [2].[1]

The workings of the right hemisphere are complicated and are not well understood, but this is because its evolution goes back further, and it is said to be more advanced than that of the left hemisphere. If logic were the highest level of intelligence, we would conclude that human beings were inferior to computers, because computers are superior in terms of logical strictness, speed, and accuracy.

We can say that the reasons artificial intelligence has made less progress than expected are as follows. The logic is too simple. There is no common

[1]That new conceptualizations cannot be obtained from logic or analysis alone, even over a great amount of time, became common knowledge early on in the field of "systems engineering," which deals with the composition of large-scale systems and their optimization, and a variety of heuristic techniques have been developed (see Ref. 3).

sense. The knowledge level is low. Data entry is complicated and difficult [4]. In order to solve these problems, high-level inferential and conjectural (including imagination, invention, and creativity) functions are necessary [5], and it is pretty much impossible to solve them using the left-hemisphere logic that has been employed up to now. In other words, with left-hemisphere logic there is strict conformity of knowledge, and if there is even one contradictory point it cannot be used. Yet the content and expressions of the knowledge that we use every day contain incompleteness, uncertainty and contradiction. Even so, human beings perform high-level inference using it. This is because people have the ambiguous knowledge known as "common sense" and in addition an ambiguous thinking ability to process it. Therefore, the realization of artificial intelligence depends on whether or not the ambiguity of thought can be processed or not [6].

The preceding technological imbalance has come about because of the posture that modern science has taken toward complete exclusion of right-hemisphere activities. In this kind of technology it is necessary to take the attitude of arriving at clear solutions for problems after being sure of the contradictions [7].

1.2 MACRO KNOWLEDGE AND MACRO THINKING

There are many levels in what we call knowledge. In general, the knowledge of specialists is seen as the most difficult of all, but from the standpoint that we have complete, well-arranged logical systems, it is the easiest to deal with. At present, this is the level of knowledge that knowledge engineering works with. But if we think about it, before we can learn to use expert knowledge, we must have expert common sense. For example, when a patient is carried into the hospital, the judgment of whether the illness belongs in the surgery department or the department of internal medicine comes first. The exact determination of the name of the illness comes later. In other words, we need a macro type of knowledge first, and after that, the expertise that only doctors have will be of use. If these are reversed, there is the danger of absurd mistakes.

Above and beyond expert knowledge is the necessity for common sense. The well-known rebuke of the famous doctor, "The operation was a success, but the patient died," shows a lack of common sense. In addition, on top of common sense there is knowledge based on noble psychological activities (such as ethics, religion, and the arts). In sum, what is called knowledge varies according to the *time*, *object*, and *person* using it, and it is difficult to discuss all of these together.

Among all these levels of knowledge, the expert knowledge in a narrow field is as accurate as it is narrow, is clear in terms of quantity, and is arranged in a logical system. Let us call this micro knowledge. On the other hand, expert common sense or everyday common sense is knowledge from an infinite number of experiences or the sum of micro knowledge, and many contradictions and illogical elements are naturally included. Therefore, it is something generalized, and if it is expressed in words only concepts come out. The contents of this kind of knowledge cannot be strictly defined, and logical conformity cannot be found. In addition, it cannot be systematized. Let us call this macro knowledge.

Not only is the expression of macro knowledge difficult, but it is also difficult to use. The reason is that the people that use it must interpret abstract meanings by themselves and apply them concretely according to the problem at hand. This is an applications problem and is one form of creativity. Human beings freely use this wide variety of knowledge in their daily lives, but this is really something surprising.

We do not know why human beings learn to use macro knowledge, but it is probably because, no matter how logically we look at it, human thinking is extremely qualitative and ambiguous. We will call this macro thinking. When a complex problem is approached by macro thinking, the essentials and minor details are divided intuitively, the details and sidelines dropped, and only the essential parts considered. In addition, macro thinking is necessary for processing large amounts of micro information, and if it did not exist, the excess information era we have now would not have come about. However, the essentials and digressions are not fixed, and they change according to the objects of human activity. In other words, people can use macro thinking because they have goals, so nothing can be gained by just spinning logic with no goal. This is why it is difficult for computers to handle macro knowledge and difficult to try and make them perform macro thinking.

1.3 MAN /MACHINE SYSTEMS

If artificial intelligence can only handle micro knowledge, the range of its usefulness becomes extremely limited. Thus, in the same way that the left and right hemispheres of the human brain cooperate, we can allow artificial intelligence to take care of the logical work such as analysis and storage, retrieval and management of past knowledge, while people handle the right-hemisphere jobs such as conception, generalization, and creation. In this way they will make up for each other's faults while making use of the good points, and have high-level functions, which is the idea behind man/machine systems. From now on it seems that research into artificial intelligence must progress in the direction of determining how to make both of these work together.

In the human brain, the corpus callosum, which links the left and right hemispheres, plays an important role, but what could be used to tie together people and artificial intelligence in man/machine systems? The authors think that both must to some extent have common (1) language, (2) knowledge, and (3) thinking processes. If these are insufficient, the composite system will not function well, no matter how great the capabilities of each part.

Since language is an important medium for communication, it would of course be ideal if natural language could be used, and in addition, one would like to use pictures if possible. Secondly, common knowledge is important even for cooperation between people themselves, and, especially when one's companion is an inanimate computer, common knowledge is indispensable for smooth communication and understanding. In addition, even though the knowledge is specialist knowledge, it would be ideal if common sense were coupled with it. Thirdly, we have thinking processes. If these are similar, accurate reasoning and inference can occur, even if the information from one of the partners is somewhat incomplete. Since human beings use qualitative, ambiguous thinking processes, artificial intelligence must deal in macro thinking as much as possible. Man/machine systems will not be complete until the people and computers understand each other's capabilities, recognize each other's simultaneous roles within the system, and communicate smoothly, as mentioned before.

1.4 FUZZY SYSTEMS

Scientific methodology requires strict logic, but one can say that not much effort goes into verification of premises and assumptions. The premises and assumptions that science and technology worry so little about are the same as axioms in mathematics, and this probably comes about because they are not logical on the whole. At present, these problems can only be processed through human perception and experience. However, if premises and assumptions are not thoroughly investigated in technical fields, there is the fear of inviting big mistakes. For example, unexpected accidents in safety systems, nonsensical conclusions in information systems, and automation systems that lack balance all occur when design premises are far from the actual circumstances.

Science and technology do their best to exclude subjectivity, but as was mentioned earlier, discovery and invention originate in right-hemisphere activities that are based on subjectivity, and objectivization and logicizing are no more than secondary processes for gaining the assent of others. Additionally, the use of subjectivity is even more effective during the process of objectivization. For example, when we are dealing with ambiguity in probability, which is often compared with fuzzy systems, truly good results cannot be obtained unless experienced people perform subjective processes such as thoroughly investigating the data in advance, assessing the general structure of the problem and discarding unreliable data. In order to establish assumptions and premises there is nothing to rely on other than subjectivity.

The problem of how to handle ambiguity that was mentioned before is the same thing as finding a way to bring human subjectivity into science and technology. This is where fuzzy sets come into play. Fuzzy sets are a mathematical method that was invented with the goal of expressing the semantic ambiguity in human language, and they are unique in that they make it possible to deal scientifically with subjectivity.

If we can express the meanings of words mathematically and deal with them logically, it means that we have a new tool that satisfies the three requirements for the realization of man/machine systems. In other words, artificial intelligence can understand the ambiguity and deep meanings of natural language by these means, and we can expect that in the future, communication will become as smooth as conversations between people. In addition, common sense, which is extremely ambiguous macro knowledge, can be expressed in terms of fuzzy propositions, if the expert field is kept narrow enough. We think this is something that will rapidly improve the usefulness of artificial intelligence. The third requirement of ambiguous thinking is more difficult. Even with fuzzy systems, it does not seem that we can come up with right-hemisphere thinking such as humans perform. However, if we limit ourselves to specialized areas, qualitative logical relations arise within macro knowledge, and it might be possible to approach human reasoning to a certain extent.

The outputs obtained from fuzzy systems are naturally ambiguous. There are probably people who think that there will be trouble if the answers from systems that support thinking are ambiguous, but only in answers concerning micro knowledge is ambiguity troublesome. Since macro knowledge concerns problems that are fundamentally unclear, ambiguous answers are natural, and conversely clear expressions could mislead people. Moreover, it can be said that ambiguous answers stimulate the right hemisphere and increase the effectiveness of man/machine systems.

In this way, we can think of many concrete examples of man/machine systems that could be developed in the future, centered on fuzzy systems, but to put it in a word, we could call it the realization of all-purpose robots to serve mankind. In other words, we can talk of such things as high-level automatic control, automatic translation, intelligent robots, preservation and safety systems, picture and voice recognition, automatic design, fault analysis, information retrieval, knowledge bases, intelligent sensors, and home automation in industrial fields. Furthermore, there are things such as medical diagnosis, treatment with Oriental medicines, artificial organ control, nursing robots, post-treatment care, health systems, and artificial limbs for medical fields. In the area of business there are things such as support for administrative decision-making, marketing, investment advising, various types of management planning, system operation, contract support, and office automation. Beyond these we can think of many others such as environmental evaluation, risk analysis, earthquake prediction, agricultural meteorology, geological surveys, produce evaluation, training systems, physical examinations, and data analysis for investigations.

Furthermore, there has been a movement toward trying to use fuzzy theory in the humanities and social sciences recently. In the near future, starting with models of human activity, thinking, psychology, reliability, and economics, it will probably be used actively in education, law, and analysis and evaluation of things such as public opinion. In sum, fuzzy methods will probably serve all fields that are connected with control, information, and systems.

Fuzzy systems show great promise for future value, but there are as many points that need to be researched and solutions to be found. For example, we can bring up the expression and acquisition of macro knowledge, methods for macro thinking (macro reasoning), the form of outputs for right-hemisphere stimulation, identification of membership functions, estimation of system functions, and systematic design methods. In some ways we can say that this research is research on humans themselves and our neverending research theme. We think the appearance of a new method for the introduction of ambiguity is a step toward a solution for these problems.

REFERENCES

1. Terano, T., *et al.*, *Introduction to Fuzzy Engineering*, Blue Backs Series, Kodansha, Tokyo (1981) (in Japanese).
2. Shinagawa, K., *Right Brain Computer*, Diamond-sha (1986) (in Japanese).
3. Terano, T., *Introduction to Systems Engineering—The Challenge of Ambiguous Problems*, Kyoritsu Shuppan (1985) (in Japanese).
4. Ueno, H., *Introduction to Knowledge Engineering*, pp. 57–62, Ohmsha, Tokyo (1985) (in Japanese).
5. Kobayashi, S., "The Current State and Future of Knowledge Systems Technology," *Measurement and Control* **27**(10), 859–868 (1988) (in Japanese).
6. Round Table on "Current State of Research on Artificial Intelligence and Its Problems," *Bulletin of the Artificial Intelligence Society* **3**(5), 14–21 (1988) (in Japanese).
7. Nakamura, V., *et al.*, *Fuzzy—A New Development in Knowledge*, Nikkan Kogyo Shinbun-sha (1989) (in Japanese).

Chapter 2

Fuzzy Set Theory for Applications

It has been a quarter of a century since University of California professor L. A. Zadeh first proposed fuzzy set theory. Therefore, the theory itself has delivered many branches, and it would take several years if one were to try and acquire knowledge about all of them. However, if concrete applications are the goal, only a small portion of this is really necessary. In this chapter we will use a rapid learning process to acquire knowledge of fuzzy set theory. In order to understand fuzzy set theory, we will first learn about crisp set theory and Boolean logic. Based on these, we will gain an understanding of fuzzy set theory and fuzzy logic. In addition, we will study fuzzy reasoning, the most important aspect for applications, centering our attention on fuzzy production rules and fuzzy relations, and we will bring our basic knowledge up to a sufficient level for understanding the chapters that follow.

2.1 CRISP SET THEORY

The "fuzz" of fuzzy means "nap" or "pile" and is a word used with textiles (something we heard from an engineer in the textile industry and probably not known by many engineers; it is basically not a technical term). From this "fuzz" comes the idea of a hazy outline, something not seen clearly, so "fuzzy" means "unclear" or "ambiguous." Fuzzy sets are used for handling attributes such as a "group of beauties," in which the meaning is understandable, but the existence of the characteristics for individual objects cannot be determined by a simple "yes" and "no" evaluation.

Conversely, the world in which characteristics have a clear two-valued determination like "man or woman?" is the *crisp* world (the original meaning being something like the clear "crispness" of fresh lettuce, something heard in the modulations of a voice). Therefore, the 0 and 1 logic that computers operate on is called crisp logic, and standard sets are called crisp sets. As extensions of these we get fuzzy logic and fuzzy sets. In order to understand these, we will first review crisp set theory.

The two large divisions of crisp set theory are axiomatic set theory and naive set theory. The former is a part of basic mathematical theory and also requires a high level of philosophical thinking. However, naive set theory is sufficient here, and this is just an extension of the set theory learned in elementary school. But the concept of characteristic functions in naive set theory is necessary.

First we will explain some basic terminology and symbols. The complete object that we are dealing with is given by a capital letter (X, for example), and the elements it is composed of are denoted by small letters (x, for example). In this case we write

$$X = \{x\}. \tag{2.1}$$

The brackets indicate a grouping. In addition, the complete object (X in this case) is called the *universe of discourse*, *total space*, or *support set*. ("Support set" is quite often used in fuzzy control systems. There are a number of differences in the meaning here and the meaning of "support" as used in functional analysis, so those familiar with this should be careful.) The individual components are known as elements or objects, and when x is an element of X, we write

$$x \in X. \tag{2.2}$$

Now we come to defining a set (crisp set) on this total space. Its name (label) is written with a capital letter, such as A, B, or C. For example, if we let the total space be all of the one-digit integers in the decimal system,

$$X = \{0, 1, 2, 3, 4, 5, 6, 7, 8, 9\}, \tag{2.3}$$

the set of even numbers A is

$$A = \{0, 2, 4, 6, 8\}. \tag{2.4}$$

In this case, the number of elements is called the Cardinal number and is indicated by the symbol #. For the foregoing example, we get

$$\#X = 10, \quad \#A = 5. \tag{2.5}$$

When $\#A = 1$, A is called a *singleton*. If $\# *$ is a *finite set*, a set with finite number of elements, it is possible to write down all of the elements as in

Eqs. (2.3) and (2.4), but if, for example, it is an *infinite set* such as all natural numbers or all real numbers, it is generally impossible. In this kind of situation, there is a method of expression that involves writing the attributes for each set on the right side of a vertical line. For example, Eq. (2.4) can be written

$$A = \{x|x \text{ being even numbers greater or equal to 0 and less than or equal to 8}\}. \tag{2.6}$$

In addition, when the concepts are expressed pictorially, *Venn diagrams* such as the one in Fig. 2.1 are often used.

In addition to these methods, there is another method for defining the concepts for crisp sets that involves defining them in terms of *characteristic functions*, and this is very important for the understanding of fuzzy sets. The characteristic function χ_A that defines set A on total space X is a mapping that gives the domain of X and a range of $\{0, 1\}$ (two-valued set of 0 or 1)

$$\begin{aligned} \chi_A\colon X &\to \{0, 1\} \\ \Psi \quad & \quad \Psi \\ x &\mapsto \chi_A(x) = \begin{cases} 0 & x \notin A \\ 1 & x \in A \end{cases}. \end{aligned} \tag{2.7}$$

In this case, when element x is attributable to set A, the value of $\chi_A(x)$ is 1, and when it is not, that value is 0. Therefore, if X is the horizontal axis and $\{0, 1\}$ the vertical axis, we get a diagram like that in Fig. 2.2.

In this way we can consider a number of sets on X, such as set A with one attribute and set B with another. What is known as the *power set* includes all of them and is denoted 2^X. If, for example, we let

$$X = \{a, b, c\}, \tag{2.8}$$

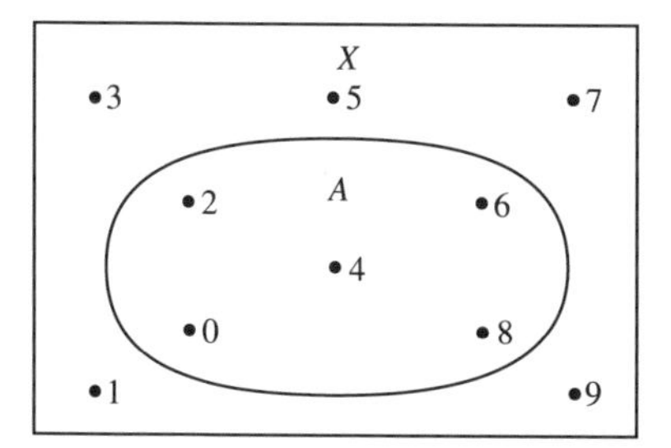

Fig. 2.1 Expressing a set using a Venn diagram.

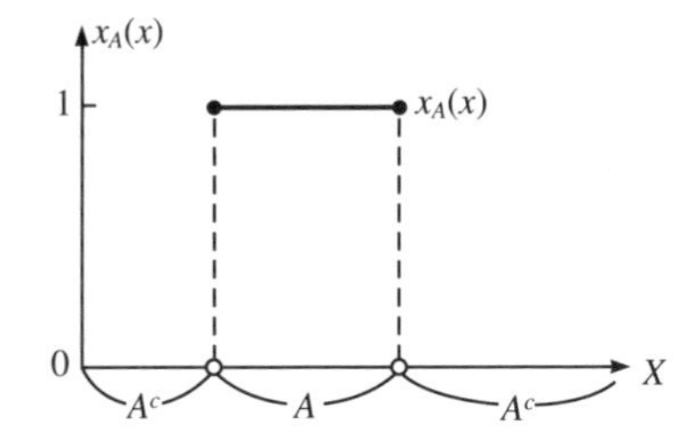

Fig. 2.2 Using characteristic functions to define a set.

the power set is

$$2^X = \{\phi, \{a\}, \{b\}, \{c\}, \{a, b\}, \{b, c\}, \{c, a\}, X\}. \tag{2.9}$$

Here ϕ is a special set that contains no elements at all and is called the *null set* or the *void set*. Its characteristic function is

$$\chi_\phi(x) = 0 \qquad \text{for } {}^\forall x \in X. \tag{2.10}$$

Here $\forall$ is called the *universal quantifier* and can be thought of as meaning "all." (There is also an *existential quantifier*, which indicates existence. The general name for $\forall$ and $\exists$ is *quantifier*, and these are commonly used in logic, artificial intelligence, and knowledge engineering.) The characteristic function for total space X is

$$\chi_X(x) = 1 \qquad \text{for } {}^\forall x \in X. \tag{2.11}$$

Looking at power,

$$\#2^X = 2^{\#X} \tag{2.12}$$

generally arises, and this equation can also be derived from (2.8) and (2.9) quite easily.

Next, we will review a number of set operations on 2^X (see Fig. 2.3). First, let us deal with inclusion relations. When, as shown in Fig. 2.3a, all elements of A are necessarily elements of B, A is said to be a *subset* of B (or B is the *superset* of A), and this is expressed by $A \subset B$. ($A \subset B$ includes $A = B$. When $A \subset B$ and $A \neq B$, A is called a *proper subset* of B.) Defining $A \subset B$ using a characteristic function, we get an inequality such as

$$\chi_A(x) \leqq \chi_B(x) \qquad \text{for } {}^\forall x \in X. \tag{2.13}$$

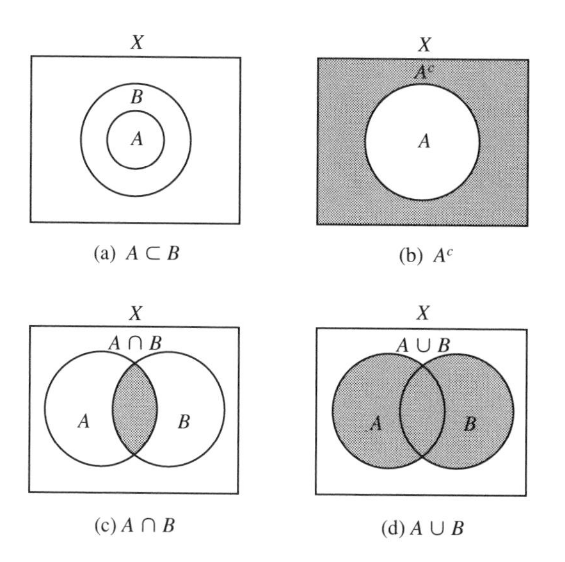

Fig. 2.3 Inclusion relation, complement, intersection, and union of sets.

When inclusion relations $\subset$ give rise to the following properties:

(1) reflexivity

$$^\forall A \in 2^X, \qquad A \subset A \tag{2.14}$$

(2) anti-symmetry

$$^\forall A, B \in 2^X \qquad A \subset B, B \subset A \rightarrow A = B \tag{2.15}$$

(3) transitivity

$$^\forall A, B, C \in 2^X \qquad A \subset B, B \subset C \rightarrow A \subset C \tag{2.16}$$

it means that $(2^X, \subset)$ is said to form a *partially ordered set* or *POSET*. (When we deal with inclusion relations, A and B do not generally give rise to convenient $A \subset B$ or $B \subset A$ relations, so we do not get a linear ordered set or a totally ordered set that can be ordered along a single line.)

Let us review one unary operation and two binary operations on POSET $(2^X, \subset)$. These are A^C, the complement of A; $A \cap B$, the intersection, and $A \cup B$, the union. These concepts are made clear in the Venn diagrams in Fig. 2.3b–c, and the characteristic functions are as follows:

$$\chi_{A^c}(x) = 1 - \chi_A(x) \qquad \text{for } ^\forall x \in X, \tag{2.17}$$

$$\chi_{A \cap B}(x) = \chi_A(x) \wedge \chi_B(x) \qquad \text{for } ^\forall x \in X, \tag{2.18}$$

$$\chi_{A \cup B}(x) = \chi_A(x) \vee \chi_B(x) \qquad \text{for } ^\forall x \in X. \tag{2.19}$$

Here $\wedge$ and $\vee$ indicate min and max operations, respectively, and they are the operations for selecting the lowest and highest values.

It is relatively easy to confirm that the following properties arise with the set operations that have been defined like this:

(1) idempotent law

$$A \cap A = A, \qquad A \cup A = A \tag{2.20}$$

(2) commutative law

$$A \cap B = B \cap A, \qquad A \cup B = B \cup A \tag{2.21}$$

(3) associative law

$$A \cap (B \cap C) = (A \cap B) \cap C,$$
$$A \cup (B \cup C) = (A \cup B) \cup C \tag{2.22}$$

(4) absorption law

$$A \cap (A \cup B) = A \cup (A \cap B) = A \tag{2.23}$$

(5) distributive law

$$A \cap (B \cup C) = (A \cap B) \cup (A \cap C),$$
$$A \cup (B \cap C) = (A \cup B) \cap (A \cup C) \tag{2.24}$$

(6) law of complements

$$A \cap A^c = \phi, \qquad A \cup A^c = X \tag{2.25}$$

In general, a POSET that satisfies the idempotent, commutative, associative and absorption laws is called a *lattice*. If, in addition, a lattice satisfies the distributive law, it is known as a *distributive lattice*, and in addition to these, a distributive lattice that satisfies the law of complements is a *complemented distributive lattice*. Furthermore, other names for complemented distributive lattices are *Boolean lattice* or *Boolean algebra*. Therefore, the concepts for crisp sets (2^X, $\subset$, $\cdot^c$, $\cup$, $\cap$) can be discussed in the framework of Boolean algebra.

Finally, the following two properties are important, so we will touch upon them, even though they arise naturally with Boolean algebra:

(1) double negation

$$A^{cc} = A \tag{2.26}$$

(2) de Morgan's law

$$(A \cap B)^c = A^c \cup B^c, \qquad (A \cup B)^c = A^c \cap B^c \tag{2.27}$$

2.2 CRISP LOGIC

As a foundation for fuzzy logic, we will now review crisp logic (standard computer logic, or two-valued Boolean algebra). This is the two-value world of "yes" and "no" or "true" and "false," and on the hardware level we can think of it as 0V and 5V. However, we can generally think of it as the two values of $\{0, 1\}$.

In general, we can think of a computer as being a *sequential circuit* like that shown in Fig. 2.4. Each of the n input variables and m output variables must be either 0 or 1, according to a given timing. In addition, these variables change in synchronization with a basic control signal called the *clock pulse*. If the input/output relations for these are given as a function, we can write

$$y_j(t+1) = f_j(x_1(t), x_2(t), \ldots, x_n(t), y_1(t), y_2(t), \ldots, y_m(t)), \quad (2.28)$$

Here $y^j(t)$ is included in the right-hand variables; that is, the equation is recursive. Therefore, each output value $y_j(t+1)$ in time frame $t+1$ depends on all past inputs up to t and is determined accordingly.

If the transfer of the values as the time frame changes is divided by a *memory* function, we can discuss the relations between fixed inputs and outputs for a given time frame. A *combinatorial circuit* (Fig. 2.5) hides the time factor and expresses the input/output relations by themselves. The equation for n inputs and m outputs can be written

$$y_j = f_j(x_1, x_2, \ldots, x_n), \qquad j = 1, \ldots, m. \quad (2.29)$$

Here we can consider each of the m output variables in order, and there is no loss of generality even if we let m be 1. In other words, we can consider

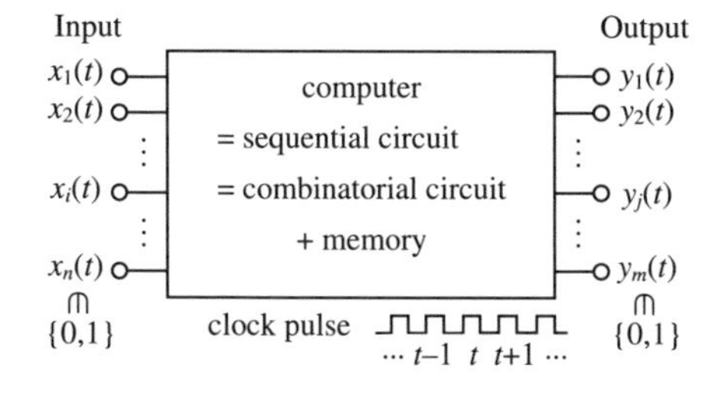

Fig. 2.4 Sequential circuit.

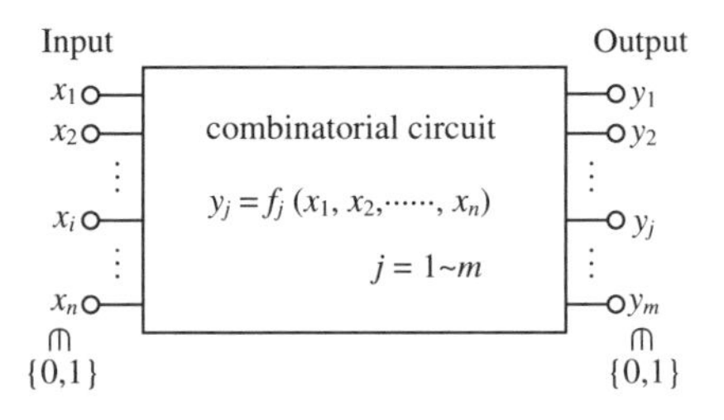

Fig. 2.5 Combinatorial circuit.

n inputs and one output. In this instance, the input/output relations can be written

$$\begin{array}{ccc} f\colon \{0,1\}^n & \rightarrow & \{0,1\} \\ \Cup & & \Cup \\ (x_1, x_2, \ldots, x_n) & \mapsto & y = f(x_1, x_2, \ldots, x_n). \end{array} \tag{2.30}$$

Relations like f are called n variable *Boolean functions*, and combinatorial circuits perform the concrete operations.

Let us first consider the simple case of $n = 1$:

$$\begin{array}{ccc} f\colon \{0,1\} & \rightarrow & \{0,1\} \\ \Cup & & \Cup \\ x & \mapsto & y = f(x). \end{array} \tag{2.31}$$

As shown in Fig. 2.6, there are four input/output relations in this case. The NOT relation is one of the most important of them, and it is performed by a device known as an *inverter* or a *NOT gate*, which is indicated by the symbol in Fig. 2.6c. In addition, Fig. 2.6d shows the order relations for the four operations in the form of a *Hasse diagram*. It is a diagram in which the 0 or 1 in each column is compared and the larger values are placed above. The places that can be compared are connected by lines. (In this case, $\bar{x}$ and x, $(1, 0)$ and $(0, 1)$, are incomparable, so they are not connected by a line.)

Let us consider two inputs next:

$$\begin{array}{ccc} f\colon \{0,1\}^2 & \rightarrow & \{0,1\} \\ \Cup & & \Cup \\ (x_1, x_2) & \mapsto & y = f(x_1, x_2). \end{array} \tag{2.32}$$

This time there are 16 input/output relations, as is shown in Fig. 2.7b. However, the AND, OR, NAND, NOR, and EXOR operations are the most important, and the gate circuit symbols shown in Fig. 2.7c are used. In addition, the Hasse diagram order expressions are given in Fig. 2.7d.

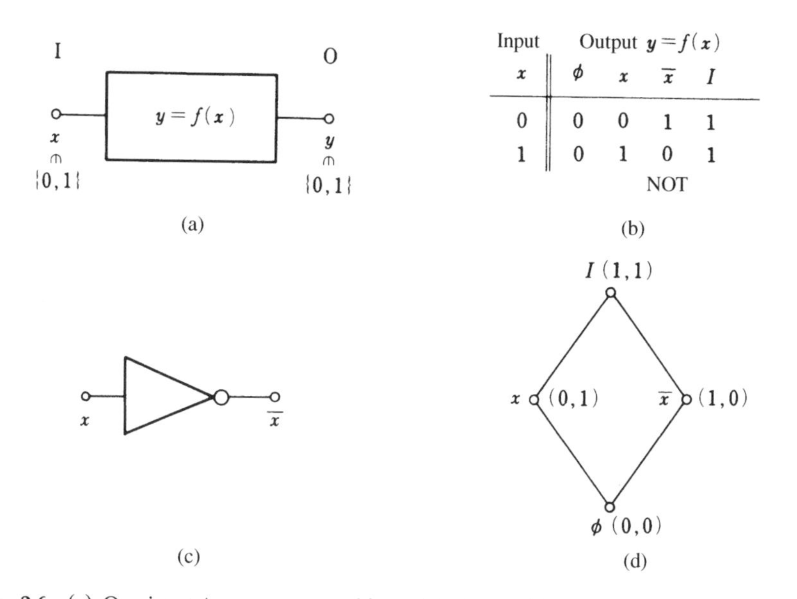

Fig. 2.6 (a) One-input/one-output combinatorial circuit. (b) Four types of response. (c) NOT gate/inverter. (d) Expression of order using a Hasse diagram.

If we continue in this way with $n = 3$ and $n = 4$, there are as many as 65,536 types of relations for $n = 4$, since there are 2^{2^n} possibilities for a combinatorial circuit with n inputs and one output. What occurs is known as a combinatorial explosion. However, it is sufficient to investigate them up to $n = 2$. This is because the following three equal propositions arise.

(1) Any Boolean function $f(x_1, x_2, \ldots, x_n)$ can be formed using NOT, AND, and OR operations with input variables $x_1, x_2, \ldots, x_n$.
(2) Any Boolean function $f(x_1, x_2, \ldots, x_n)$ can be formed using NAND operations with input variables $x_1, x_2, \ldots, x_n$.
(3) Any Boolean function $f(x_1, x_2, \ldots, x_n)$ can be formed using NOR operations with input variables $x_1, x_2, \ldots, x_n$.

If this is expressed in hardware terms we get

(a) Any given combinatorial circuit can be constructed using only inverters, AND gates, and OR gates.
(b) Any combinatorial circuit can be constructed using only NAND gates.
(c) Any combinatorial circuit can be constructed using only NOR gates.

In addition, when we consider order circuits, taking into account the time factors, it means considering the basic *flip-flop* circuits of memory ele-

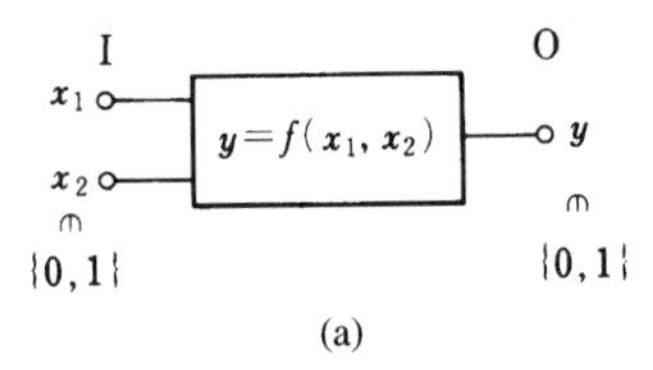

(a)

Input		Output $y=f(x_1, x_2)$															
x_1	x_2	ϕ	$x_1 \cdot x_2$	$x_1 \backslash x_2$	x_1	$x_2 \backslash x_1$	x_2	$x_1 \oplus x_2$	$x_1 + x_2$	$\overline{x_1 + x_2}$	$\overline{x_1 \oplus x_2}$	$\overline{x_2}$	$\overline{x_2 \backslash x_1}$	$\overline{x_1}$	$\overline{x_1 \backslash x_2}$	$\overline{x_1 \cdot x_2}$	I
0	0	0	0	0	0	0	0	0	0	1	1	1	1	1	1	1	1
0	1	0	0	0	0	1	1	1	1	0	0	0	0	1	1	1	1
1	0	0	0	1	1	0	0	1	1	0	0	1	1	0	0	1	1
1	1	0	1	0	1	0	1	0	1	0	1	0	1	0	1	0	1
			AND					EXOR	OR	NOR						NAND	

(b)

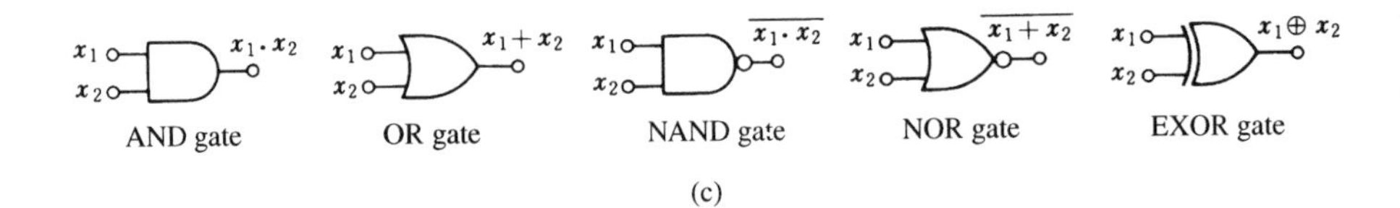

(c)

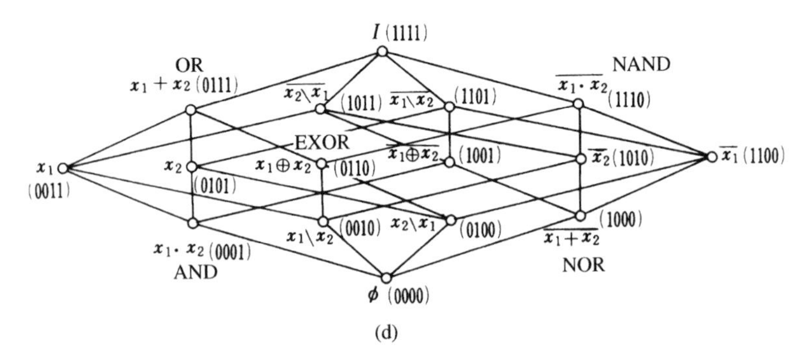

(d)

Fig. 2.7 (a) Two-input/one-output combinatorial circuit. (b) Sixteen types of response. (c) Symbols for two-input gates. (d) Expression of order using a Hasse diagram.

ments, but this can be done using the gate circuits mentioned earlier. This means that NOT, AND, and OR; NAND; and NOR are each capable of executing a complete system in Boolean algebra, and this is a sufficient understanding of computer theory for the moment.

However, fifth-generation computers are the first with *artificial intelligence* (*AI* for short). In addition, it is best to give a table of the operations that differ from these basic ones and are used with *fuzzy computers*, which are based on *fuzzy inference* or *approximate reasoning*. These correspond to those under $x_1 \setminus x_2$ in Fig. 2.7b, and they are called *implications*. These are usually shown with an arrow as in "$x_1 \rightarrow x_2$," and this is read "If x_1 then x_2." Here x_1 is called the *antecedent*, *precondition*, *condition part* or *assumption*, and x_2 the *conclusion* or *operation part*. If we summarize the values for implication operations, we get something like Table 2.1. (This type of table is generally known as a *truth table*.) The area inside the dotted lines—which is the rule that if condition x_1 is false ($= 0$), the truth value of $x_1 \rightarrow x_2$ is 1, regardless of conclusion x_2— is somewhat difficult for those who are learning logic for the first time.

In addition, we can confirm (easily, if we use truth tables) that if we show the complete NOT, AND, OR implication system, we get

$$x_1 \rightarrow x_2 = \bar{x}_1 + x_2, \tag{2.33}$$

and we should learn this property, which is one of the most basic ones. In general, *syllogisms*, which are based on this type of implication operation, are performed in the *reasoning* or *inference* for AI systems. There are a number of forms that syllogisms can take, such as

$$\begin{array}{c} x_1 \rightarrow x_2 \\ x_2 \rightarrow x_3 \\ \hline x_1 \rightarrow x_3 \end{array} \tag{2.34}$$

for reasoning like "if it is a bird, then it can fly" and "if it can fly, then it can go to the island you can see over there" with the conclusion "if it's a

Table 2.1. Truth Value Table for Implication $x_1 \rightarrow x_2$

x_1	x_2	$x_1 \rightarrow x_2$
0	0	1
0	1	1
1	0	0
1	1	1

bird, then it can fly to the island you can see over there"; or

$$\begin{array}{l} x_1 \rightarrow x_2 \\ x_1 \\ \hline \qquad x_2 \end{array} \tag{2.35}$$

for reasoning like "if it is a bird, then it can fly" and "this animal is a bird" with the conclusion "this animal can fly"; or

$$\begin{array}{l} x_1 \rightarrow x_2 \\ \qquad \overline{x_2} \\ \hline \overline{x_1} \end{array} \tag{2.36}$$

for reasoning like "if it is a bird, then it can fly" and "this animal can't fly" with the conclusion "this animal is not a bird." (However, we do not take into account exceptional cases here.) If we want to learn fuzzy reasoning for applications, the equation form known as the *modus ponens* (2.35) is very important. In addition, $x_1 \rightarrow x_2$ or "if ~ then ... " knowledge like "if it is a bird, then it can fly" is sometimes called *procedural knowledge*, and the knowledge in statements like "this animal is a bird" is called *declarative knowledge*.

2.3 FUZZY SET THEORY

If we take a look at the definition equation (2.7), which is based on a characteristic function $\chi_A(x)$ for crisp set A on total space X, the rule is that if x has the characteristics of A, the value is 1 and if not it is 0. Naturally this is the crisp world in which the existence or absence of properties can be divided into 1 and 0 (or yes and no).

However, things are not so black and white in the real world, or at least there are many matters that we do not wish to make so. This situation is more common. For example, questions like "We're thinking of bringing Mr. X in on this next project; how do you feel about it?" with answers like "I really don't think it's very good, but if there's no one more appropriate, that's the way it goes" are the kind found in everyday events. It is inexpedient to discuss this "yes" on the same level with the "yes" of "Mr. X is certainly the most appropriate." For this reason, University of California Professor L. A. Zadeh extended the two-valued evaluation of 0 or 1, $\{0, 1\}$, to the infinite number of values from 0 to 1, $[0, 1]$, and advanced the concept of *fuzzy sets* at the beginning of the 1960s, publishing a paper in *Information and Control* entitled "Fuzzy Sets" in 1965. (Here { } brackets are used to indicate sets, but square [ ] brackets and parentheses ( ) are used to denote real-number closed intervals and open intervals, respectively. For example, they are used as follows:

$$[0,1] = \{x \mid 0 \leqq x \leqq 1\}, \tag{2.37}$$

$$(0,1) = \{x \mid 0 < x < 1\}, \tag{2.38}$$

$$[0,1) = \{x \mid 0 \leqq x < 1\}. \tag{2.39}$$

In addition, Zadeh used the term *membership function* instead of characteristic function.

In other words, fuzzy set A on total space X is defined by membership function $m_A(x)$. (In his paper, he says "characterized," but we use "defined" here:

$$\begin{array}{rcl} m_A\colon X & \rightarrow & [0,1] \\ \Downarrow & & \Downarrow \\ x & \mapsto & m_A(x). \end{array} \tag{2.40}$$

("Ambiguous sets" and "attribution functions" are sometimes used instead of "fuzzy sets" and "membership functions." In addition, symbols such as μ_A and f_A are sometimes used instead of m_A. Furthermore, fuzzy set A is sometimes written $\bar{A}$ in order to make the distinction from crisp sets clear.) This alone forms the definition of a fuzzy set, and we will leave the interpretation up to the reader. However, interpretations such as "the value for $m_A(x)$ expresses a subjective value for the degree of A-ness of x, and $m_A(x) = 0.8$ means x has an A-ness of about 80%," are common. Therefore, it follows that we will have "my membership function" and "your membership function," as well as "specialist x's membership function." In addition, when we come to graphic descriptions, the Venn diagram of Fig. 2.1 is replaced by the contour line figure in Fig. 2.8a, and in order to get the feeling of the function, the rectangular waveform of Fig. 2.2 is often replaced by the bell-shaped curve of Fig. 2.8b.

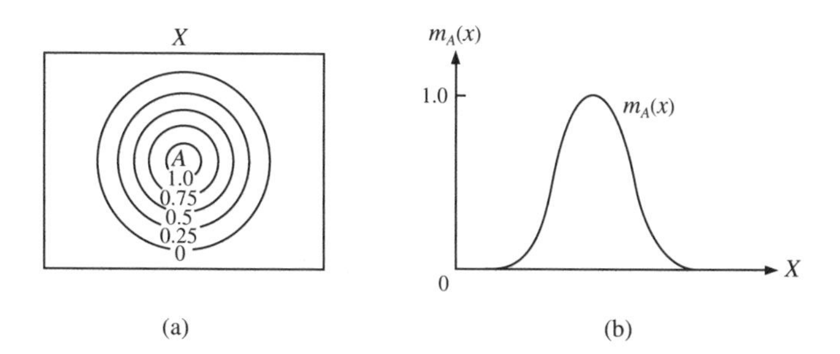

Fig. 2.8 Conceptual diagrams of fuzzy sets. (a) Expression of membership function using contour lines. (b) Bell-shaped membership function.

Let us now discuss the care that must be taken about the relationships between fuzzy and crisp sets. The two values of $\{0, 1\}$ are included in closed space $[0, 1]$. Therefore, crisp sets are special instances of fuzzy sets, or put another way, the concept of fuzzy sets is an extended concept that includes the concept of crisp sets. (In other words, crisp sets like those in Figs. 2.1 and 2.2 are also fuzzy sets. "Crisp sets" are sometimes referred to as "nonfuzzy sets," but this invites the misinterpretation of taking "nonfuzzy" to mean "not fuzzy," so we will not use the term "nonfuzzy" here.)

Fuzzy sets are clearly (crisply) defined by the membership function in Eq. (2.40), and there is nothing ambiguous about the definition itself. That is, the logic (meta-logic) for the conceptual definition is crisp and absolutely unambiguous. (When the meta-logic is ambiguous, we call it "nonsense" or "haphazardry.") It is clearly defined by an evaluation $[0, 1]$ on X, that is, by the membership function. When it comes to applications, cases in which X is a finite set are most often considered, but in these instances the expression for the membership function is mostly written using *separators* "/" and OR "+" symbols, as is shown in the following example. For example, when X is one-digit decimal integers as shown by Eq. (2.3) and A is "small numbers," we get an expression like

$$A = 1/0 + 1/1 + 0.8/2 + 0.5/3 + 0.1/4 \tag{2.41}$$

Here, for example, $0.8/2$ means $m_A(2) = 0.8$. When the value for the membership function is 0, it is left out.

We can consider the same set operations for fuzzy sets as we did for crisp sets. The definitions for the most basic ones—inclusion relation, fuzzy complement, fuzzy intersection, and fuzzy union—are usually given as follows:

$$A \subset B \leftrightarrow m_A(x) \leqq m_B(x) \qquad \text{for } {}^\forall x \in X, \tag{2.42}$$

$$m_{A^c}(x) = 1 - m_A(x) \qquad \text{for } {}^\forall x \in X, \tag{2.43}$$

$$m_{A \cap B}(x) = m_A(x) \wedge m_B(x) \qquad \text{for } {}^\forall x \in X, \tag{2.44}$$

$$m_{A \cup B}(x) = m_A(x) \vee m_B(x) \qquad \text{for } {}^\forall x \in X. \tag{2.45}$$

If we use bell-shaped membership functions to express these concepts graphically, they come out as in Fig. 2.9.

When the all-fuzzy sets on X are given by $P(X)$, the $(P(X), \subset, \cdot^c, \cap, \cup)$ system unfortunately does not give us Boolean algebra. However, reflexivity, anti-symmetry, and transitivity are assured for $\subset$, and furthermore, it can easily be shown that the idempotent, commutative, associative, absorption, distributive, double negative, and de Morgan's laws all

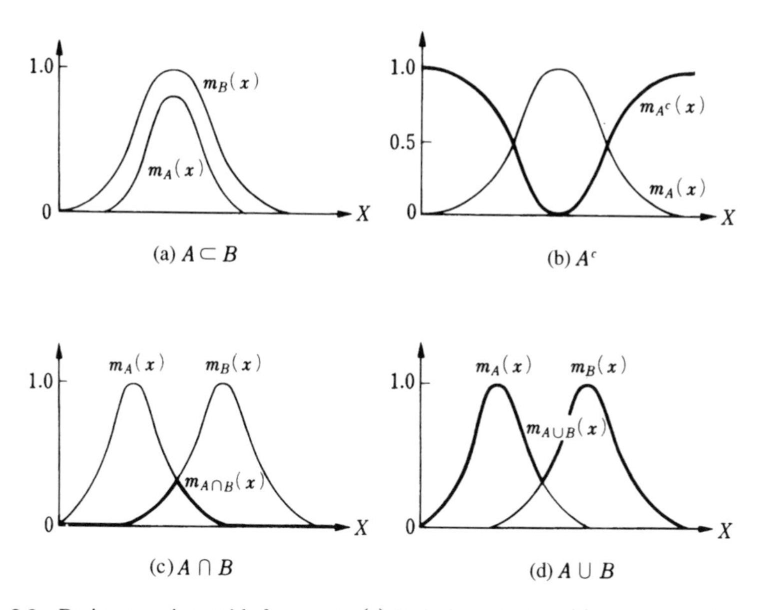

Fig. 2.9 Basic operations with fuzzy sets. (a) Inclusion relation. (b) Fuzzy complement. (c) Fuzzy intersection. (d) Fuzzy union.

arise. The only one that does not arise is the law of complements. In other words, we have

$$A \cap A^c \supset \phi, \qquad A \cup A^c \subset X \tag{2.46}$$

with fuzzy sets, and there is no equality. This should be clear if one looks at Fig. 2.10. However, all of the others arise, and $(P(X), \subset, \cdot^c, \cap, \cup)$ give rise to a system known as *complete pseudo-Boolean algebra*.

We can define an unlimited number of other operations for fuzzy sets, but we will just explain those that are important from the point of view of practical applications.

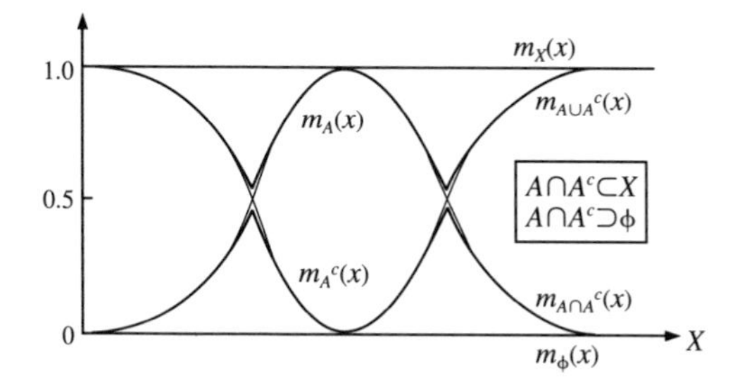

Fig. 2.10 The law of complements does not arise.

First, there is a monomial operation, the α power for fuzzy set A (where α is a positive parameter), and it is defined by

$$m_{A^\alpha}(x) = \{m_A(x)\}^\alpha \qquad \text{for } {}^\forall x \in X. \tag{2.47}$$

If we express the most commonly used 2 power and $\frac{1}{2}$ power in graphic form, we get Fig. 2.11. When some ambiguous information is expressed by fuzzy set A, A^2 corresponds to the narrow, sharp "very A, and $A^{1/2}$ to the wide, unclear "somewhat A." In addition, 4 power and $\frac{1}{4}$ power are often used with analogous interpretations, like "A^4 = very very A = (very)^2A). However, if A is a crisp set, the 0 and 1 for the 0 and 1 powers do not change, so A and A^α are the same. The definition is possible, but it is meaningless.

Next we will talk about binomial operations. Among the most used that are related to intersection $A \cap B$ are the *algebraic product* $A \cdot B$, *bounded product* $A \odot B$, and drastic product $A \wedge B$:

$$m_{A\cdot B}(x) = m_A(x) \cdot m_B(x) \qquad \text{for } {}^\forall x \in X, \tag{2.48}$$

$$m_{A\odot B}(x) = (m_A(x) + m_B(x) - 1) \vee 0 \qquad \text{for } {}^\forall x \in X, \tag{2.49}$$

$$m_{A \wedge B}(x) = \begin{cases} m_B(x) & \text{if } m_A(x) = 1 \\ m_A(x) & \text{if } m_B(x) = 1 \\ 0 & \text{otherwise} \end{cases} \qquad \text{for } {}^\forall x \in X. \tag{2.50}$$

Parallel to this are the commonly used ones related to union $A \cup B$; the *algebraic sum* $A \dotplus B$, *bounded sum* $A \oplus B$, and *drastic sum* $A \vee B$:

$$m_{A\dotplus B}(x) = m_A(x) + m_B(x) - m_A(x) \cdot m_B(x) \qquad \text{for } {}^\forall x \in X, \tag{2.51}$$

$$m_{A\oplus B}(x) = (m_A(x) + m_B(x)) \wedge 1 \qquad \text{for } {}^\forall x \in X, \tag{2.52}$$

$$m_{A \vee B}(x) = \begin{cases} m_B(x) & \text{if } m_A(x) = 0 \\ m_A(x) & \text{if } m_B(x) = 0 \\ 1 & \text{otherwise} \end{cases} \qquad \text{for } {}^\forall x \in X. \tag{2.53}$$

Expressing these graphically we get Fig. 2.12a and b.

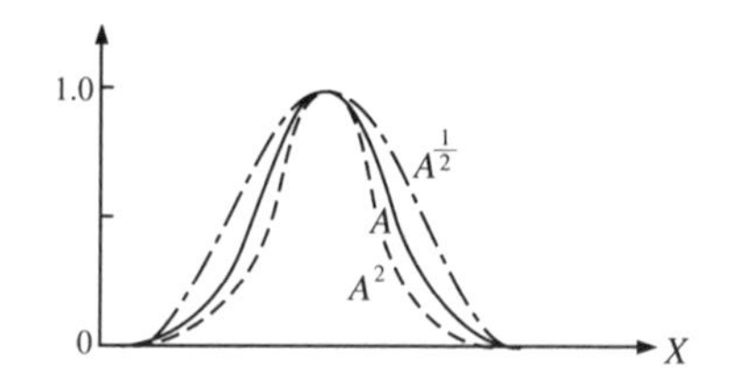

Fig. 2.11 α power of a fuzzy set ($\alpha = 2, 1/2$).

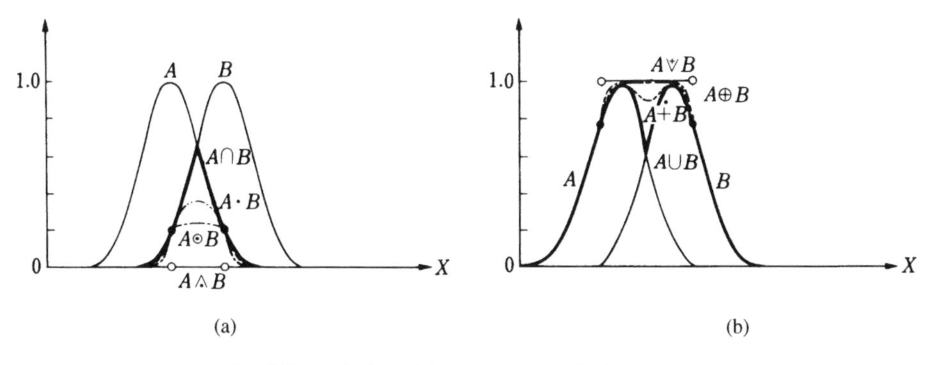

Fig. 2.12 (a) Fuzzy intersections and (b) fuzzy unions.

Other important operations include the *difference* between fuzzy sets $A - B$ and the *absolute difference* $A \triangle B$:

$$m_{A-B}(x) = (m_A(x) - m_B(x)) \vee 0 \qquad \text{for } {}^\forall x \in X, \tag{2.54}$$

$$m_{A \triangle B}(x) = |m_A(x) - m_B(x)| \qquad \text{for } {}^\forall x \in X. \tag{2.55}$$

All of the binomial operations we have mentioned so far have corresponding ones for crisp sets, but there are many binomial operations that can only be defined after fuzzification. Here we will only introduce the frequently used λ-sum $A \overset{+}{_\lambda} B$ (λ being a parameter between 0 and 1):

$$m_{A \overset{+}{_\lambda} B}(x) = \lambda m_A(x) + (1 - \lambda) m_B(x) \qquad \text{for } {}^\forall x \in X. \tag{2.56}$$

In other words, this is the weighted mean (or the convex linear combination of A and B) for the weights λ and $(1 - \lambda)$ for A and B, respectively. When $\lambda = 0.5$, it is the arithmetic mean of A and B. (For this operation, $\{0, 1\}$ is not closed, so a crisp definition is impossible.)

One can also easily confirm that the following inclusion relations arise with the operations related to the intersection and the union through an investigation of the membership function inequalities:

$$\begin{aligned} \phi &\subset A \wedge B \subset A \odot B \subset A \cdot B \subset A \cap B \subset A \overset{+}{_\lambda} B \subset A \cup B \\ &\subset A \dotplus B \subset A \oplus B \subset A \veebar B \subset X. \end{aligned} \tag{2.57}$$

2.4 EXTENSION OF THE CONCEPT OF FUZZY SETS

The basic fuzzy sets discussed in the previous section are called *type* 1 *fuzzy sets*, and many further extensions have been investigated. However, the greater portion of those considered from the point of view of practical applications are essentially type 1 fuzzy sets, so this section is probably not necessary. But since there are several important items when viewed from the point of view of expressing ambiguous information, we will give a simple explanation.

If we take a look at the basic definition equation (2.40) for fuzzy sets, there is a $[0, 1]$ real number evaluation $m_A(x)$ given for every element x of total space X. In other words, it is a pointwise definition for each element x. Therefore, if we consider x separately as being fixed, the main point is evaluating ambiguous information using real numbers $[0, 1]$.

Just as the two crisp values $\{0, 1\}$ were fuzzified to the infinite number of evaluation values $[0, 1]$, we can obtain corrected fuzzy concepts by replacing $[0, 1]$ with other things. Here we will introduce some of the more important ones.

For example, let us say that the A-ness of x is about 80%, $m_A(x) = 0.8$. If in this instance, we ask "Does $m_A(x)$ always have to be 0.8 and not 0.7 or 0.9?", cases are common in which the answer is "That's OK. 0.8 is the most probable value, but it has some width." Thus, the concept of *interval-valued fuzzy sets* has been proposed, in which the fixed value from the $[0, 1]$ real number interval is replaced by upper and lower bounds that give a width to the evaluation, as when $m_A(x)$ is indicated by interval values like [0.7, 0.9] (Fig. 2.13b). However, here we can ask, "Is it impossible for the upper bound of 0.7 to go to 0.69 or in the same way the lower bound of 0.9 . . . ?" Hence, the concept of *type 2 fuzzy sets*, in which the evaluation values for $m_A(x)$ are fuzzy sets on $[0, 1]$ instead of being values on a crisp interval (Fig. 2.13c). In this way, fuzzy sets like that of Eq. (2.40) are known as basic type 1 fuzzy sets, and type 2 fuzzy sets are expressed by

$$m_A: X \to [0, 1]^{[0, 1]}. \tag{2.58}$$

(In general, when U and V are sets, U^V expresses the set of mappings where V is the domain and U the range:

$$U^V = \{f \mid f: V \to U\}.) \tag{2.59}$$

Therefore, we get

$$[0, 1]^{[0, 1]} = \{f \mid f: [0, 1] \to [0, 1]\}. \tag{2.60}$$

Next we have the case in which we cannot determine $[0, 1]^{[0, 1]}$ and it undergoes fuzzification once more, and so on; the generalized concept of *type n fuzzy sets* has been proposed. Thus, type n fuzzy set A is defined as

$$m_A: X \to \overbrace{[0, 1]^{[0, 1]^{\cdot^{\cdot^{\cdot^{[0, 1]}}}}}}^{n} \tag{2.61}$$

On the other hand, another point of view has produced a proposal for *probabilistic sets* in which the randomness of evaluation values is brought in

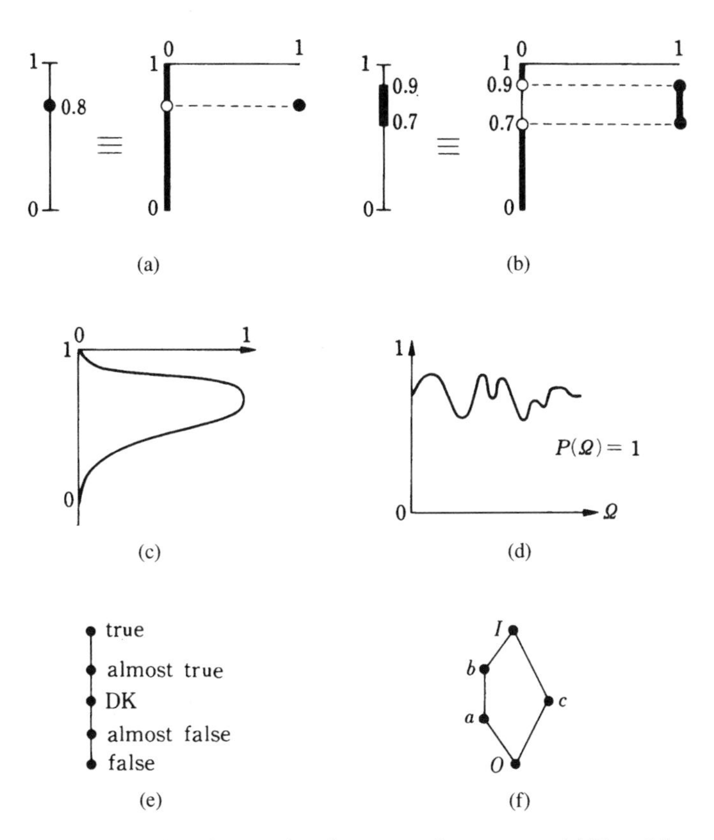

Fig. 2.13 Evaluation values from various fuzzy extension concepts. (a) Type 1 fuzzy set. (b) Interval-valued fuzzy set. (c) Type 2 fuzzy set. (d) Probabilistic set. (e) Linguistic-valued fuzzy set. (f) Lattice-valued fuzzy set.

using the concept of probability. In other words, the value for $m_A(x)$ is 0.8 in a certain case and 0.7 in another. In this way $m_A(x)$ can be thought of as being indicated by a probability distribution on [0, 1] (Fig. 2.13d).

In contrast to the foregoing ideas, the concept of *linguistic-valued fuzzy sets* does not make limiting the evaluation values necessary (Fig. 2.13e). To begin with, the results of human thinking are given in words such as, "There is some disappointment, but in general it's OK." These are not forced into numbers, but rather an attempt is made to use language-based evaluations (*linguistic variables*) just as they are. This idea is good for *knowledge representation* and *knowledge acquisition*, during which the knowledge of on-site experts (people) is converted into data bases, during the construction of expert systems.

If we consider linguistic evaluation from a crisp point of view, it means paying attention to the order structure of the evaluation. This brings in the idea of evaluations being given by various ordered sets, and many types have been proposed, such as *Boolean-valued fuzzy sets*, *lattice-valued fuzzy sets* (*L fuzzy sets*), and *poset-valued fuzzy sets*.

However, from the point of view of applications, the fuzzy sets within mechanical systems are essentially expressed in terms of type 1 fuzzy sets. Even if linguistic expressions are used at the knowledge acquisition level, system designers most often implement systems by interpreting the linguistic information in membership functions with [0, 1] evaluations. This means that in what follows we will essentially be discussing type 1 fuzzy sets with [0, 1] evaluations. The reason we use the expression "essentially" [0, 1] evaluations is that on the hardware level, [0 μA, 5 μA] currents and [0 V, 5 V] voltages are used, and on the software level, integers from 0 to 1,000 and dispersions of the 16 values for 4 bits or the 256 values for 8 bits are often used to speed up calculations.

2.5 FUZZY LOGIC

Just as the basis for crisp set theory is crisp logic, fuzzy logic forms the basis for the various operations mentioned in Section 2.3, in the case of fuzzy sets. In the case of crisp two-valued logic, there were the complete NOT · AND · OR, NAND and NOR systems, and all of the other logical operations could be expressed using these alone. However, since there are an unlimited number of operations in the case of fuzzy logic, we cannot express all operations in terms of a few basic operations. Therefore, we will limit our explanation to the important ones.

First we will discuss the fuzzy extensions that correspond to NOT, AND, and OR. These are called *fuzzy negation*, *t norm*, and *s norm*. Since there are an infinite number of cases with fuzzy systems, it is impossible to give explanations in terms of truth tables, as was the case with two-valued systems. Explanations will be made in terms of functions and a number of axioms, and furthermore, graphs will be used to get an image of the concepts.

Fuzzy negation, which is the fuzzification for NOR, is a monomial operation in which a [0, 1] evaluation is negated in a fuzzy way, resulting in another [0, 1] evaluation. The axioms that are usually given are

$$ⓝ : [0,1] \to [0,1], \tag{2.62}$$

$$\text{(N1)} \quad 0^{ⓝ} = 1, \tag{2.63}$$

$$\text{(N2)} \quad (x^{ⓝ})^{ⓝ} = x \qquad \text{for } ^\forall x \in [0,1], \tag{2.64}$$

$$\text{(N3)} \quad x_1 < x_2 \to x_1^{ⓝ} > x_2^{ⓝ}. \tag{2.65}$$

Here, (N1) is to preserve the two-valued NOT properties, so that fuzzy negation of 0 is 1; in other words, it is a boundary condition. Next, (N2) is the double negation rule that says that if two negations are performed in sequence, the original evaluation is reproduced. This is a very important requirement, and if this is eliminated a variety of good properties will fail to arise, so at present it is commonly one of the axioms for fuzzy negation. The final one, (N3), is an essential requirement for the concept of "negation." In other words, it is the requirement that fuzzy negation reverses (meaning no equality) the order of evaluations (or their goodness and badness). This can be interpreted as meaning that a graph with x plotted along the horizontal and $x^{ⓝ}$ along the vertical will give a narrowly defined monotonically decreasing function.

Everything that satisfies (N1), (N2), and (N3) is a fuzzy negation. A representative fuzzy negation is "variance from 1." In other words, we have

$$x^{ⓝ} = 1 - x \tag{2.66}$$

and this is the fuzzy operation that corresponds to the concept of a fuzzy complement shown in Eq. (2.43). We can easily confirm that this equation satisfies the three axioms in (2.66).

For (N1):

$$0^{ⓝ} = 1 - 0 = 1. \tag{2.67}$$

For (N2):

$$(x^{ⓝ})^{ⓝ} = 1 - x^{ⓝ} = 1 - 1(1 - x) = x. \tag{2.68}$$

For (N3), $1 - x$ is a strictly monotonic decreasing function.

Other properties of Equation (2.66) include the fact that when $x = 0.5$ ($=$ don't know), there is no change and $x^{ⓝ} = 0.5$. We can see that in general, x and $x^{ⓝ}$ have symmetrical values centered on 0.5. From the preceding we can see that the difference-from-1 operation is a representative fuzzy negation, and it would not be going too far to say that it is the only fuzzy negation that is used in industrial applications.

However, let us enumerate a few other facts that derive from (N1)–(N3) in order to gain a greater understanding of the meaning of these axioms:

(1)

$$1^{ⓝ} = 0. \tag{2.69}$$

This can be obtained immediately by using (N1) when $x = 0$ in (N2). Therefore, Eq. (2.69) can be employed in place of (N1); or, even though it is redundant, an equivalent axiom system is created even if both (2.63) and (2.69) are employed for (N1).

(2) A graph of $x^{ⓝ}$ with x on the horizontal and $x^{ⓝ}$ on the vertical has axial symmetry across a 45° line going through the origin. This is made clear by the fact that if the point $(x, x^{ⓝ})$ exists, $(x^{ⓝ}, x^{ⓝ ⓝ}) = (x^{ⓝ}, x)$ also exists, and they are axially symmetrical to each other across a 45° straight line.

(3) The $x^{ⓝ}$ graph is a continuous strictly monotonically decreasing graph. But since the proof of the continuity of $x^{ⓝ}$ is somewhat difficult, we will leave it out here.

We can see from the foregoing that an infinite number of fuzzy negations generally exists, as shown in Fig. 2.14. At the risk of being repetitive, the most commonly used of these is the line in Eq. (2.66). In addition, a variety of fuzzy logic circuits that perform this operation and are among the basic devices for fuzzy computers have been developed. For convenience, they are shown by the symbol at the top of Fig. 2.14.

Next we come to the fuzzy extension of AND, the *t norm* or *triangular norm*. The t norm itself, along with the s norm discussed next, is a concept developed by mathematicians, but when fuzzy logic is under discussion, it is convenient, so it is often employed. A t norm circuit has two inputs and

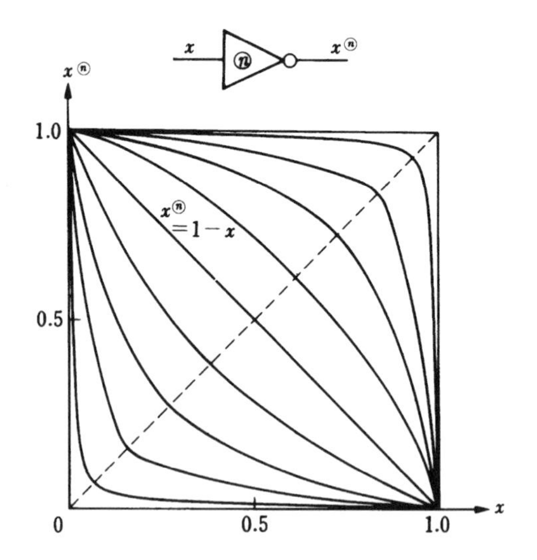

Fig. 2.14 Various fuzzy negations.

one output, and the function has two variables. There are four axioms:

$$\textcircled{t}:[0,1]\times[0,1]\rightarrow[0,1], \tag{2.70}$$

$$\text{(T1)}\quad x\textcircled{t}1=x,\, x\textcircled{t}0=0 \qquad \text{for } ^{\forall}x\in[0,1] \text{ is left out}, \tag{2.71}$$

$$\text{(T2)}\quad x_1\textcircled{t}x_2=x_2\textcircled{t}x_1, \tag{2.72}$$

$$\text{(T3)}\quad x_1\textcircled{t}\left(x_2\textcircled{t}x_3\right)=\left(x_1\textcircled{t}x_2\right)\textcircled{t}x_2, \tag{2.73}$$

$$\text{(T4)}\quad x_1\leqq x_2\rightarrow x_1\textcircled{t}x_3\leqq x_2\textcircled{t}x_{33}. \tag{2.74}$$

(T1) is a boundary condition and arose with the crisp AND. (T2) and (T3) are the commutative and associative laws; the former can be interpreted on the hardware level as meaning that the input pins are the same and there is no need to distinguish between them, and the latter as meaning that when a construction with three or more inputs is used with the two-input devices, the order of combination does not make any difference. (T4) requires the preservation of order, and it guarantees that the evaluation order cannot be reversed at the third evaluation.

The *logical product* produced by a min operation is a representative t norm operation.

$$x_1\textcircled{t}x_2=x_1\wedge x_2 \tag{2.75}$$

and this corresponds to the concept of the intersection of fuzzy sets in Eq. (2.18). It is almost completely self-evident from this that Eq. (2.57) satisfies (T1)–(T4), so we will not go into details here. If this is expressed graphically, we get Fig. 2.15.

Let us consider the meaning of the geometry when this is expressed graphically as in Fig. 2.15. From (T1) and (T2), the domain of $x_1\textcircled{t}x_2$, values within a simple square on the (x_1, x_2) plane, is fixed. From (T1) we can see that we have the line $x_1\textcircled{t}x_2=x_1$ at the $x_2=1$ boundary and the line in the $x_1\textcircled{t}x_2=0$ plane at the $x_2=0$ boundary, and in addition if we use the symmetry of (T2), the line $x_1\textcircled{t}x_2=x_2$ at the $x_1=1$ boundary and the line in the $x_1\textcircled{t}x_2=0$ plane at the $x_1=0$ boundary. The four vertices of the $x_1\textcircled{t}x_2$ square are the values for the crisp AND operation (Fig. 2.16). In addition, from (T2) we can see that the graph shows symmetry across the plane made by the $x_1=x_2$ diagonal line.

What other kinds of operations can this t norm produce? The most important ones for practical applications are the *algebraic product* $x_1\cdot x_2$,

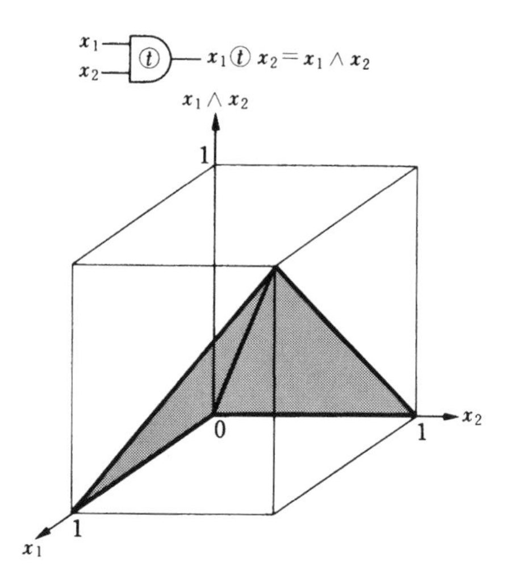

Fig. 2.15 Logical product ($x_1 \wedge x_2$).

the *bounded product* $x_1 \odot x_2$, and the *drastic product* $x_1 \mathbin{\dot\wedge} x_2$:

$$x_1 \cdot x_2 = x_1 x_2, \tag{2.76}$$

$$x_1 \odot x_2 = (x_1 + x_2 - 1) \vee 0, \tag{2.77}$$

$$x_1 \mathbin{\dot\wedge} x_2 = \begin{cases} x_2 & x_1 = 1 \\ x_1 & x_2 = 1 \\ 0 & \text{otherwise}. \end{cases} \tag{2.78}$$

The fuzzy set operations that correspond to these are given in Eqs. (2.48)–(2.50). In addition, the graphs look like those in Fig. 2.17a, b and c.

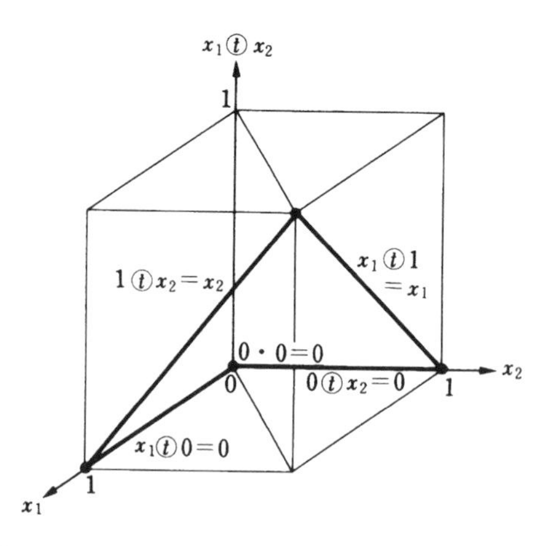

Fig. 2.16 Boundary conditions for a general t norm.

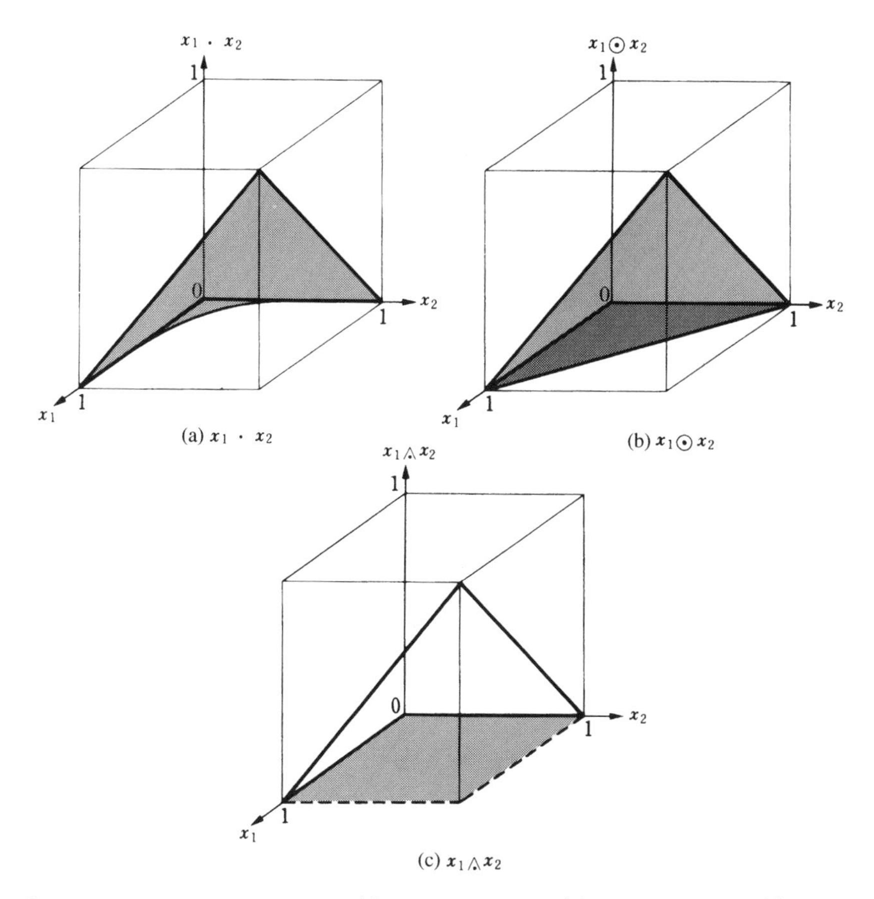

Fig. 2.17 Frequently used t norms. (a) Algebraic product. (b) Bounded product. (c) Drastic product.

It can also be proved from these graphs that

$$0 \leqq x_1 \mathbin{\dot\wedge} x_2 \leqq x_1 \odot x_2 \leqq x_1 \cdot x_2 \leqq x_1 \wedge x_2 \tag{2.79}$$

arises.

The t norm can produce an infinite number of other operations. For example, Yager, Frank, Weber, and Schweizer, and others have each suggested t norm equations that give an infinite number of t norms using a single real number parameter. However, no matter how t norms are produced (satisfying the four axioms), they can always be placed in order between the drastic product and the logical product.

The *s norm*, which is the extension of OR, is also called the *t conorm*, and it is discussed relative to the t norm. Only the boundary condition axiom differs from the t norm, and the rest are exactly the same:

$$ⓢ : [0,1] \times [0,1] \to [0,1], \tag{2.80}$$

$$\text{(S1)}\quad x ⓢ 1 = 1, \quad x ⓢ 0 = x, \tag{2.81}$$

$$\text{(S2)}\quad x_1 ⓢ x_2 = x_2 ⓢ x_1, \tag{2.82}$$

$$\text{(S3)}\quad x_1 ⓢ (x_2 ⓢ x_3) = (x_1 ⓢ x_2) ⓢ x_3, \tag{2.83}$$

$$\text{(S4)}\quad x_1 \leqq x_2 \to x_1 ⓢ x_3 \leqq x_2 ⓢ x_3. \tag{2.84}$$

A representative s norm is the *logical sum* produced by a max operation,

$$x_1 ⓢ x_2 = x_1 \vee x_2, \tag{2.85}$$

and others include the *algebraic sum* $x_1 \dot{+} x_2$, the *bounded sum* $x_1 \oplus x_2$ and the drastic sum $x_1 \mathbin{\dot{\vee}} x_2$:

$$x_1 \dot{+} x_2 = x_1 + x_2 - x_1 x_2, \tag{2.86}$$

$$x_1 \oplus x_2 = (x_1 + x_2) \wedge 1, \tag{2.87}$$

$$x_1 \mathbin{\dot{\vee}} x_2 = \begin{cases} x_2 & x_1 = 0 \\ x_1 & x_2 = 0 \\ 1 & \text{otherwise.} \end{cases} \tag{2.88}$$

The properties of these are given in Fig. 2.18a–d. As is obvious from these figures,

$$x_1 \vee x_2 \leqq x_1 \dot{+} x_2 \leqq x_1 \oplus x_2 \leqq x_1 \mathbin{\dot{\vee}} x_2 \leqq 1, \tag{2.89}$$

so the order is the reverse of the t norm.

As with the t norm shown in Fig. 2.16, a general s norm must meet the boundary conditions shown in Fig. 2.19. In addition, it can be shown that the smallest s norm is the logical sum and the largest s norm the drastic sum.

In this way, various fuzzy negations, t norms and s norms have been proposed, but it is convenient to choose to employ the ones that meet the following conditions:

$$(x_1 ⓢ x_2)^{ⓝ} = x_1^{ⓝ} ⓣ x_2^{ⓝ}, \tag{2.90}$$

$$(x_1 ⓣ x_2)^{ⓝ} = x_1^{ⓝ} ⓢ x_2^{ⓝ}, \tag{2.91}$$

These correspond to de Morgan's laws for crisp operations and are called *fuzzy de Morgan's laws*. Since we can derive one from another with these

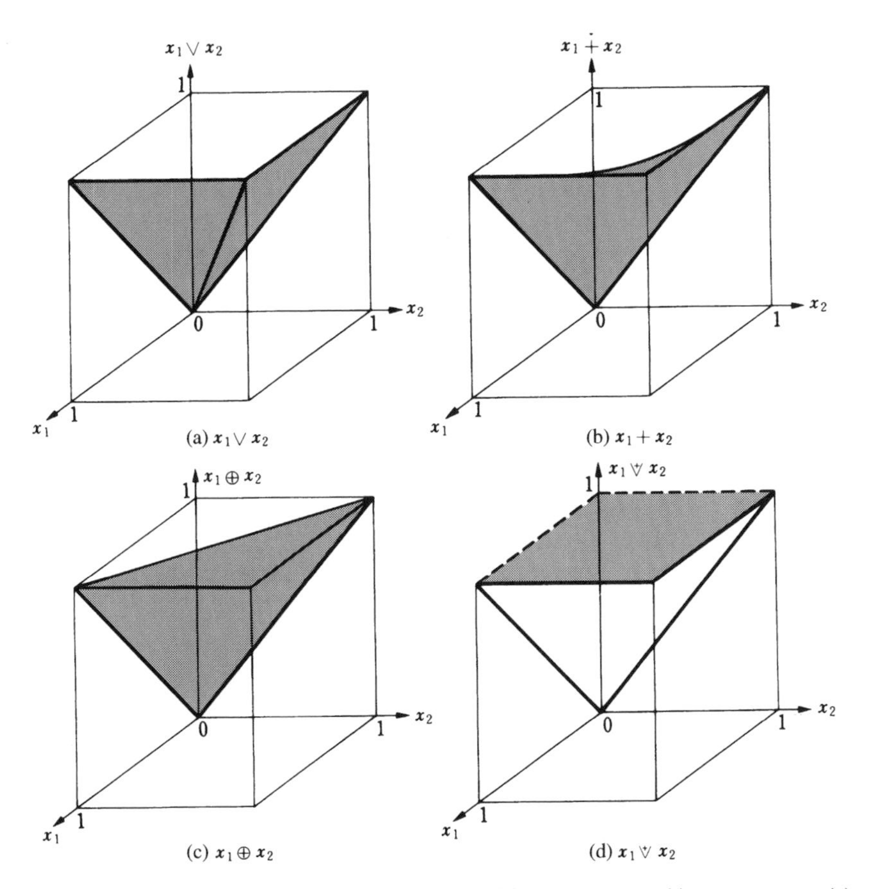

Fig. 2.18 Representative s norms. (a) Logical sum. (b) Algebraic sum. (c) Bounded sum. (d) Drastic sum.

two equations, it is enough to show one side, if we use the axioms for fuzzy negation, t norms, and s norms. When Eq. (2.90) or Eq. (2.91) (therefore both) arises, the t norm and the s norm are *dual* in reference to the fuzzy negation. It can be shown that the logical product and logical sum, algebraic product and sum, bounded product and sum, and drastic product and sum all show duality for the variance from 1 of the fuzzy negation. In practical applications, the logical pair is standard, and the algebraic and bounded pairs are used at times. The drastic pair has the property of being discontinuous and is important in terms of the lower bound for t norms and the upper bound for s norms, but it is not often used in practical applications.

The reason that the logical pair is the standard is that it has an explicit physical meaning and that the $([0,1], \leqq, 1-, \wedge, \vee)$ system gives com-

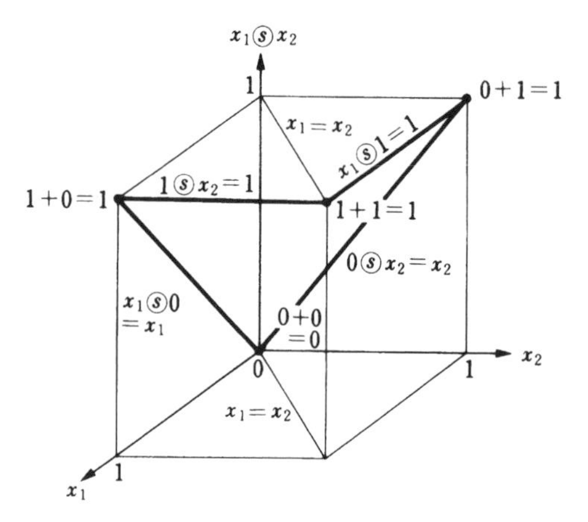

Fig. 2.19 General s norm boundary conditions.

plete pseudo-Boolean algebra, and thus good mathematical characteristics. Only complements do not arise:

$$x \wedge (1 - x) \geqq 0, \qquad x \vee (1 - x) \leqq 1, \tag{2.92}$$

and all the other properties that arise with crisp logic arise just the same. (For example, if we employ the algebraic product as a t norm, the idempotent law does not arise, so we have nothing but a lattice.) With crisp logic, the equality of Eq. (2.92) arises, but the complements are called the laws of contradiction and exclusion. The law of contradictions—that is, that no property and its negation can exist at the same time—and the exclusion law—that is, that both the property and its negation exist with no ambiguous intermediates—are properties peculiar to the crisp world of two-valued evaluations. It can be considered natural that they do not arise in fuzzy logic, which takes into account ambiguous intermediates.

Fuzzy implications are also important in fuzzy logic, but we will take them up under the subject of fuzzy reasoning.

2.6 FUZZY REASONING

Fuzzy inference or *fuzzy reasoning*, often called *approximate reasoning* recently, is the most important technique for practical applications of fuzzy theory. Since all *reasoning* or *inference* for crisp AI is included as special cases of fuzzy reasoning, the *expert systems* constructed using crisp AI techniques can be viewed as special cases of *fuzzy expert systems*. At the research level for fuzzy expert systems, studies investigating the extension to fuzzy reasoning of almost all of the crisp reasoning techniques that have been developed, such as rule- and frame-based systems, have been done. However, among all the real systems that actively use ambiguous evaluations in industrial applications, at present almost all use rule-based systems that employ *fuzzy production rules* or relational systems that employ *fuzzy relations*. Both of these can be explained theoretically using *compositional rules of fuzzy inference*, but since actual engineering algorithms appear very different at first, we will divide up our explanation of them.

We will explain rule-based fuzzy inference using the following very simple example. Let us say that there is on-site expert knowledge for "when the water level is high, the valve must be opened." This is expressed in an "if. . .then" form in what is called a *fuzzy production rule*:

$$\text{If the water level is high then the valve must be opened.} \tag{2.93}$$

Here the "if" clause is called the *antecedent*, *assumption*, or *condition part*, and the "then" clause is called the *conclusion* or *operation part*. The important thing here is that the descriptions in the antecedent and the conclusion are *fuzzy propositions*. In other words, direct expressions such as "if the water level is x meters" or "the valve is opened y degrees" are not present. However, the on-site experts should have a rough idea of this. For example, when asked "At what level do you judge the water to be high?" they might reply "about 2 meters." In this instance, an interpretation using a fuzzy set like

$$\begin{aligned}\text{HIGH} &= 0.1/1.5\text{ m} + 0.3/1.6\text{ m} + 0.7/1.7\text{ m} + 0.8/1.8\text{ m}\\ &\quad + 0.9/1.9\text{ m} + 1.0/2.0\text{ m} + 1.0/2.1\text{ m} + 1.0/2.2\text{ m}\end{aligned} \tag{2.94}$$

is obviously much easier for the experts to actually get used to than a crisp interpretation such as "1.9 m is not high, but 2 m is." In the same way, if for example we let a turn of 90° open the valve completely, the angle of rotation of the valve can be described by this membership function:

$$\begin{aligned}\text{OPEN} &= 0.1/30^\circ + 0.2/40^\circ + 0.3/50^\circ + 0.5/60^\circ\\ &\quad + 0.8/70^\circ + 1.0/80^\circ + 1.0/90^\circ\end{aligned} \tag{2.95}$$

The people that construct the system produce the membership functions in (2.94) and (2.95) from the linguistic expression rules in (2.93). Basically these are determined by the following method:

(1) Determination by question and answer (Q & A) or by apprenticeship.
(2) On-site experts actually perform the operation, and the system is constructed by monitoring the conditions.
(3) The function values are corrected to obtain good results from simulations.

If the membership function is determined through these operations, it can be recorded in computer memory as a knowledge base. For example, Eqs. (2.94) and (2.95) can be recorded as one-dimensional array information that makes each element of the total space an index. For simplicity, we will say that there is only one fuzzy production rule for Eqs. (2.93) stored in the knowledge base. Under these circumstances, we observe the present water level, and our observation is

$$\text{Present water level is rather high.} \tag{2.96}$$

If for the present we can make an accurate enough observation of the water level, we can obtain crisp information such as "the present water level is 1.7 m." However, if the characteristics of this system are such that we cannot obtain good information with crisp accuracy no matter what we do—that is, if the expected observational error overlaps the observed 1.7 m or if there is no way to set up a water level measurement system (for example, as is the actual case in many places at present, if the water level has to be guessed by tapping on the upper part of the tank and guessing the level from the sound)—in such cases, we can get observational information by putting (2.96) into the form of a fuzzy set such as

$$\text{rather HIGH} = 0.5/1.6\ \text{m} + 1.0/1.7\ \text{m} + 0.8/1.8\ \text{m} + 0.2/1.9\ \text{m} \tag{2.97}$$

How should the valve be operated in these circumstances? In other words, we get a problem setup like

$$\begin{array}{l} \text{If HIGH then OPEN} \\ \text{rather HIGH} \\ \hline \hfill ? \end{array} \tag{2.98}$$

in which we want to know what to put down for the "?" mark. It is obvious that this means matching the antecedent HIGH with the observation "rather HIGH." With crisp logic there would be no match, so there can be no inference. However, human beings can perform *approximate matching*,

$$\begin{array}{l} \text{If HIGH then OPEN} \\ \text{rather HIGH} \\ \hline \hfill \text{a little OPEN} \end{array} \tag{2.99}$$

and perform the operation of opening the valve a little. This is the way fuzzy reasoning works. (This approximation process accounts for the development of the term "approximate reasoning" in English, but in Japanese it is better to use "fuzzy reasoning.")

If we are dealing with human thought, that is, with explanations on the linguistic level, (2.99) is a typical example of fuzzy reasoning, but how are the calculation processes for the membership functions that come in here performed in programs and fuzzy reasoning chips? In one report it has been said that there are more than 100 methods for converting this linguistic-level fuzzy reasoning into calculations, but if we limit ourselves to the methods most frequently used in applications, we can explain them as shown in Fig. 2.20. The total space of the antecedent is water level, and the total space of the conclusion is the angle of valve operation. If we let these be X and Y, respectively, we can depict the fuzzy production rule for (2.93) graphically as shown in the top row of Fig. 2.20, using (2.94) and (2.95). (These are drawn with interpolated curves between data points.) In addition, we have simplified the expressions by representing the fuzzy set HIGH on antecedent X and A and the fuzzy set OPEN on conclusion Y as B. From Eq. (2.97) we know that the "rather HIGH" fuzzy set, which represents the observed data, on X (written A') can be described graphically as in the figure in the second row of Fig. 2.20. The third row in the same figure shows a typical fuzzy inference process. The figure on the left determines the matching for the antecedent of the rule A and the observed data A' through $A \cap A'$. We then take the maximum value of $A \cap A'$, α, to be the degree of matching, and the conclusion of the rule B is reduced by degree of matching α, giving us the inference result B' (the right-hand figure on the third row). This subtraction method means taking degree of matching α off the top of B. (Here αY means

$$m_{\alpha Y}(y) = \alpha \qquad \text{for } {}^{\forall}y \in Y.) \tag{2.100}$$

In this way we obtained B' (= a little OPEN) from the current observational data A' (= rather HIGH) using $A \to B$ (= If HIGH then OPEN). Inference result B' is a fuzzy set on Y like that shown in the lower right-hand position in Fig. 2.20. However, no work can be done with things as they are. In other words, we must reduce this operation value to a single point based on the characteristic function of B', $m_{B'}$. This process is called *defuzzification.* In Fig. 2.20, the *center of gravity method* (*CG method*) is employed for defuzzification, and the operation value CG = 70(°) is finally determined. The valve should be opened 70°.

Now let us consider things in a little more detail using the shorthand just given. First of all, the cause and effect relationship between the antecedent and the conclusion for expert knowledge $A \to B$ is called a *fuzzy relation* and is denoted by R:

$$R = A \to B. \tag{2.101}$$

R can be thought of as a fuzzy set on the *Cartesian product* $X \times Y$ of the total space for the antecedent X and the total space for the conclusion Y. Also, the process for obtaining (fuzzy) inference result B' from observed data A using $A \to B$ is expressed in the form

$$B' = A' \circ R = A' \circ (A \to B). \tag{2.102}$$

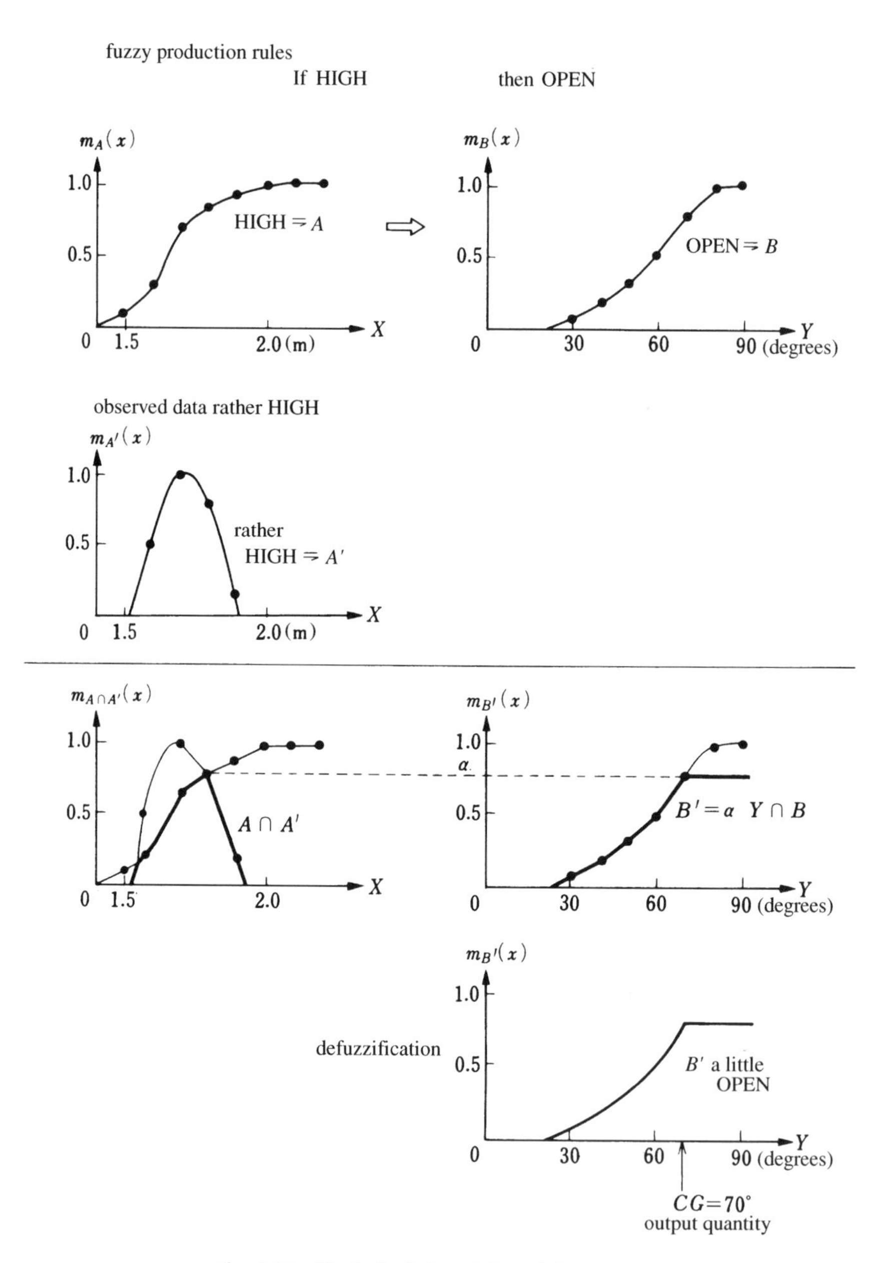

Fig. 2.20 Typical rule-based fuzzy inference.

Here $\circ$ is called the *compositional rule of fuzzy inference*. In addition, the $\rightarrow$ in $A \rightarrow B$ is known as a *fuzzy implication*. What happens if we write the preceding expressions using membership function labels? The fuzzy inference in Fig. 2.20 actually has a max–min composition for the compositional rule of fuzzy inference, and fuzzy implications can only be performed using min operations. In other words, we get

$$m_{B'} = \bigvee_{x \in X} (m_{A'}(x) \wedge m_R(x, y)) \tag{2.103}$$

$$= \bigvee_{x \in X} (m_{A'}(x) \wedge (m_A(x) \wedge m_B(y))) \tag{2.104}$$

$$= \left(\bigvee_{x \in X} (m_{A'}(x) \wedge m_A(x)) \right) \wedge m_B(y) \tag{2.105}$$

$$= \bigvee_{x \in X} m_{A' \cap A}(x) \wedge m_B(y) \tag{2.106}$$

$$= \alpha \wedge m_B(y) \tag{2.107}$$

$$= m_{\alpha Y \cap B}(y) \tag{2.108}$$

$$CG = \int_Y y \cdot m_{B'}(y)\, dy \Big/ \int_Y m_{B'}(y)\, dy \tag{2.109}$$

(This is not really important, but the conversion of Eq. (2.104) into Eq. (2.105) employs the generalized distribution rule for max–min operations.) In this way, it can also be said (in the author's [K. Hirota's] opinion) that the max–min composition center of gravity method is a representative fuzzy inference method.

Given the explanation up to now, well-informed readers will be able to think of various alternative forms for Fig. 2.20. For example, the *median method*, which employs the *median*, or the *height method*, which makes use of the y that gives the maximum membership value, could be used instead of the center of gravity method for defuzzification. The α reduction for B, αB, could be used for B' instead of cutting off the top by α, $\alpha Y \cap B$. We could enumerate more than 100 versions that have been invented. At the risk of repeating, we can say that among all of these, experience tells us that best one is the max–min center of gravity method in Fig. 2.20, and at present almost all the VLSI chips for fuzzy inference and other parts employ this system.

Now let us explain fuzzy implications from the standpoint of fuzzy logic. This is the basis for expressing the fuzzy relations in Eq. (2.101), and it is extremely important. If the elements of the total space are fixed and we confine our discussion to evaluations within [0, 1], *fuzzy implications* are two-variable functions or two-item relations of [0, 1]:

$$\begin{array}{ccc} \rightarrow : [0,1] \times [0,1] & \rightarrow & [0,1] \\ \Uparrow & & \Uparrow \\ (x_1, x_2) & \mapsto & x_1, x_2 \end{array}. \tag{2.110}$$

Using the crisp case as a model, implications for Boolean algebra are given in Table 2.1. Using Eq. (2.33) we know that $x_1 \rightarrow x_2$ is described by $\bar{x}_1 + x_2$ in crisp cases. Therefore, if we replace NOT with "variance from 1" and OR with max, the most standard fuzzy logic operations, we get

$$x_1 \rightarrow x_2 = (1 - x_1) \vee x_2. \tag{2.111}$$

In the case of fuzzy inference, if we try making the following substitution with Eq. (2.104), we see that we do not obtain very good results:

$$\begin{aligned} m_R(x, y) &= m_{A \rightarrow B}(x, y) \\ &= (1 - m_A(x)) \vee m_B(y). \end{aligned} \tag{2.112}$$

In order to look into the meaning of Eq. (2.111), let us graph the t norm and the s norm in the same way, as in Fig. 2.21a. The four crisp points (black dots in the figure) are preserved, and the figure is composed of two triangular planes. However, given just the coordinate axis and the four crisp points, many readers might more readily imagine Fig. 2.21b for the same graph of two triangular planes, if asked to interpolate. If we express

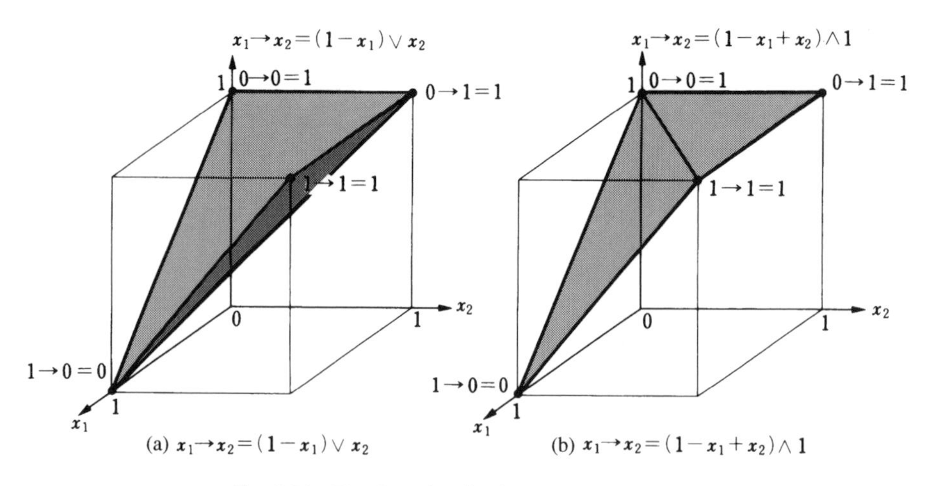

(a) $x_1 \rightarrow x_2 = (1 - x_1) \vee x_2$ (b) $x_1 \rightarrow x_2 = (1 - x_1 + x_2) \wedge 1$

Fig. 2.21 Two fuzzy implication operations $x_1 \rightarrow x_2$.

Fig. 2.21b as an equation, we get

$$x_1 \to x_2 = (1 - x_1 + x_2) \wedge 1. \qquad (2.113)$$

This is the limited sum operation for fuzzy negation of the variance from 1 of x_1 and x_2, and it is an equation known as the *Lukasiewicz implication* in *multiple logic*. In the early stages, Zadeh explicated a fuzzy inference that used this operation and made the following substitution with Eq. (2.103):

$$\begin{aligned} m_R(x, y) &= m_{A \to B}(x, y) \\ &= (1 - m_A(x) + m_B(y)) \wedge 1. \end{aligned} \qquad (2.114)$$

If actual tests are performed with primary delay and other control systems, very good results are obtained. However, Mamdani, who gave the perspective for actually controlling the automatic operation of a steam engine, obtained excellent results by employing the max–min composition in Eqs. (2.104)–(2.109).

$$x_1 \to x_2 = x_1 \wedge x_2 \qquad (2.115)$$

In this case, the fuzzy implication is a min operation, and the properties are those of the logical product t norm shown in Fig. 2.15. Yet the crisp implication table, 2.1, is not preserved. In other words, in Table 2.1, when antecedent x_1 is false = 0, the crisp implication is 1, but with the min we get 0. With min operation fuzzy implications, the idea is that answers are not given when the antecedent breaks down.

There are many other concrete proposals for fuzzy implication operations, but min operations are most often used, because experience has proven that they give good results. In addition, since the crisp implication operations are not preserved, an axiomatically organized system has not been set down for fuzzy implications, as in the case of t norms.

We have already explained the basics of fuzzy inference using fuzzy production rules, but a small supplementary explanation is necessary on the level of practical applications. In the first place, there are usually a number of rules housed in a knowledge base. In addition, the number of items in the fuzzy propositions that make up the descriptions in the antecedents and conclusions is usually more than one. Therefore, we generally consider knowledge bases in the form of

$$\begin{aligned} \{\text{If } A_{i_1} = \bigcirc, A_{i_2} = \bigcirc, \ldots, A_{i_m} = \bigcirc, \text{ then} \\ B_{i_1} = \triangle, B_{i_2} = \triangle, \ldots, B_{i_n} = \triangle\}_{i=1}^{I} \end{aligned} \qquad (2.116)$$

Here I is the number of rules, m the number of items in the antecedent and n the number of items in the conclusion. In actual applications, I is most often around 30 (up to around 100), and the ratio of m to n is around 2:1 ranging up to 5:2. In crisp rule-type expert systems, there are usually several hundred to several thousand rules and in large-scale systems more than 10 thousand. In comparison, fuzzy systems have fewer rules by a factor of 10. This is not because crisp systems break things down minutely and generate rules accordingly, but because the whole is expressed through similarities using a small number of essential rules. We often hear the opinion that "When compared with the number of crisp rules, fuzzy rules allow a reduction in the number of rules by factors of 10 to 100." The small number of rules is tied to things such as ease of knowledge acquisition from experts, ease of overall system adjustment and the building of cost-effective systems, and it is one of the major factors spurring practical applications of fuzzy systems.

When operations are actually performed in the case of the standard max–min composition center of gravity calculation method, the conclusion for each rule is found using a synopsis in which B' in Fig. 2.20 is calculated, and a method that uses the fuzzy union of these as the final result of inference is often employed. In other words, it takes the form of a *parallel firing system*, in which all of the rules are used simultaneously.

Let us get a better understanding of the situation here by taking a look at a simple example. Let there be three rules with one item in the antecedent and one item in the conclusion of each rule, given as follows:

$$\begin{aligned} &R_1\text{: IF } A \text{ is } R \text{ then } B \text{ is } L,\\ &R_2\text{: IF } A \text{ is } C \text{ then } B \text{ is } C,\\ &R_3\text{: IF } A \text{ is } L \text{ then } B \text{ is } R. \end{aligned} \tag{2.117}$$

Here, the membership functions for fuzzy labels R, C, and L are given as in Fig. 2.22. In other words, R, C, and L are right, center, and left, respectively, and Eq. (2.117) is a reversing rule of the "if right then left"

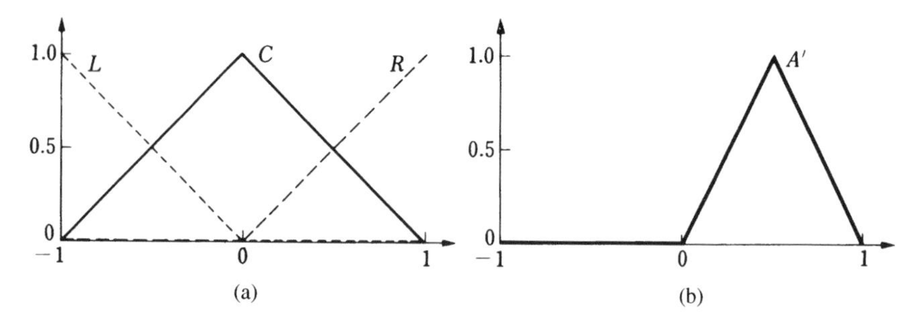

Fig. 2.22 Basic data for rule-based fuzzy inference problems. (a) Membership functions for fuzzy labels L, C, and R. (b) Membership function for observed data A'.

type. If A' in Fig. 2.22b is input as observed data, what kind of answer would we get from a max–min composition center of gravity calculation? Since the A' observational information is "somewhat right," having a peak at 0.5 on the right side, we can expect an answer of "a little left" from the reversing rule. In actual fact, we obtain a result in which the operation value CG is a little to the left, as in Fig. 2.23.

In standard control processes there is a repeated process of implementation in which A' is observed on a given schedule, the operation value CG is determined as in Fig. 2.23, and the control operation is performed. In the next time frame a new A' is observed, the operation values determined once again as in Fig. 2.23, and the control operation is performed. When a high-speed response is required of the system, the time for performing the

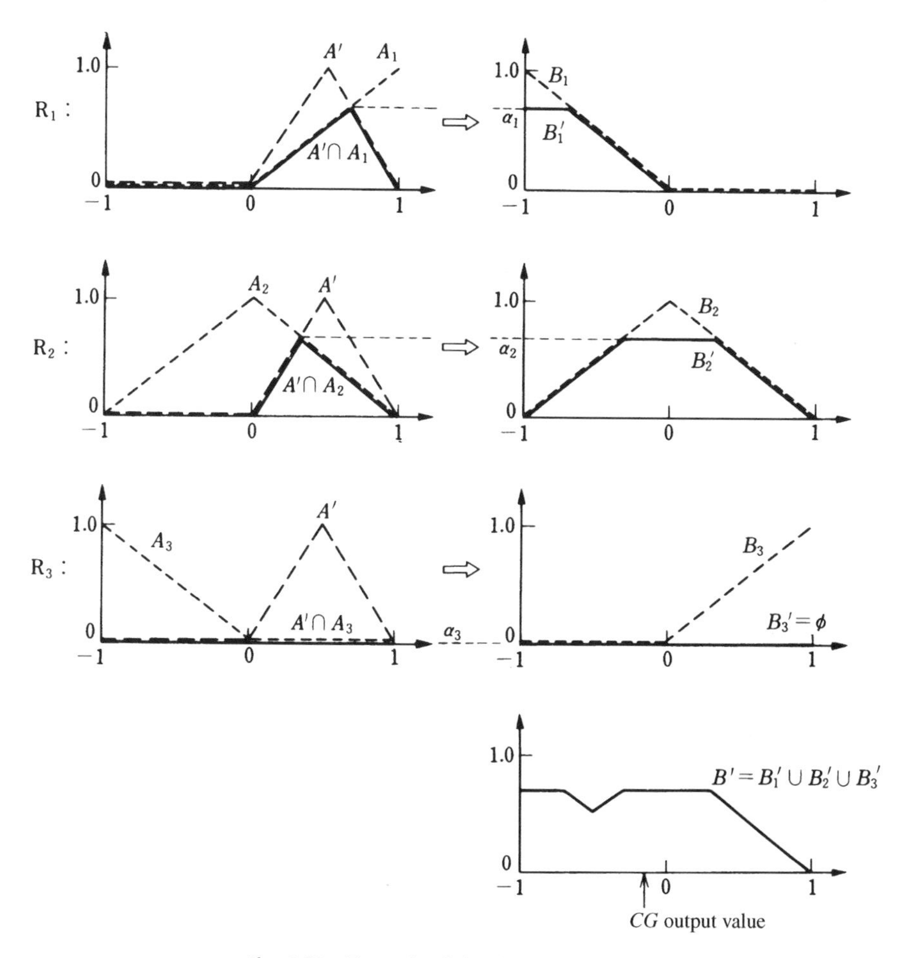

Fig. 2.23 Example of simple fuzzy inference.

single fuzzy inference after A' has been input must be short. The units most often used for processing speed are FLIPS (fuzzy logical inference per second). This depends on the number of rules and items and on the Cardinal number of the total spaces, but normally a program using a single-chip microprocessor can work with quite a large system. For example, the fuzzy control for the Sendai subway system performs 10 FLIPS with an 8-bit processor and 24 rules, bringing about comfortable operation in which the condition of the train three seconds later is predicted every 100 ms. As can be seen from this one example, applications of fuzzy control technology for reduction of labor are proceeding at a rapid pace. In addition, in cases in which even more rapid responses are required, the making of fuzzy inference hardware for this can be considered, and already several types are coming onto the market. At present the greatest speed possible is about 40 million FLIPS (completion of one fuzzy inference in the time it takes for light to move 7.5 m in free space). Investigations of applications of this kind of *fuzzy inference chip* in things such as robots and aircraft and rocket control are being done.

The fuzzy reasoning that has been discussed so far has been inference from the antecedent side to the conclusion side or *forward reasoning*, and it is the most frequently used in practical applications. However, since *backward reasoning* has also come into frequent use, especially in diagnostic-type fuzzy expert systems, we will discuss it. This is a method for modeling using *fuzzy relational equations*.

Let us explain diagnostic systems. The total space X for the antecedents is made up of m causes, and the total space for conclusions is made up of n systems:

$$X = \{x_1, x_2, \dots, x_m\}, \tag{2.118}$$

$$Y = \{y_1, y_2, \dots, y_n\}. \tag{2.119}$$

For example, in a simplified automotive fault diagnosis, let $m = 2$ and $n = 3$; we can consider these items: x_1: battery deterioration, x_2: engine oil deterioration, y_1: poor starting, y_2: extent of bad coloration of exhaust, and y_3: degree of insufficiency of power. Therefore, a cause and effect relation exists between the antecedent items x_i and the conclusion items y_j. This cause-and-effect relation $x_i \to y_j$ is simply written r_{ij}, and it is known as a *fuzzy relation* of x_i and y_j. If we bring together all of the fuzzy relations for x_i and y_j, we obtain matrix R with m rows and n columns, and this is called a *fuzzy relational matrix*:

$$R = [r_{ij}]_{i=1\sim m,\, j=1\sim n}. \tag{2.120}$$

With (type 1) fuzzy systems, the grade for the cause and effect relation for each r_{ij} is expressed as a $[0, 1]$ real number value. In addition, if we think of the antecedents as the *inputs* and the conclusion as *outputs* in the foregoing circumstances, we get something in the form of the *fuzzy system* in Fig. 2.24. The actual antecedents (inputs) and conclusions (outputs) can be taken to be fuzzy sets A and B on X and Y, respectively. If we write the relation

$$B = A \circ R, \tag{2.121}$$

we have the same formulation as for rule-type systems. Here $\circ$ is the compositional rule of fuzzy inference. However, when this is actually used, the reasoning is in the opposite direction from that of rule-type systems. In other words, a method in which R is identified using the knowledge of experts, the output B (or symptoms) observed and the input A (or causes) found is used in diagnostic systems. In the preceding example of automotive fault analysis, let the knowledge of the repair technician be identified as

$$R = \begin{bmatrix} 0.9 & 0.1 & 0.2 \\ 0.6 & 0.5 & 0.5 \end{bmatrix}. \tag{2.122}$$

A car is brought in and after testing it is found that it starts poorly, but the power and exhaust are fine, so the judgment is

$$B = 0.9/y_1 + 0.1/y_2 + 0.2y_3. \tag{2.123}$$

This means we want to find the cause,

$$A = a_1/x_1 + a_2/x_2. \tag{2.124}$$

In a case like this, Eqs. (2.123) and (2.124) are often expressed as *fuzzy row vectors*:

$$B = [0.9, 0.1, 0.2], \tag{2.125}$$

$$A = [a_1, a_2]. \tag{2.126}$$

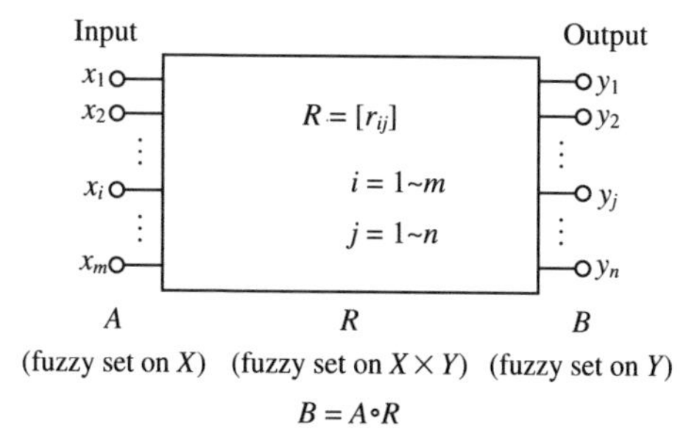

Fig. 2.24 Modeling by means of fuzzy systems.

Thus, Eq. (2.121) can be expressed as

$$[0.9, 0.1, 0.2] = [a_1, a_2] \circ \begin{bmatrix} 0.9 & 0.1 & 0.2 \\ 0.6 & 0.5 & 0.5 \end{bmatrix}. \tag{2.127}$$

Alternatively, we can transpose this and express it as a *fuzzy column vector*,

$$\begin{bmatrix} 0.9 \\ 0.1 \\ 0.2 \end{bmatrix} = \begin{bmatrix} 0.9 & 0.6 \\ 0.1 & 0.5 \\ 0.2 & 0.5 \end{bmatrix} \circ \begin{bmatrix} a_1 \\ a_2 \end{bmatrix}. \tag{2.128}$$

which may be easier to follow (for readers who have studied matrices or determinants). At this point there are numerous fuzzy compositional rules of inference $\circ$ that have been investigated, but the max–min composition is the standard one. In this case, we get

$$\begin{cases} 0.9 = (0.9 \wedge a_1) \vee (0.6 \wedge a_2), & (2.129) \\ 0.1 = (0.1 \wedge a_1) \vee (0.5 \wedge a_2), & (2.130) \\ 0.2 = (0.2 \wedge a_1) \vee (0.5 \wedge a_2), & (2.131) \end{cases}$$

for Eq. (2.127) or Eq. (2.128). We can think of this as being just like the replacement of addition with the max $\vee$ and multiplication with the min $\wedge$ in modeling that uses simultaneous linear equations. Let us solve this. Because of Eq. (2.129), the two terms on the right-hand side do not matter for the left-hand side, so we get

$$0.9 = 0.9 \wedge a_1, \qquad a_1 \geqq 0.9. \tag{2.132}$$

From this and Eq. (2.130) we obtain

$$0.1 \geqq 0.5 \wedge a_2, \qquad a_2 \leqq 0.1. \tag{2.133}$$

Equations (2.132) and (2.133) satisfy Eq. (2.131). Thus, we get

$$0.9 \geqq a_1 \geqq 1.0, \qquad 0 \leqq a_2 \leqq 0.1. \tag{2.134}$$

as a solution, and changing the battery is sufficient (a_1 is the degree of battery deterioration, and a_2 the degree of engine oil deterioration).

In reality there are usually several to several dozen items for m and n and several types of compositional rules of fuzzy inference employed, and we usually deal with a fuzzy system with two- or three-stage cascade connection. In addition, the solution to this example is determined in terms of spatial values as in Eq. (2.134), and we obtain results with a maximum value of $\{1.0, 0.1\}$ and a minimum value of $\{0.9, 0\}$. In general we can see that with max–min compositions, there is usually a single maximum solution and several minimum solutions. Of course there are instances in which there are no solutions. In comparison with the rule method there are still few samples of actual applications of this method of solving fuzzy relational equations, and applications software for solving them has yet to come out.

As was mentioned in the opening paragraph of this section, many other methods for fuzzy inference, such as frame-based fuzzy inference, have been investigated, and in the near future, there is a good possibility that these will be tied to practical applications, but given the current situation in applications technology, we will leave off any further explanations.

2.7 PROBLEMS FOR PRACTICE

We have laid out the basics of fuzzy theory from the point of view of applications technology, but for those who wish to deepen their understanding and raise their ability to the standards for beginning research into the development of concrete applications, we provide a list of practice problems to the extent that space allows. Since all of the answers can be inferred from the preceding discussion, we will not include them here.

1. Show that set inclusion equation (2.13) satisfies Eqs. (2.14)–(2.16) and forms a POSET.
2. Show Eqs. (2.20)–(2.25) for crisp set operations and confirm that they are Boolean algebra.
3. Confirm that the double negation equation (2.26) and de Morgan's equation (2.27) arise for crisp set operations.
4. Think about why there are 2^{2^n} combinatorial circuits when there are n inputs and one output.
5. Give the proof for (1) on page 17, and give the proofs for (2) and (3) based on this. (Hint: standard forms for the main addition and multiplication).
6. Give the proof for Eq. (2.33) using the truth table.
7. Show that reflexivity, anti-symmetry, transitivity, double negation and the idempotent, commutative, associative, absorption, distributive and de Morgan's laws all arise for the fuzzy set operations in Eqs. (2.42)–(2.45), but that the law of complements does not.
8. Try drawing Fig. 2.12 for yourself.
9. Using the same configuration as in Fig. 2.12, draw graphs for Eqs. (2.54) and (2.55). In addition, graph Eq. (2.56) for $\lambda = 1/2$.

10. Give the proof for Eq. (2.57).

11. Give the proof for (3) on page 30.

12. Give the proof for Eq. (2.79).

13. For positive parameter p, the Yager's t norm is defined as

$$x_1 \textcircled{t} x_2 = 1 - \left(1 \wedge \left((1 - x_1)^p + (1 - x_2)^p\right)^{1/p}\right). \tag{2.135}$$

Show that this satisfies t norm axioms (T1)–(T4).

14. Show that the maximum value for the t norm is the logical product and that the minimum value for the t norm is the drastic product.

15. Give the proof for Eq. (2.89).

16. For positive parameter p, the Yager's s norm is defined as

$$x_1 \textcircled{s} x_2 = 1 \wedge (x_1^p + x_2^p)^{1/p}. \tag{2.136}$$

Show that this satisfies s norm axioms (S1)–(S4).

17. Show that the maximum s norm is the drastic sum and that the minimum s norm is the logical sum.

18. Using the fuzzy negation and the axioms for t norms and s norms, derive Eq. (2.91) from Eq. (2.90), and then in reverse derive Eq. (2.90) from Eq. (2.91).

19. When we let the fuzzy negation be "the difference from 1," show that the logical, algebraic, bounded, drastic, and Yager (Eqs. (2.135) and (2.136)) pairs are reciprocal.

20. Write a program to implement Fig. 2.20.

21. Write a program in which the defuzzification section uses the median method and one in which it uses the height method. In addition, write a program for the α contraction when finding B'.

22. Write a program for fuzzy reasoning using the configuration in Eq. (2.116).

23. Think up a circuit for performing Problem 22.

Chapter 3

Applications in Industry

The most advanced applications of fuzzy sets have been made in industrial areas. The most notable are applications in automatic control and pattern recognition in which there is a great deal of interest and where a large amount of research has been done. It is commonly known that the first research into applications of fuzzy theory was that done by Mamdani on automatic boiler control. It is said that he started out studying adaptive control and learning control, but these were too difficult using current methods, so he took the bold step of employing fuzzy theory. In this chapter we will give a number of actual examples of high-level automation that was difficult using older theories and first made possible using fuzzy control. These are all things that go hand in hand with the larger trend toward making machines intelligent, and they can be thought of as things that will be developed more and more in the future. The things taken up in each section are different, but their aims are the same in this aspect, and they should provide effective suggestions to those researching fuzzy applications.

3.1 OUTLINE

Application of fuzzy theory is most active in industrial areas, and this is connected to the fact that the idea of fuzzy theory came from engineering. Zadeh, the originator of fuzzy sets, was a specialist in control theory. The major industrial areas that can be considered for applications are control, fault analysis, pattern recognition, image processing, safety analysis, system design, and computers, but the applications best known at present are mainly control applications. Recently, however, research into applications in pattern recognition and computers has been flourishing. If we were to say why out of all these areas control was the one in which applications were carried out, it would be because fuzzy theory itself is easily brought into line with the ideas behind control. In addition, control is carried out in most industrial processes and plants, and it seems there was a great need for fuzzy theory and many opportunities for application.

Zadeh has said that one of the major points in the creation of the idea of fuzzy sets was the principle of incompatibility. This comes from the fact that when a system gets extremely large and complicated, it becomes difficult to carry out modeling using equations, as in the past. In other words, the number of variables and parameters grows dramatically, but the determination of each of the parameters and measurement of the variables becomes difficult, so a model with good predictability and operability cannot be built. Instead of this, what Zadeh proposed was linguistic modeling using qualitative words rather than equations. Using words means a loss of precision when compared with mathematical modeling, but it is possible to build a qualitative model for a system with good predictability. This kind of model is actually used in systems other than industrial systems, such as economic and social systems. What becomes a problem here is the ambiguity of the words that make up the descriptive language of the system. But instead of this ambiguity being a defect in this model, Zadeh claimed that it was one of the good points. Zadeh took the parking of one of two automobiles as an example and clearly explained the merits of ambiguous description.

If, following the methods used in standard control theory, this problem is solved by creating an equation for the motion of the car, it is almost impossible to park the car. The tire and surface conditions cannot be accurately assessed, and even if we take for granted the construction of the equation, we cannot determine the parameters. However, students at a driving school can learn to park a car correctly just from what the instructor says, without knowing any equation for describing the motion of the car. What the students learn is something like "While turning the wheel full right, ease forward, turn back to the left, and stop. Next, back up while turning full right, and turn back to the left. If you miss, try again." People can drive cars given this kind of general instruction in words. However, this is impossible for a computer. Computers must be given instructions using deterministic equations. However, as mentioned earlier, an equation for the driving of a car cannot be made. Say we have an equation, and let it be a simplified one with a structure for which the characteristics of the car and its environment are determined beforehand. If we use the equation to control the car, the motion of the car will be jerky, because the reality and the model do not match.

The qualitative processes people can think of to solve problems are called fuzzy algorithms, and the idea behind fuzzy control is having a computer carry out fuzzy algorithms using fuzzy logic. Among all the areas for applications of fuzzy theory, fuzzy control is the one, as mentioned in the opening paragraph, where research started and one that has produced big results—the results being a rise in the level of interest in fuzzy theory in society due to the success of fuzzy control. When we consider applications of fuzzy theory in fields other than control, the reasons for its success are only intensified in the ideas for fuzzy control. No matter what area of application we consider, the methods used for fuzzy control will be instructive. Here we will take a simple look at some methods for fuzzy control and their characteristics.

Normally, fuzzy control is not the successive carrying out of single groups of instructions such as those in fuzzy algorithms, but rather the carrying out of a number of rules in parallel. The use of rules in actual conditions employs a technique from fuzzy theory that is known as fuzzy inference. As an example of a rule and inference, we can consider the following schema.

Rule: Reduce speed if the distance between cars gets small.
Condition: The distance between cars is 20 m.
___
Action: Reduce speed very much.

The "Action" in the schema is the result of fuzzy inference. In actuality, there would be a number of these rules, so each of the inference results would be intensified, brought together into one, and converted into numerical values for inputting to the plant or whatever is the object of the control. Out of all the methods for fuzzy control, the one that is most appropriate is chosen, and that one is employed. Rules, such as the one in the schema in the example, that use descriptions employing ambiguous expressions, are called fuzzy control rules.

The following four methods are used for devising fuzzy control rules—in other words, there are four methods for designing a fuzzy controller. These methods are (1) basing the rules on the knowledge and experience of experts, (2) building operator models, (3) deriving them through learning, and (4) basing them on a fuzzy model of the plant. Method (1) is the same as what is known as an expert system, and it is a method in which the experience of skilled operators or the knowledge of control engineers is assimilated and brought together in the form of "if/then" rules. Method (2) is a method for times when rules cannot be obtained from experts in linguistic form. For example, the skills of experienced operators, such as when they have learned a manual operation, are especially difficult to express in words. In such cases, if we can model operator actions in the same if/then form as before, we can use this model as is for control rules. Method (3) is effective when there is a plant model or when experiments can be carried out using an actual plant, and it is appropriate for robot control. Since the fuzzy control rules can be created by learning from a situation in which there are no control rules and can be improved according to environmental changes and the learning of control rules, it can be done without experts. Method (4) takes this even further and is a method in which we try to build a fuzzy model of the plant. If at this time we build a model of the plant in if/then form, it will be theoretically convenient for deriving fuzzy control rules from the control objectives and the plant model.

Fuzzy control, which was first tested by London University's Mamdani in the first half of the 1970s, was given its first practical application in controlling a cement kiln in Denmark in the 1980s. Research into applications in Japan really got going around 1985, but already there are various successful applications. If we conjecture about the reasons for the success of fuzzy control looking at actual examples, the three following characteristics of fuzzy control come to the surface: (1) It is logical control. (2) It is parallel (distributed) control. (3) It is linguistic control.

(1) comes from the fact that fuzzy control rules are logical because of the if/then form of the conditional propositions. The employment of the rules is of course carried out by means of inference. Being logical means that it is easy to express the logic of expert control, and furthermore, that it can work with various antecedents. In real plants, the information in clean forms that could be used with past forms of control is not all that is interpreted for use in control, and control modes change with various conditions, such as time and the values for variables. Logical control is also appropriate for out-of-the-ordinary processes. In most real plants special considerations are necessary at times such as during start-up, but since with fuzzy control both the time of start-up and that of regular operation are described by rules using the same if/then form, it is convenient for automation.

(2) represents a big difference between the control methods used up to now and fuzzy control. Whether the older form of control be classical or modern, all of the control rules are expressed in the form of a single equation, but with fuzzy control, multiple control rules of a partial nature are employed. One control rule is effective in a specific area of the information space that is used for control, and it is sufficient to create control rules for each sub-area of the information space. In addition, when there is more than one output, it is possible to make control rules for each output individually. Furthermore, when there is more than one control

objective, control rules can be created for each objective. In older forms of control all objectives have to be summarized in one function, so the kinds of objectives that can be handled logically are extremely limited. With fuzzy control, there is no need for objective functions, and furthermore there is no need to solve optimum control problems, so it can cope with various objectives, and even objectives that are mutually competitive.

(3) comes from the fact that the if/then rules are written using words. This means that interactive control can be carried out with an operator and that the controller is a gray box rather than the black box it was with older forms of control. Therefore, it is easy for operators to understand the operation of the controller. Furthermore, an operator can learn and improve the performance of the fuzzy controller. The higher levels of human learning are not skills involving the movement of limbs brought about through repetition, but learning on the level of language. Acquisition of new knowledge is made easy through learning that uses language as its intermediate.

What about the overall quality of the ideas behind fuzzy theory that show up in a straightforward manner in fuzzy control? In a word, it can express areas in which human thinking appears very well, but more concretely, it gives rise to methods for modeling of complex objects and human judgements and can aid the expression of knowledge. People maintain a grip on their judgments by dividing them according to circumstances, and they can cope with complex circumstances by using a form in which one set of rules for judgment are used in one case and another set in another case. They do this skillfully using experience in the process of determining characteristics, and this is helpful for overall judgments. Judgments are not made according to uniform values; rather, there are times when they are made based on several competing value systems. Furthermore, inference is a help when information is incomplete. Fuzzy control takes the form of rules divided according to circumstances and goals and fuzzy inference, and these take in human methods for judgment. In everyday life, people do nothing that approaches modeling done with equations. In addition, they do not summarize their governing rules in a single rule. The language that people use for modeling is ambiguous natural language. The models that arise are not uniform; rather, they describe the fragmentary characteristics of their objects or are a collection of conditional partial models. When we say partial models, this does not mean nonmathematical micromodels, but rather models that have a certain amount of range and that are easily understood qualitatively. In accordance with this, fuzzy control uses words and develops operator models using the if/then formula.

People do not express knowledge by memorizing unprocessed data or by memorizing each of their own deeds or actions separately. Rather, data and activities are summarized and organized as knowledge in a way that makes them easy to understand and refer to later. This is of course expressed in natural language. If things are expressed in and intensified by words with ambiguous divisions among small and medium-sized units instead of being arrangements of numbers, there is a large reduction of memory used, and in addition, recall is easy when necessary. Cases in which data are incomplete can naturally be expressed well using ambiguous language. The use of words for ambiguity and the fact that those words contain ambiguity are essential characteristics of the expression of human knowledge.

The quality and characteristics of fuzzy theory and its good characteristics can be captured in this way. If we take a new look at problems, even if they are problems outside the area of control, from the point of view of the usefulness of fuzzy theory that can be seen in the area of control and with the idea that there might be some commonality, we might be able to make some steps toward the solutions that were unthinkable using past methods. In other words, there are many things that come up in various control problems that can be seen as being common to other areas as well, and in addition, among the successful examples of the use of human intelligence in fuzzy control, there are important hints for industrial applications of theory.

3.2 BLAST FURNACE CONTROL

Control of blast furnaces, which occupy the most important position in integrated iron and steel works, was always carried out based on the working knowledge and experience of operators and operations engineers. At the NKK Fukuyama Iron Works, a sensor-based expert system, which performs on-line, real-time automatic control of the thermal condition of the blast furnace (hereafter "blast furnace temperature") using knowledge engineering methods, was developed for blast furnace thermal condition control.

Blast furnaces are very large and complex, and in addition, the in-furnace thermal condition cannot be measured directly. To add to this, the sensors that indirectly measure the in-furnace thermal condition include noise, so there is inevitable fuzziness in the interpretation of sensor data.

This system makes use of membership functions, which are used in fuzzy control, and employs methods for determining certainty factors (CF values) as a means for expressing the fuzziness associated with the blast furnace. In addition, there is a learning mechanism for membership functions, and it can easily and rapidly respond to differences in the blast furnace process, changes in facilities over time, and operational changes.

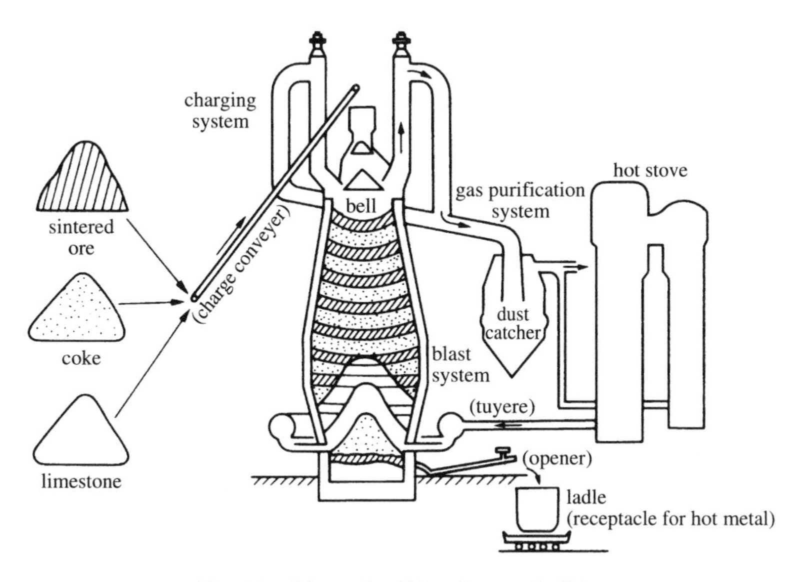

Fig. 3.1 Schematic of blast furnace facilities.

In this section, we will discuss the application of fuzzy control in this expert system for blast furnace control.

3.2.1 Blast Furnace Operation and Measurement Control System

(a) Blast Furnace Facilities and Operations

Figure 3.1 gives a summary of the equipment for a blast furnace. A blast furnace consists of the main furnace, equipment for charging the furnace with raw materials, hot stoves and blowers for creating the hot blast, gas cleaning apparatus for purifying the gas produced in the furnace, and iron and slag channels for discharging the molten pig iron (hereafter hot metal) from the furnace.

With a blast furnace, sintered ore, which is lumps formed by baking powdered ore, and coke are charged alternately into the top of the furnace. In addition, a 1,100° C blast is continually blown into the bottom of the furnace through 20–30 tuyeres set up in a ring around it. As the sintered ore descends from the top to the bottom of the furnace (at about 6 m/hour), it is continuously reduced and melted by the reductive gases produced by the burning of the coke.

The furnace is cylindrical in shape, with a diameter of about 10 m at the top and about 15 m at the bottom. In addition, the distance from where the raw materials are charged to the tap holes is approximately 40 m. Large furnaces have an internal volume of more than 5,000 m^3 and can produce as much as 10,000 tons per day. This kind of large blast furnace has the following characteristics.

(1) There is a high-temperature, high-pressure atmosphere inside the furnace, and reactions in all three phases—gas, solid, and liquid—progress simultaneously. In addition, the reactions inside the furnace are extremely complex.
(2) The hot blast is turned into a reductive gas, and the time it takes to travel from the bottom to the top of the furnace is several seconds, but it takes 6–7 hours before the raw material charged into the top of the furnace turns into molten iron. A blast furnace involves processes with this kind of long delay time.
(3) Once a blast furnace is started up, it is not stopped until the end of its campaign (approximately 10 years).

On the other hand, the mission of a blast furnace is to respond swiftly and flexibly to changes in production plans and operational changes and stably produce high-quality molten iron. Because of this, proper maintenance of temperature is indispensable, and minute control of the thermal condition is usually required. Also, in order to maintain the furnace thermal condition, irregular phenomena such as furnace cooling due to operational mistakes, "slipping"[1] which can cause it, and "channeling,"[2] must be avoided.

The thermal condition is controlled by the quantity of raw material charged, charging pattern, quantity of blast, temperature, and moisture level. Because of this, furnace technicians infer the thermal condition and other conditions (conditions that lead to slipping and channeling) from the information obtained from actual everyday operational results, real-time signal changes in patterns from a multitude of sensors and statistical models, and they work toward optimizing methods for charging raw materials and introducing the hot blast.

[1]Slipping: This is a phenomenon in which the raw materials slip from their suspended condition (in which the descent of the charge is stopped) and drop suddenly. When slipping occurs, there is generally a change in the amount of raw material that is supposed to be reduced and melted, and a destabilization of in-furnace thermal condition.

[2]Channeling: This is a phenomenon in which the pressure pushing the reducing gas upward becomes greater than the weight of the charge and blows out the top of the furnace all at once. When channeling occurs, the reducing gases hardly reduce any of the raw materials within the furnace, so this is a cause of furnace cooling.

(b) Blast Furnace Measurement Control System

The conditions inside a blast furnace are constantly changing, and slipping and channeling occur when the balance between charged material and gas pressure is destroyed. In addition, there are large changes in the thermal condition at these times. Because of this, sensors and models for inferring the thermal condition and other conditions are necessary for blast furnace control.

Figure 3.2 shows the sensor installation. The sensors are mainly thermometers and pressure gauges, but they number more than 1,000 for a single furnace. The information from the sensors is processed by a process computer, and trend graphs and distribution maps are at the disposal of the operators at all times. In addition, numerous in-furnace thermal control models have been developed and tested under actual operating conditions. However, since reactions in all three phases—gas, solid and liquid—progress inside the blast furnace simultaneously and the reactions interact in a complex manner, there is no model that has proven useful

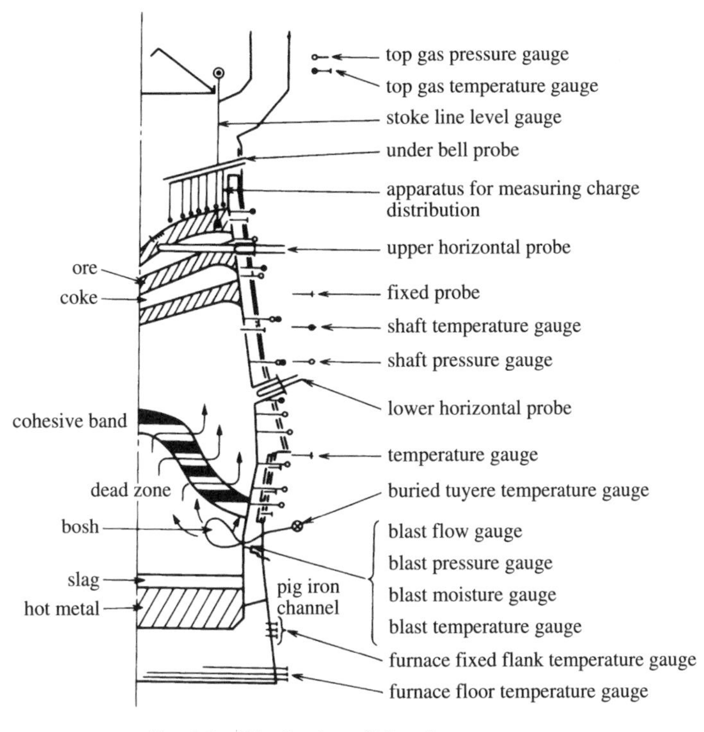

Fig. 3.2 Distribution of blast furnace sensors.

enough. Thus, blast furnace control uses the operational knowledge of experienced operators to a great extent.

3.2.2 In-Furnace Thermal Condition

The raw materials that are charged into the top of the furnace are reduced by the reductive gases produced by the combustion of coke, are liquefied, drip down as molten pig iron and slag and accumulate at the bottom of the furnace. The in-furnace thermal condition is the temperature of the accumulated molten iron or the atmosphere. If the thermal condition is too high, there is generally an increase in the impurities in the hot metal, and its quality deteriorates. If the thermal condition is too low, melting of the sintered ore is delayed, and there is a reduction in the amount of molten iron produced. Thus, the thermal condition is an important index that shows the operating conditions. However, no method for its direct measurement has been developed. For this reason, this index has been managed using the temperature of the hot metal when it is tapped.

(a) Hot Metal Temperature as Representative of Furnace Temperature

The hot metal and slag that gathers at the bottom of the furnace is tapped alternately from three tap holes set up around the circumference of the bottom of the furnace. After the tapped hot metal and slag are separated by means of the difference in specific gravity by a skimmer that is installed along the trough, the hot metal is recieved by a ladle (volume: 200–300 tons) and carried to the steel plant by a transfer car. In addition, the slag is disposed of in a slag pit. The hot metal temperature of the iron is measured intermittently at the skimmer shown in Fig. 3.3 just after the slag and hot metal have been separated. Because of wear on the refractory brick of the top hole, the duration of a tap is usually 2–6 hours. Since the tapping of the hot metal is this kind of intermittent process, the hot metal temperature turns out to be information with a fuzziness peculiar to the process, as is shown in Fig. 3.4.

In general, even if the in-furnace thermal condition is stable, the hot metal temperature produces a lower measurement because (1) the hot metal cools because it has been held in the bottom of the furnace for a long time, and (2) the iron trough that serves as the flow path absorbs the apparent hot metal temperature. With the passage of time, the holding time of the iron within the furnace decreases, and the cooling from the base of the furnace is reduced. In addition, the temperature of the trough rises, and the hot metal temperature gives information more representative of the in-furnace condition.

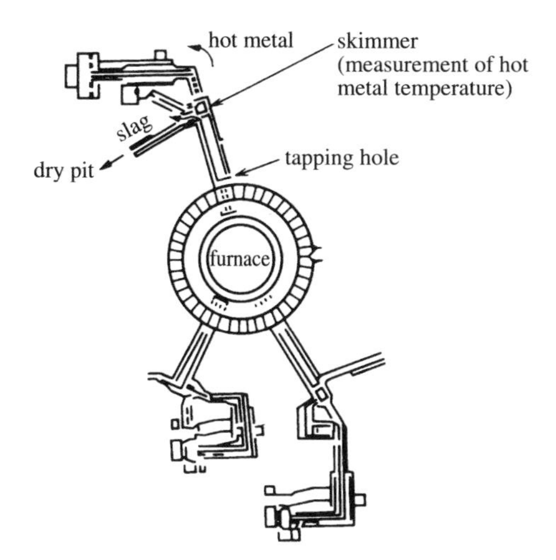

Fig. 3.3 Furnace and casting floor cross section.

In this way, information that shows the thermal condition fairly well can only be obtained after time, and in addition, the hot metal temperature during a tap changes according to operating conditions rather than being fixed. In this way the hot metal temperature normally contains fuzziness as a source of information on thermal condition. Because of this it is necessary to keep the elapsed time from the beginning of the tap and the operating conditions in mind, and in addition to consider the fuzziness peculiar to the system.

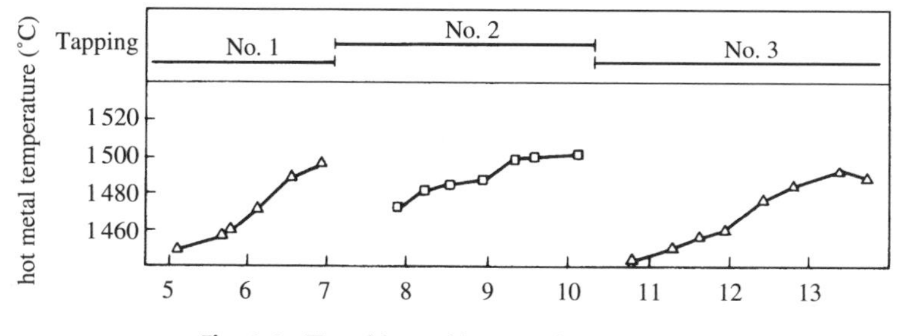

Fig. 3.4 Transitions of hot metal temperature.

3.2.3 Expert System for Furnace Temperature Control

(a) System Structure

Figure 3.5 shows the system structure. This system is made up of a process computer that performs collection of sensor data and the preprocessing for the data used for inference, an AI processor that uses the knowledge base for inferring thermal condition, and a distributed digital controller for carrying out thermal condition control from the inference results. In addition to performing sensor data processing, the process computer carries out the analysis of data that has been developed and used in the past and houses the thermal condition diagnostic functions that make use of mathematical models.

Using sensor information and the knowledge base, this system infers the current thermal condition and the direction in which the temperature is

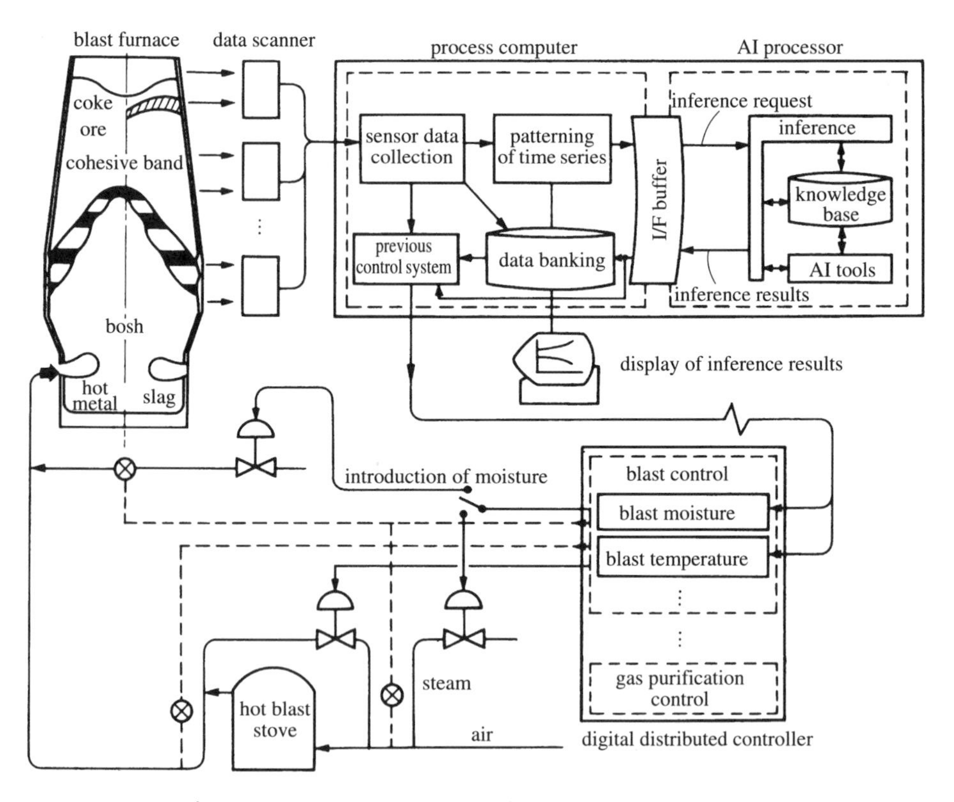

Fig. 3.5 Structure of the expert system for blast furnace control.

moving on a 20-minute cycle, uses control rules with the results, and determines the amount of work, such as amount of moisture in the blast, to be done.

(b) Inference Mechanism

The inference mechanism is shown in Fig. 3.6. The knowledge sources (KS) form units according to the attributes of sensors and each functional unit of the rules and are housed in the AI processor. The sensor data that accumulates in the computer is transformed into real data for inference and written on the common blackboard (BB). The inference engine is activated by the real-time scheduler construction of the process computer, and after that the "inference control knowledge source" begins to be activated by the inference engine. For example, when an intermediate result concerning thermal condition is derived from a certain "sensor level knowledge source," this intermediate result is written on the blackboard along with the "furnace level knowledge source" to be used next. Inference control is usually carried out by determining the knowledge source

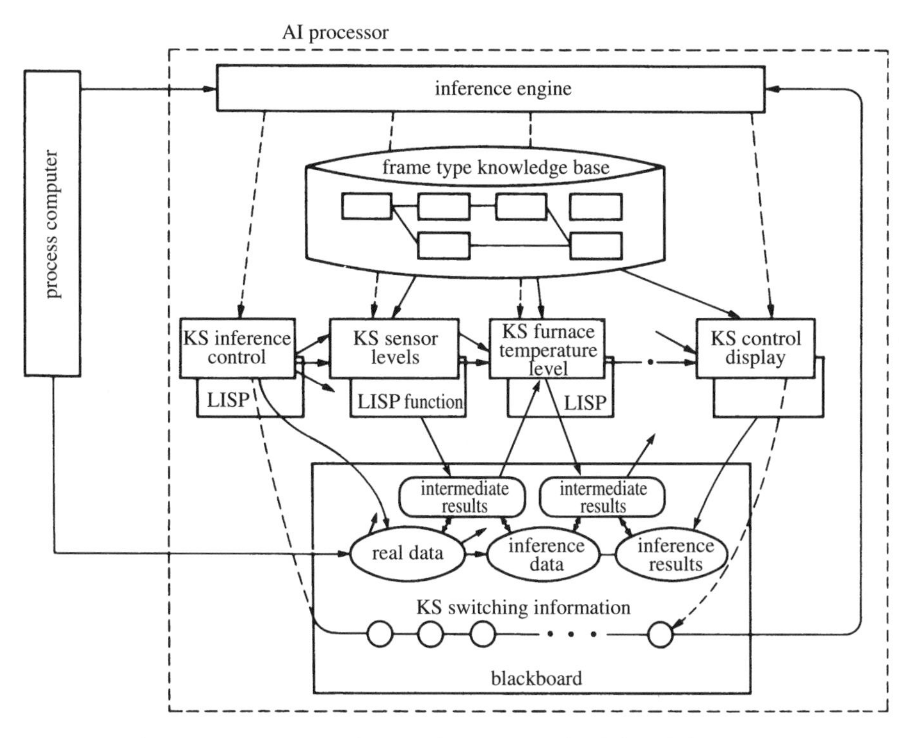

Fig. 3.6 Inference flow for blast furnace control. KS: knowledge source.

that should be put into motion next, based on the information on the blackboard.

(c) Operation Knowledge of Furnace Control

Figure 3.7 gives the knowledge engineering development of the operational knowledge that has been used in blast furnace control up to the present. Operational knowledge includes diagnostic knowledge of thermal condition, basic knowledge about blast furnaces, and knowledge that is common sense to blast furnace operators.

Diagnostic knowledge of thermal condition includes knowledge used for diagnosing furnace temperature that comes from sensory data from operators (thing like coke combustion conditions at the tuyeres or the color of slag during tapping) and sensor information, as well as knowledge for substantiating suppositions about thermal condition from the behavior of particular pieces of information after the supposition has been made. Operators instantly bring together these two kinds of knowledge and make judgments on thermal condition, and from these results, their knowledge of furnace control, and the use of operation manuals, figure out what actions to take. Basic knowledge of blast furnaces includes knowledge

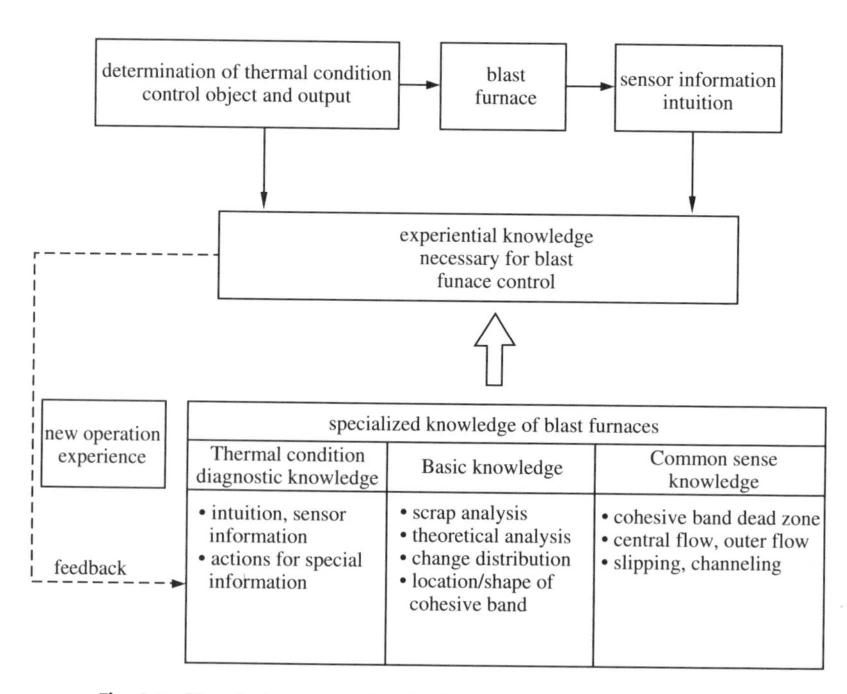

Fig. 3.7 Knowledge engineering development of blast furnace operation.

from scrap analysis, specialist knowledge obtained from theoretical analyses such as metallurgical reaction theories and long years of experience, and this is the support behind operations technology. Furthermore, common sense that can only be used by blast furnace experts is behind all of this knowledge. For example, this includes things like the fact that the cohesive zone is the 1,250° C zone, the mollification temperature for the ore, or that hanging is a temporary stagnation in the descent of the raw materials within the furnace.

(d) Knowledge Expressions and Knowledge Bases

In this system, the knowledge necessary for inference is described in three forms: production rules, frames, and LISP functions. The majority of the knowledge is expressed in terms of production rules, divided according to the functional units of the rules and sensor attributes, unitized, and placed in a hierarchical structure.

Figure 3.8 shows an example of a knowledge base and rules. The knowledge source group for determining the thermal condition level, for example, is constructed from four knowledge source subgroups, and of these the "hot metal temperature knowledge source" is the source of the knowledge for inferring the thermal level from measurements of the hot metal temperature, and it is made up of approximately 10 rules. The hot metal temperature in inferred with reference to the elapsed time since the beginning of the tap and the number of ladles taken, using the idea of fuzzy sets that will be discussed later. Next, the thermal level is inferred from the knowledge sources concerned with values for Si, S, and other components of the hot metal. After this processing is over, the "sensor level knowledge sources" and "human judgment knowledge sources" are activated, and an overall furnace temperature level determination is made from the inference results of all of the knowledge source subgroups.

Static knowledge such as constants, blast moisture levels and temperatures for thermal condition control, coke ratios, and the amounts of work to be done are housed in frames. In addition, since each frame contains knowledge in common with the others, knowledge sequencing is used to enhance the effectiveness of knowledge expression.

Knowledge in the form of certainty factors for the expression of the fuzziness within knowledge, output quantities from the past, and processes for the compensation calculations for output actions are described using LISP functions.

(e) Expression of Fuzziness

The major problem with an expert system such as this one that is mainly based on experimental rules is the expression of the fuzziness contained in

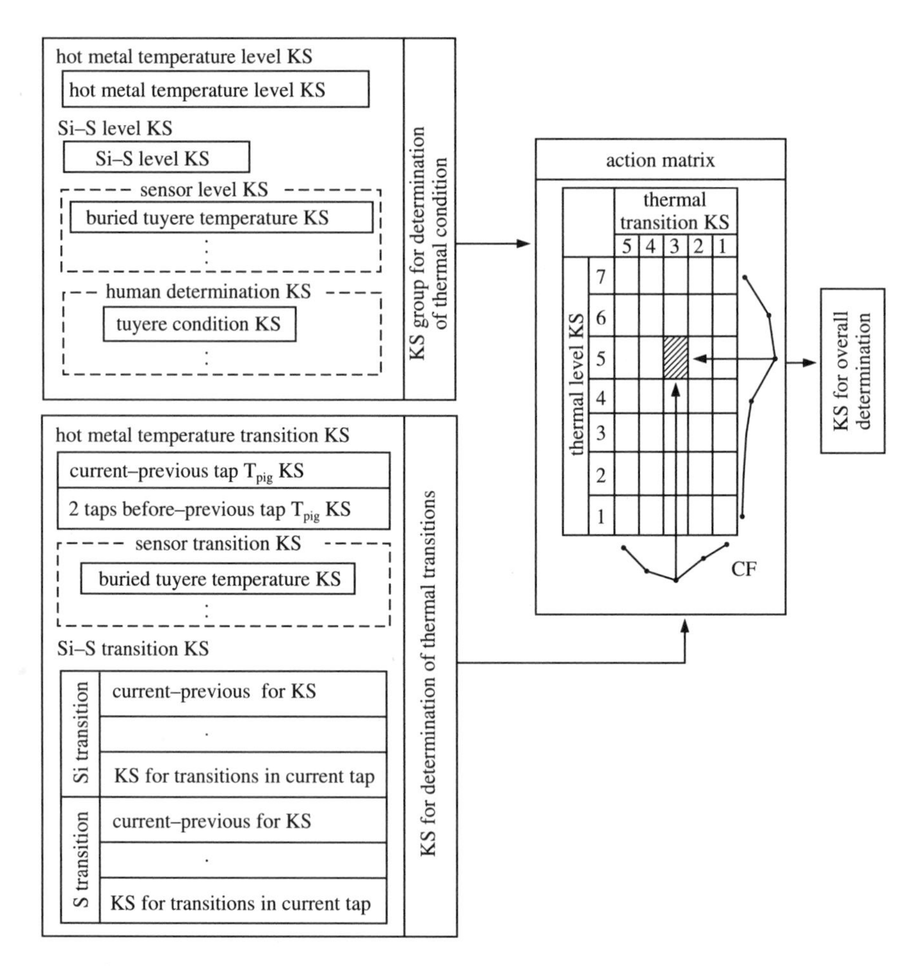

Fig. 3.8 Structure of knowledge base. CF: certainty factor; KS: knowledge source.

the knowledge. In general, the methods for expressing fuzziness include the introduction of certainty factors (CF: a value that expresses the reliability of each inference result) into the rules and the use of fuzzy sets. The "fuzzy control" based on fuzzy sets uses membership functions that can freely express even the most subjective "fuzzy concepts" people have, and it sets out to express the working knowledge of experienced operators and perform control on the same level as they do. On-line, real-time control has already been developed and introduced into the sintering processes in the iron-making industry.

However, since the blast furnace process is an extremely complicated process in which gas, solid, and liquid phase reactions all progress simultaneously, it is very difficult to express the operational knowledge of experienced operators with membership functions alone and carry out control using a single control system. Because of this, this system uses production rules to express the abundance of operation knowledge and introduces the idea of fuzzy sets as a method for expressing fuzziness. This is connected with the fact that simplification of knowledge expression avoids excessive expansion of the number of rules and contributes to the shortening of inference time. The result is that on-line, real-time control by an expert system is made possible.

(f) Introduction of Membership Functions

When thermal conditions are inferred from measurements of the hot metal temperature, operators do not make unconditional determinations of thermal condition. Because of this, thermal condition, which is primarily a continuous quantity, is divided into the seven levels shown in Table 3.1 in this system, in accordance with the working experience of operators.

Normally, operators do not make declarations such as "The current thermal condition is level 7," using only one piece of data, but more often make judgments such as "The current level is 6–7." Thus, rather than make uniform determinations concerning actual data (hot metal temperature), this system uses fuzziness expressions that are given certainty factors at each level for any single piece of data. Furthermore, the maintainability of the certainty factor level is taken into consideration, and a method of expression that employs three-dimensional extended membership functions that use actual data (hot metal temperature, for example), conclusions (thermal level, for example), and certainty factors as their three elements was developed.

Figure 3.9 shows the membership function for hot metal temperature, thermal condition, and the certainty factor for the Nth ladle within one tap. The axes here show the following coordinates:

X: hot metal temperature,
Y: thermal level,
Z: certainty factor.

Table 3.1. Furnace Temperature Level and Condition

Level	Furnace Temperature Condition
7	Very high
6	High
5	Somewhat high
4	Normal
3	Somewhat low
2	Low
1	Very low

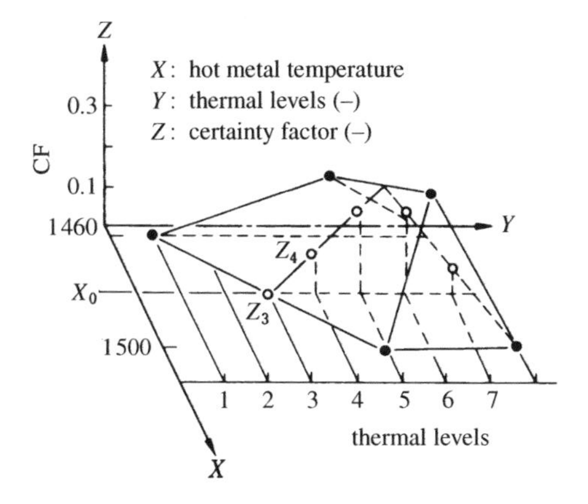

Fig. 3.9 Extended membership function.

For example, when the measured value of the hot metal temperature X is X_0, the Y–Z plane for X_0 in this three-dimensional figure gives certainty factor X_j for thermal condition at each level ($j = 1$–7).

Since the meaning of the information and the fuzziness of the hot metal temperature change with the elapsed time from the beginning of the tap, the number of ladles taken per tap was taken as one of the parameters here. In addition, there are membership functions not only for the hot metal temperature, but also for the information from each of the sensors connected with thermal condition.

(g) Derivation of Membership Functions

In the following, we will discuss the derivation of the membership functions used in this system. Figure 3.10a is a plot of measured values for the hot metal temperature (X axis) and the maximum hot metal temperature during the tap (Y axis), which is the figure that comes the closest to expressing the thermal condition. Figure 3.10b shows the expression of membership functions in which the Z axis has been plotted as the rate of inclusion within a certain temperature range, when the foregoing are divided by temperature range. Furthermore, since each of the membership functions resembles a triangular function, we connect the vertices, so Fig. 3.10c shows the extended membership function. In addition, since the hot metal temperature is information that depends on the time elapsed since the beginning of the tap and the operational conditions, 30 or more extended membership functions are prepared inclusive of these conditions and set up so that they can be chosen among automatically.

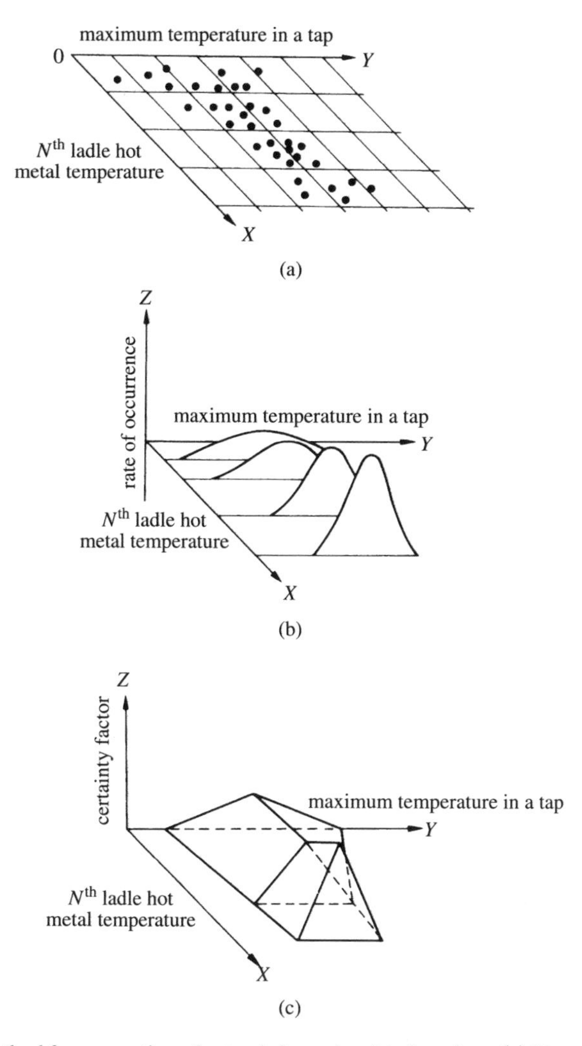

Fig. 3.10 Method for generation of extended membership functions. (a) Hot metal temperature of hot metal in Nth ladle and maximum temperature in a tap. (b) Expression of membership function. (c) Expression of extended membership function.

(h) Learning of Membership Functions

In cases such as this, in which a process is controlled based on the working knowledge of operators, the success or failure of the system depends on whether or not the membership functions that express fuzziness show sensitivity to the sensory perceptions of operators. In addition, the sensory perceptions of operators change minute by minute, along with changes in the blast furnace facilities and operation.

Because of this, whether or not membership functions keep up with changes in facilities and operational conditions has a great effect on the life of the system, when it comes to actual operation. In consideration of these points, extended membership functions can be learned through the process shown in Fig. 3.11. In this system, sensor and operational data for the preceding several months are stored in the process computer, and the data are retrieved routinely. Irregular data, such as (1) data just after blowing down, (2) data during the occurrence of irregular conditions (slipping, channeling, etc.), and (3) data when there is a large difference between the data just before and after switching tap holes, are removed and divided into the four cases shown in Table 3.2, and a cause analysis is performed. Membership functions are then rederived as described in Section 3.3.3(h) when case (2) occurs.

(i) Effects of Fuzzy and Learning Control

Effects of Fuzzy Control. In this system, ideas from fuzzy set theory were introduced as a means for expressing the fuzziness included in operator working knowledge. When compared with the case in which this working knowledge is described by production rules alone, the results are (1) simplification of knowledge expression, (2) reduction of the number of rules, and (3) improvement in inference efficiency. On-line, real-time control is made possible. In addition, maintainability is vastly improved.

Effects of Learning. The results of having a learning control function for membership functions are that the system is able to react effectively to

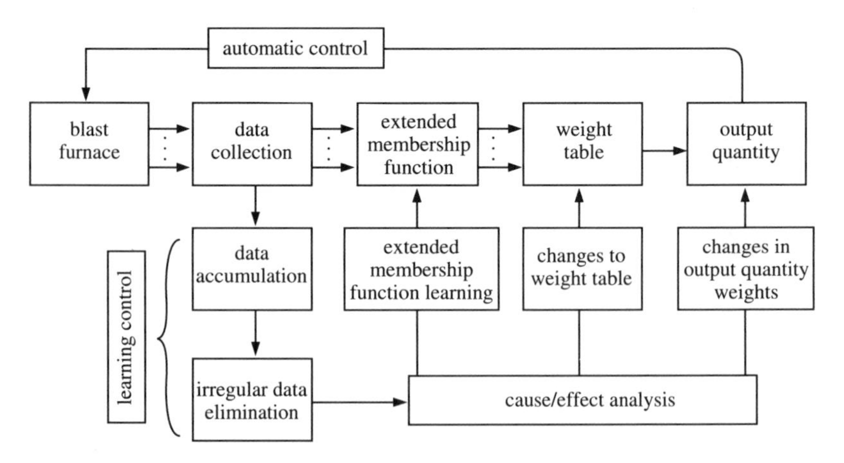

Fig. 3.11 Extended membership learning control.

Table 3.2. Cause Analysis and Data Processing

Case	Cause Analysis	Data Processing
1	Deviation from hot metal temp. objective value small, and variations small	Maintain present conditions
2	Deviation from hot metal temp. objective value large, and variations large	Membership function learning
3	Blast furnace operation and conditions change, control worsens	Revise weight assignment table
4	Hot metal temp. increases or decreases and there is stable operation (low output quantities)	Revise output weight assignment

differences in the blast furnace process and changes in operating conditions, and that results such as (1) improvements in blast furnace controllability, (2) lengthening of system life, and (3) extension of utilization to other blast furnaces can be expected.

3.3.4 Actual Operational Use

This system has been working smoothly since it was first employed for actual operations in January 1987, and the rate of utilization of automatic control is more than 95%.

The opportunities for application of the concepts of fuzzy control in process control will probably increase in the future. Systems will vary according to the method of application and the nature of the process, but the methods described in this chapter can be considered effective for complex processes such as blast furnaces, or cases of use as a part of the structure of an expert system.

3.3 COLD-ROLLING CONTROL

3.3.1 Introduction

The job of the cold rolling done during steel production is to perform the finish rolling of strip steel that has been rolled to a thickness of several millimeters or less during hot rolling in order to make the thin sheet steel used in things such as automobiles and cases for consumer electronics products. It is the last process in the production of the material and sizes required by regular consumers. Because of this, a high level of precision is required for properties such as thickness and shape. Initial set-up control of the cold-rolling apparatus is done using a process computer and sequencer, and DDC control is carried out. The model used for set-up control is called the set-up model, and control is carried out using methods based on the cold-rolling theories used up to now. The set-up model provides the basis for calculations of operational conditions (such as the roll gap necessary for obtaining the final dimensions and shape of the product, rolling speed, and the tensile force applied to the material being rolled) from the type of steel, size, and other characteristics of the hot-rolled strip. This control is the most important thing affecting productivity and stability of cold rolling as well as product quality.

In other words, if the precision of these calculations is poor, prescribed product characteristics such as thickness and shape cannot be met, quality deteriorates, and a drop in the pace of production is invited. In addition, the worst possibility is an inability to maintain mass flow balance among the roll stands, which gives rise to production troubles such as ruptures. On the other hand, the requirements for improvements in the precision of set-up models have increased because of the increasing frequency of set-up changes due to a tendency toward small lot production of a variety of products.

The problem is that since variables that are difficult to measure, such as coefficient of friction between roll and material being rolled and the deformation resistance of the material, are required by the equations in the cold-rolling theories used up to now, hypothetical values must be used in the calculations. Because of this, large errors sometimes arise in the calculation results, and this is a large barrier to automatic set-up. In other words, the current state of things is that in order to avoid deterioration in quality and operational problems, experienced operators usually observe the results of the calculations from the set-up model and, in cases of irregularity, correct the calculated values for roll gap, rolling speed, etc., to ensure operational stability.

However, it is natural that the manual adjustments made by these operators lengthen the time taken for set-up changes. Normally, rolling speed is reduced when set-up changes are made in cold-rolling operations, so productivity is sacrificed for the production stability derived from manual adjustments made by operators.

In order to solve these problems, a method in which the control performed by operators based on their past experience is expressed using fuzzy reasoning has been developed, and rolling load, which is particularly important for the set-up model mechanism and the precision of which has a large effect on other control systems (sheet temperature, lubrication, etc.), is predicted. Good results have been obtained. In this section, we will discuss this rolling load model, which is based on fuzzy theory, and furthermore, we will explain the results of verifications of the accuracy of the predictions that make use of actual rolling data.

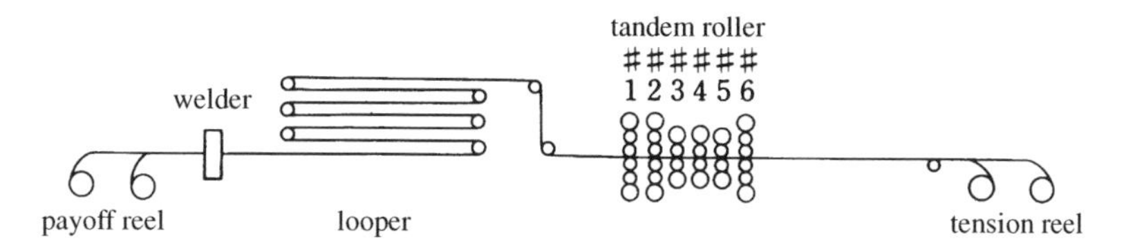

Fig. 3.12 Example of cold-rolling mill layout.

3.3.2 Plant Outline

Cold rolling is a process for rolling strip to prescribed thicknesses and shapes at low temperatures, and the facilities are usually constructed of a number of rolling stands. Figure 3.12 is an example of the layout of a plant with a six-stand tandem roller.

The material is first attached to the payoff reel on the feed-in side and unwound. Continuous rolling can be carried out by welding the beginning of this coil to the end of the previous coil, which is still being rolled. The coil that has been welded on goes through the looper, and after it has been rolled to a fixed thickness by the roller, we have the final cold-rolled product, which is then taken up by the tension reel on the output side. The set-up changes mentioned earlier take place over the period of transition before and after the weld.

The cold-rolling mill that is our object this time produces the wide range of sizes and qualities shown in Table 3.3, and model-based set-up changes are frequent. The normal pattern for rolling speed is the trapezoidal pattern shown in Fig. 3.13, and as was mentioned earlier, the speed is raised to the top capacity of the roller after completion of set-up adjustments and then lowered again for set-up changes as the next weld approaches. As is shown in Fig. 3.14, the rolling process is carried out through the application of tension from the front and rear and pulling on the material between the rolls as the roll pressure is applied. At this time, it is imperative that the previously mentioned roll pressure (known as rolling load) and tension be accurately estimated for the production of

Table 3.3. Facility Specifications

Production capacity	120,000 tons/month
Rolling speed	2,170 mpm (MAX)
Product thickness	0.15–1.0 mm
Sheet width	500–1,300 mm
Steel types	Normal steel, electromagnetic steel

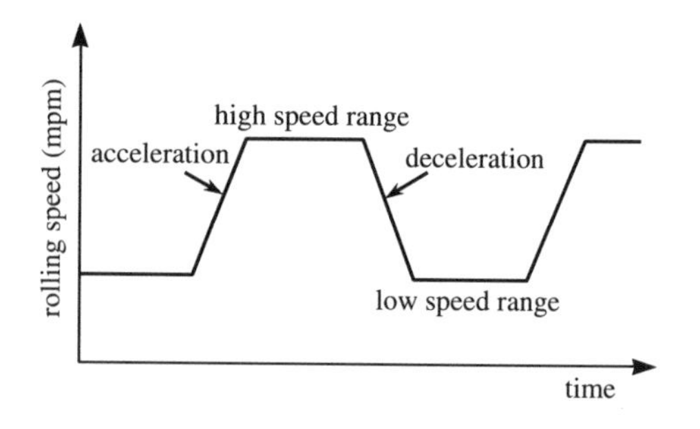

Fig. 3.13 Normal rolling speed pattern.

material of the desired thickness. In addition, the tension applied to the two ends of normal strips is standardized according to type of steel, quality, and dimensions. Therefore, the precision of the thickness of the sheet is determined by the accuracy of prediction of rolling load, based on a given tension.

As was mentioned earlier, the set-up model is a model for carrying out the initial set-up of the roller during the rolling process. It is a model that calculates the load for each stand for production of the desired thickness and the roll position necessary to apply this load to the strip from information about the dimensions of the material to be rolled, the type of steel, and the characteristics of the product to be made from that material. It is a model for passing the results on to the roller beforehand.

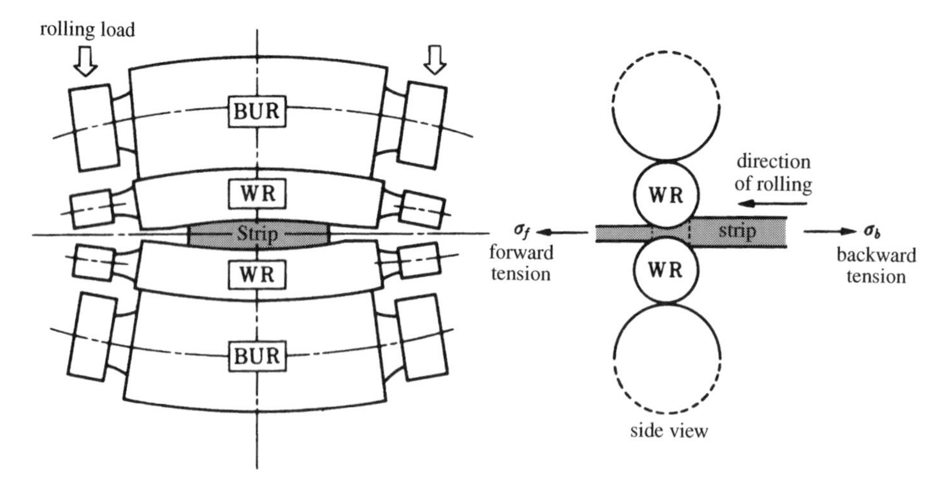

Fig. 3.14 Schematic of rolling.

As was mentioned before, the problem is that present set-up models sometimes produce large errors in their results, and because of this, manual adjustments of the roll positions and speed set by calculations are frequently performed by operators.

3.3.3 Control Algorithm

As was mentioned earlier, operators have the know-how to determine what kind of corrections should be made to the set-up resulting from the model in order to obtain the prescribed quality of product and make stable operation possible. Thus, a control method that improves the precision of the model by quantifying this know-how using fuzzy theory and integrating it into the set-up model was proposed.

There are several proposals for control methods that use fuzzy theory for the quantification of the control knowledge of experienced workers and that perform control, and based on these there have been publications of examples of various types of applications. However, in this case, the operators have almost no consciousness of rolling load as they work; they only have knowledge concerning roll position (adjustment of the gap between rolls). The point is that roll position can be precisely determined by taking the elastic displacement of the rollers into consideration, if the rolling load can be accurately predicted; so, when we consider applicability, rolling load is better. Thus, in this section we produced a model by identifying the fuzzy model for predicting rolling load based on operational data obtained as results of operators' adjusting roll position. Because of this, a model in which the operation part was a linear equation was employed with the goal of simplifying identification, as is mentioned in the next section.

3.3.4 Fuzzy Model

The fuzzy model used for this rolling load prediction is made up of the two mechanisms shown in Fig. 3.15: the rolling load prediction model described by fuzzy relational equations, and a learning function that performs point-by-point correction of the coefficients of this model. The inputs of the rolling load prediction model are information about the material to be rolled, orders concerning that material that are determined beforehand (product width and thickness), and other conditions (tension between stands and the rolling reduction for each stand); the model predicts the rolling load for each stand using fuzzy inference.

Next, after rolling of that coil is completed, point-by-point learning is carried out for the coefficients of the rolling load prediction model based

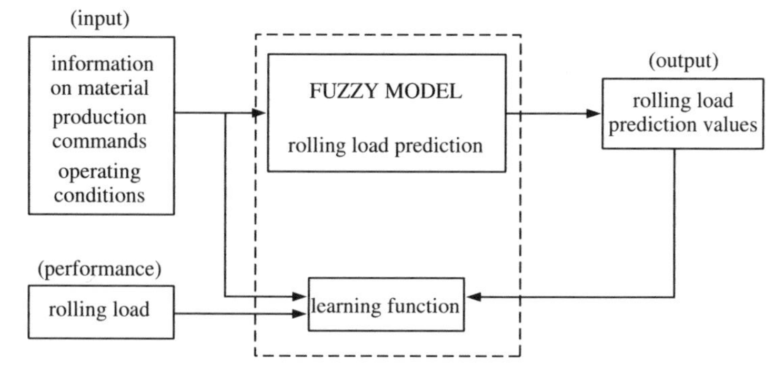

Fig. 3.15 Fuzzy model system diagram.

on the actual performance of the rolling, using the learning model. The learning model is a function necessary for the maintenance of a high level of predictive accuracy in the model, and especially in a mill that produces many types of products in many sizes like this one, it is not too much to say that it is indispensable.

The rolling load prediction model expresses rolling loads, its output, as sets of if/then rules that perform inference, and each rule is expressed in the form shown below:

$$R_i \quad \text{If } X_1 \text{ is } A_1^i,\ X_2 \text{ is } A_2^i,\ \ldots,\ X_m \text{ is } A_m^i$$
$$\text{then } y^i = a_0^i + a_2^i X_1 + a_2^i X_2 + \cdots + a_m^i X_m,$$

Here R_i is the ith rule, A_j^i the fuzzy set, X_j the input variable, y^i the ith rule output and a_j^i the coefficient.

When input values $X_1, \ldots, X_m$ are given the inference value $\hat{y}$ for output, y is found as follows:

$$\hat{y} = \sum_{i=1}^{n} g^i \cdot y^i \Big/ \sum_{i=1}^{n} g^i.$$

Here n expresses the number of rules, and y^i is the output of the calculations using the rule. Weight g^i expresses the truth value of the ith rule and is calculated as follows:

$$g^i = \min\left\{A_1^i(X_1), \ldots, A_j^i(X_j), \ldots, A_m^i(X_m)\right\}$$

$A_j^i(X_j)$ is the grade of the membership function of fuzzy set A_j^i. In other words, the final result of inference is the weighted mean of the values for y^i asserted by each rule, with the truth value g^i for each rule used for the weighting. For example, Fig. 3.16 shows a four-rule case.

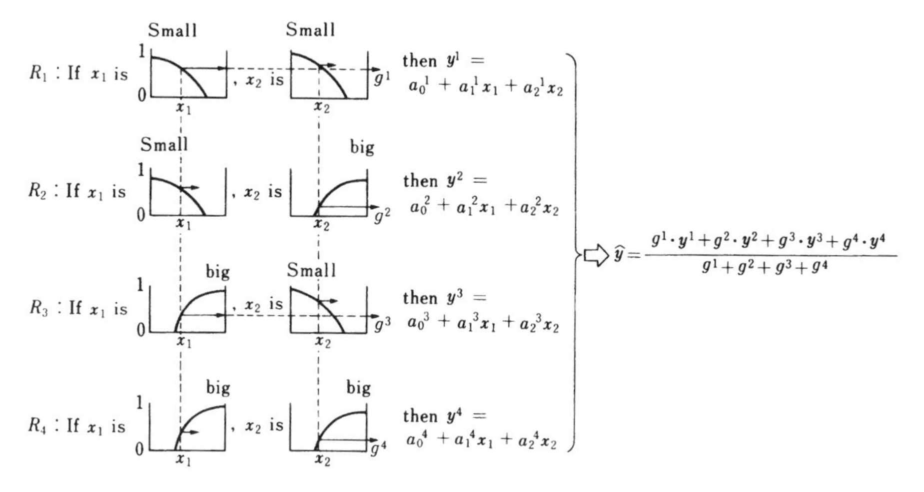

Fig. 3.16 Fuzzy relational equations and inference image.

3.3.5 Derivation of the Fuzzy Model

One of the key points in coming up with a fuzzy model is the choosing of variables for fuzzy divisions. In the present case, three of the parameters that are known to have an effect on rolling load and that were thought appropriate were chosen. The following two items were taken as the basis for this at the time:

(1) There should be a clear, confirmed correlation with the output variables (in this case rolling load) of the object.
(2) The tendency of this correlation should change according to whether the areas for the values of descriptive variables are large or small.

(1) is obvious, but with (2), we can expect the effect to be greater depending on the division of the input space.

For example, one of the ambiguous variables that were chosen this time, accumulated length of rolling of the work roll, can be considered an index of changes in roll roughness, but when the rolls are rearranged and the rolling length is a small, there is a sudden drop in rolling load, and if it goes beyond a certain value, changes are easy. Therefore, if we center the division of areas on the point where the tendency of the correlation changes, we can expect that the effects of division will be maximized.

Based on the preceding standards, three variables, including accumulated rolling length of work rolls, were chosen, and they were divided into two parts: "small" and "big." Here, we will let these variables be x_1, x_2, and x_3. In addition, the linear equation for the conclusion includes four parameters besides these, which are considered important for predicting the rolling load—for example, carbon equivalent, which is an index of hardness and forward tension on the sheet. So it is constructed of a total of seven variables.

In addition, the fuzzy model is constructed from the eight rules shown in Table 3.4, through the combination of x_1–x_3 mentioned earlier. Here we have

$$S_j: \text{ small},$$
$$B_j: \text{ big } (1 \leqq j \leqq 3),$$

$$f_i(x_1, x_2, \ldots, x_7) = a_0^i + a_1^i x_1 + a_2^i x_2 + \cdots + a_7^i x_7, \qquad i\text{: rule number.}$$

In addition, the following functions were used as the membership functions of the fuzzy subsets of x_1, x_2, and x_3:

$$\mu_{\text{big}}(x_i) = (1/\pi) \cdot \tan^{-1}\{a_i(x_i - b_i)\} + 0.5,$$
$$\mu_{\text{small}}(x_i) = (-1/\pi) \cdot \tan^{-1}\{c_i(x_i - d_i)\} + 0.5 \qquad i = 1,2,3.$$

Appropriate values that correspond to the correlation tendencies of x_1, x_2, and x_3 are chosen for coefficients a, b, c, and d. Here equivalent choices are made for a and c and b and d, so this is a symmetrical function across grade 0.5 as shown in Fig. 3.17.

Based on actual rolling data, the coefficients for the linear equations for conclusions are found using the weighted minimum square, with the truth value of the antecedent as the weighting factor. In other words, coefficients are determined for each rule by finding the minimum for the following equation as an evaluation value:

$$J = [\mathbf{Y} - \mathbf{Da}]^{\text{t}}\, \Lambda [\mathbf{Y} - \mathbf{Da}],$$

Table 3.4. Structure of Fuzzy Model

R1: If x_1 is S_1, x_2 is S_2, x_3 is S_3 then P is $f_1(x_1, x_2, \ldots, x_7)$
R2: If x_1 is S_1, x_2 is S_2, x_3 is B_3 then P is $f_2(x_1, x_2, \ldots, x_7)$
R3: If x_1 is S_1, x_2 is B_2, x_3 is S_3 then P is $f_3(x_1, x_2, \ldots, x_7)$
R4: If x_1 is S_1, x_2 is B_2, x_3 is B_3 then P is $f_4(x_1, x_2, \ldots, x_7)$
R5: If x_1 is B_1, x_2 is S_2, x_3 is S_3 then P is $f_5(x_1, x_2, \ldots, x_7)$
R6: If x_1 is B_1, x_2 is S_2, x_3 is B_3 then P is $f_6(x_1, x_2, \ldots, x_7)$
R7: If x_1 is B_1, x_2 is B_2, x_3 is S_3 then P is $f_7(x_1, x_2, \ldots, x_7)$
R8: If x_1 is B_1, x_2 is B_2, x_3 is B_3 then P is $f_8(x_1, x_2, \ldots, x_7)$

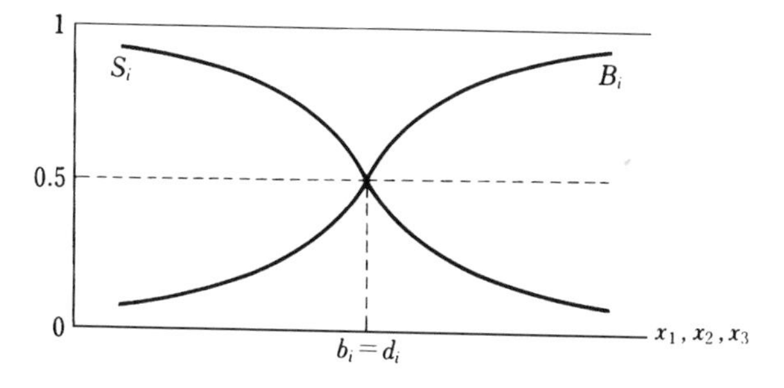

Fig. 3.17 Membership function.

where

$$\mathbf{D} = [\mathbf{Z}_1 \quad \mathbf{Z}_2 \quad / \quad \mathbf{Z}_n]^{\mathrm{t}},$$

$$\mathbf{Z}_i^{\mathrm{t}} = [1 \quad x_{1i} \quad x_{2i} \quad \cdots \quad x_{7i}]: \quad i\text{th product data vector},$$

$$\mathbf{a} = \left[a_0^k \quad a_1^k \quad a_2^k \quad \cdots \quad a_7^k\right]: \quad k\text{th rule coefficient vector},$$

$$\Lambda = \begin{bmatrix} \eta_1^k & & & \\ & & 0 & \\ & \eta_2^k & & \\ & & \ddots & \\ & 0 & & \eta_N^k \end{bmatrix}: \quad \text{weight matrix},$$

where n_i^k is the standardized weight for the ith piece of data for the kth rule:

$$\eta_i^k = g_{ki} \Big/ \sum_{i=1}^{8} g_{ji}.$$

In addition, g_{ij} is the grade for the ith piece of data for the jth rule:

$$\mathbf{Y} = [y_1, y_2, \cdots, y_N]^{\mathrm{t}}: \quad \text{actual rolling load vector}$$

In addition, as was mentioned earlier, in order that the model include the changes peculiar to the roller and operational changes for things like the sudden changes in roll roughness when rolls are rearranged, corrections to the model that make use of actual data are indispensable. Therefore, the weighted minimum square operation mentioned earlier was transformed point by point, and a learning function that makes corrections through point-by-point learning for the coefficients of linear equations for conclusions was employed:

$$\mathbf{a}_N = \mathbf{a}_{N-1} - \eta_N \mathbf{P}_N \left[z_N \quad z_N^{\mathrm{t}} \quad \mathbf{a}_{N-1} - \mathbf{z}_N y_N \right],$$

$$\mathbf{P}_N = \mathbf{P}_{N-1} - \eta_N \mathbf{P}_{N-1} \mathbf{z}_N (1 + \eta_N \quad \mathbf{z}_N^{\mathrm{t}} \quad \mathbf{P}_{N-1} \quad \mathbf{z}_N)^{-1} \mathbf{z}_n^{\mathrm{t}} \mathbf{P}_{N-1},$$

where

η_N = standardized weight of antecedent
$\mathbf{a}_N$ = vector of learning coefficient
$\mathbf{z}_N$ = Nth data vector
$\mathbf{P}_N$ = correction gain matrix

3.3.6 Results of Use

A fuzzy model for predicting rolling load was constructed using the preceding methods. The data used were actual data from the plant, and as shown in Fig. 3.18, the data accumulated in the existing process computer were transferred to a microcomputer for technical analysis. The microcomputer performed lower bound and other checks for the exclusion of abnormal data, carried out direct engineering transformations, and stored

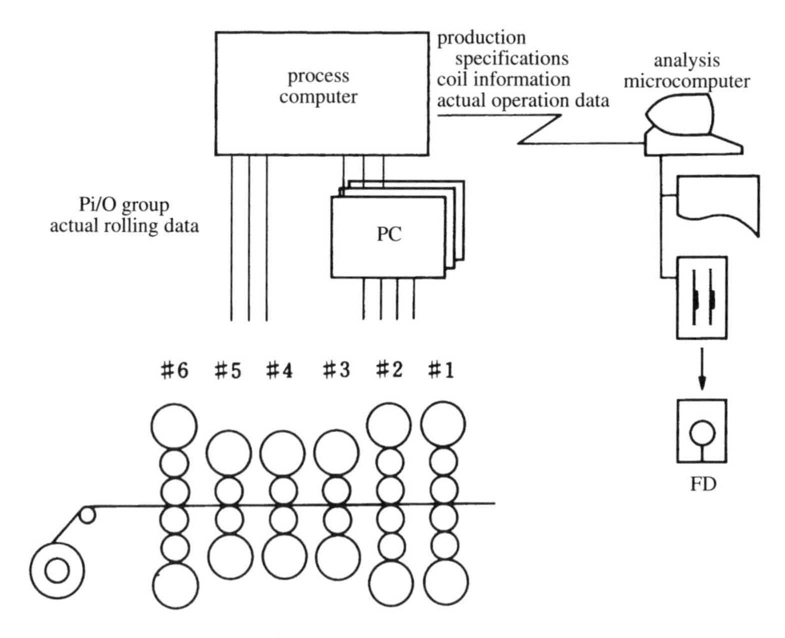

Fig. 3.18 System structure.

the data on a floppy disk. Analysis was done off-line, based on this data. The rolling materials taken as objects were thin rolled materials for which the estimation accuracy of older models had been extremely poor because of extra-small rolling lots and wide ranges in size and quality. In addition, the stand taken as a object was stand No. 1, the one said to have the greatest effect on the thickness of the product.

(a) Performance of the Model

In this case, evaluations from two viewpoints are necessary, in order to evaluate the performance of the model. The first is prediction accuracy, that is, the predictive error separating the actual load and the predicted load, and this was evaluated using a statistical index, standard deviation. The other one is the frequency of generation of large inference errors that have an effect on the stability of operation. In terms of operation, an improvement in the standard deviation of prediction error can bring about improvements in product quality and productivity, or for example, even if the standard deviation is good, there can be large errors in the results from the model, so that automatic operation under the model is filled with possibilities for troubles with rolling, and the model cannot replace people.

Here we considered the limit of prediction error to be $\pm 20\%$ from the actual rolling load based on operational experience, an actual load difference of ± 150 tons.

The results of using the model generated in the previous section on 361 sets of data for different thin cold-rolled materials are shown in Fig. 3.19.

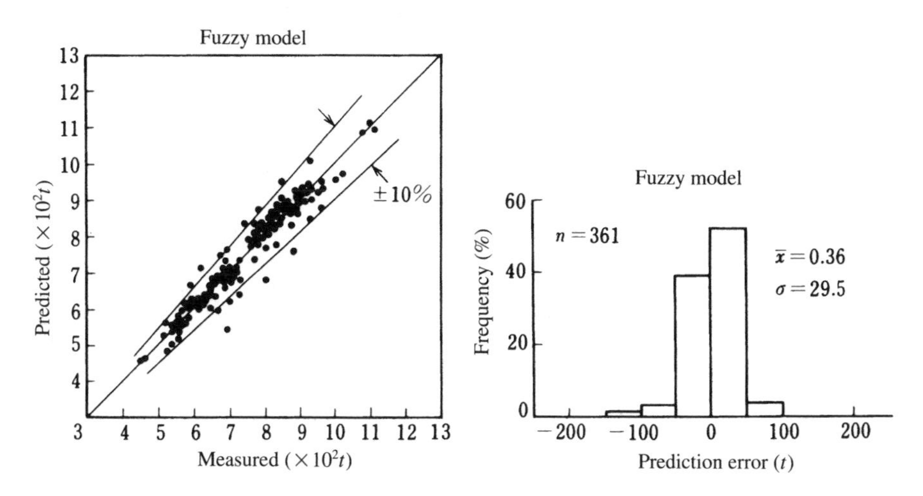

Fig. 3.19 Precision of fuzzy model rolling load predictions.

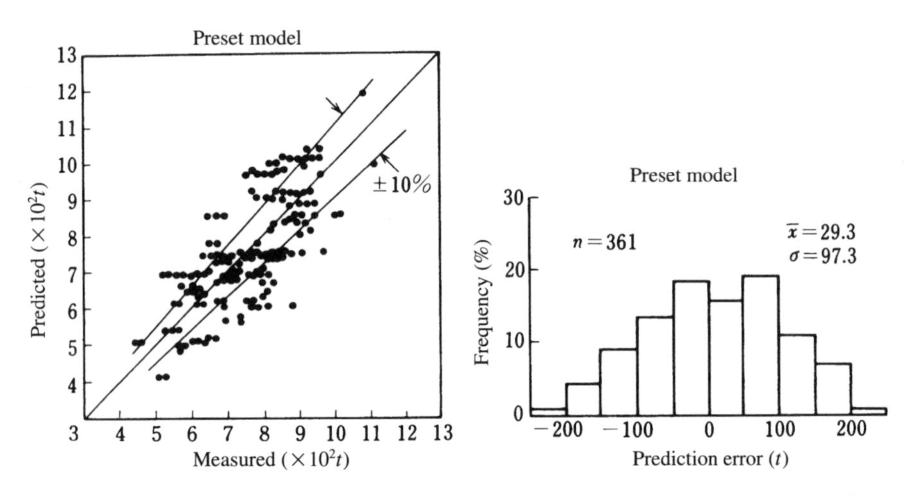

Fig. 3.20 Precision of rolling load predictions for cold-rolling set-up model based on older methods.

The predictive error for rolling load was held to roughly ±10%, so there was good predictive accuracy. In addition, there were no inference errors above the 150 ton level that can cause trouble with the rolling, so we can see that there are good possibilities for actual use.

Figure 3.20 shows inference errors in rolling load for a set-up model based on older methods, and the fuzzy model is much better in all areas: average inference error, standard deviation, and frequency of errors over 150 tons. In addition, the effects of learning for each coil are shown in Fig. 3.21. Figure 3.21a shows the results with learning, and 3.21b shows them without learning.

As learning was carried out with each coil, the inference error distribution came close to a clean normal distribution, and the frequency of errors of 20% or more diminished to almost nothing. We can see that learning makes for a large improvement in accuracy of inference.

Figures 3.22 and 3.23 are a comparison of the fuzzy model and multiple regression, a standard statistical model, in order to take a look at the effectiveness of the area divisions. As is shown in Table 3.5, the fuzzy model agrees more closely with the results of control by operators, and this shows that the fuzzy model does a good job of expressing the nonlinearity of the system.

(b) Summary

The following are clear from the results of the use of the fuzzy model for predicting rolling load, which is the most important function of a set-up model for cold rolling:

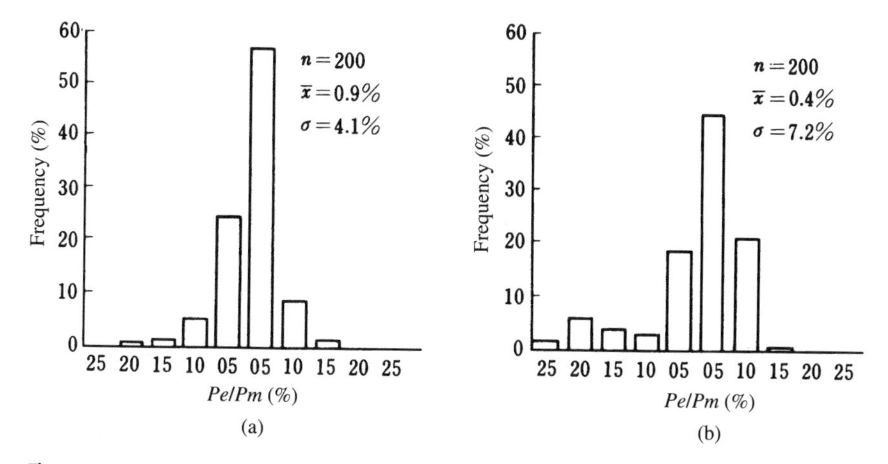

Fig. 3.21 Frequency distribution of ratio of rolling load predictive error P_e and observed value P_m. (a) With learning. (b) Without learning.

(1) The employment of fuzzy control can improve the accuracy for rolling load prediction to $\pm 10\%$, and eliminate extreme prediction errors of over 150 tons. This is a threefold improvement when compared to previous set-up models. Steady-state errors were also roughly zero.

(2) Because learning was employed that weighted the conclusions of the fuzzy model with the truth values of the antecedents between coils, the standard deviation of inference error was reduced by almost half, and inference misses of 20% or more were virtually eliminated.

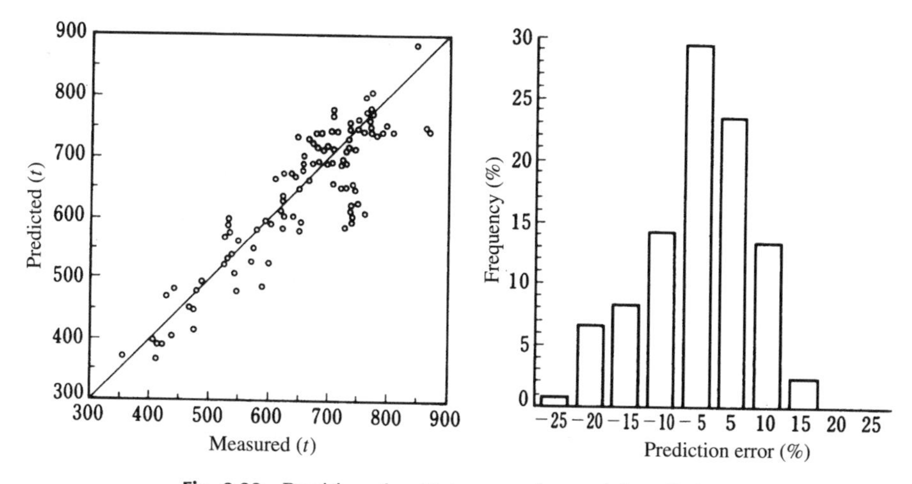

Fig. 3.22 Precision of multiple regression model prediction.

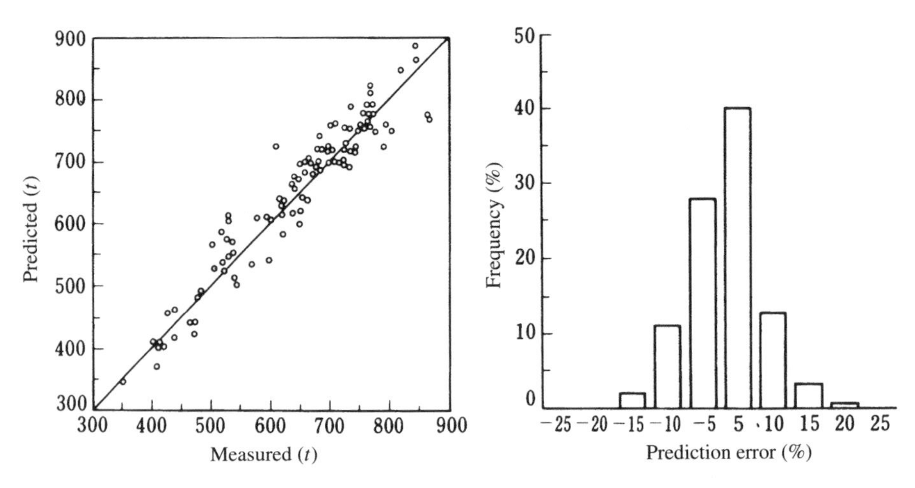

Fig. 3.23 Precision of fuzzy model prediction.

(3) Even when compared to multiple regression, the fuzzy model was in closer agreement with the results of operator control.

3.3.7 In Conclusion

In this section we have discussed the use of fuzzy theory in the set-up control for the rollers used in cold rolling. Set-up control for rollers depends heavily on the experiential knowledge of operators. The control rules were expressed using fuzzy reasoning, and a fuzzy controller for predicting rolling loads was constructed based on actual operational data, the results of operator control. In addition, a learning function that judges the inference results and corrects the rules was set up in order to incorporate the changes that occur in roller characteristics over time and changes in operational conditions. The results of simulations using actual data show that the prediction accuracy of the fuzzy model matches closely with that of operator control, and it is clear that there is a much higher prediction accuracy than with existing models.

The control method described in this section is raising great expectations for improvements in productivity and product quality, as well as the

Table 3.5. Comparison of Predictive Accuracy

	Multiple Regression Model	Fuzzy Model
Standard deviation	41.6 tons	29.5 tons

realization of stable operation. At present there are plans for the extension of this model beyond stand No. 1 to all stands, as well as for uses with materials other than thin cold-rolled materials. With uses in the set-up models in other industrial processes, the range of application is growing.

3.4 AUTOMOTIVE SPEED CONTROL

3.4.1 Introduction

In recent years there has been rapid progress in automotive electronics. Electronic fuel injection, engine control that makes use of microcomputers, 4WD and 4WS control, and fixed-speed cruise controls are all in use, and we now see many cars that are equipped with cruise computers capable of calculating distance traveled, fuel consumption, etc.

In the 1960s, top-of-the-line American cars were equipped with cruise controls. At that time, almost all of the control methods were mechanical. After that, cruise control devices that made use of electronic control methods such as PID control, adaptive control, nonlinear control, and optimal control appeared, and that brings us up to the present. The conditions for these kinds of control are that (1) the dynamic characteristics of the car are known, (2) the dynamic characteristics do not vary, and (3) there is little disturbance. However, the dynamic characteristics of automobiles vary with gear changes and loads. In other words, with each gear, load and road condition, the dynamic characteristics differ. In addition, the dynamic characteristics are greatly affected by hilly roads (inclines). Because of this, we need to develop control rules that will carry out the best control, even if the dynamic characteristics are changed.

The characteristic points of fuzzy control are the possibility of a construction made up of drivers' knowledge and skills using linguistic control rules (LCR) and the lack of a requirement for a quantified model of the control object. Because of this, a fuzzy logic controller (FLC) constructed of qualitative relations between the car's carburetor opening and speed—that is, control knowledge rules—was developed in order to have a controller capable of handling a car's changes in dynamic characteristics. With an FLC, fuzzy inference is carried out following control rules, and the manipulated variable (carburetor opening) is calculated. The FLC inputs when this is done are speed deviation and first- and second-order differences in speed deviation.

Since we constructed a hardware fuzzy controller that uses indirect fuzzy inference using a microcomputer, installed it in an automobile (Toyopet Crown, 1970 model, automatic), and conducted a cruise test on a city expressway (Kitakyushu Expressway, speed limit 60 km/hr), we will describe the methods and results and compare them to the results of using PID control.

In addition, an FLC controller with a learning function was designed, in order to increase the robustness of the FLC for speed control. In other words, we will discuss the design of an FLC that is equipped with a function in which the control response and performance of the FLC is evaluated repeatedly in real time and the FLC parameters adjusted. We will investigate the effectiveness of this FLC by means of a simulation.

3.4.2 Outline of the Automotive Speed Control System

The hardware automotive speed control system that was constructed using a microcomputer is shown in Fig. 3.24. Speed information is received by the speed sensor as a voltage. In the microcomputer, the voltage for the target speed and the detected speed are first converted to digital values by an A/D converter, and then speed deviation, etc., is calculated. Next, based on fuzzy control rules (LCR), the manipulated variable is calculated by means of fuzzy inference, passed through a D/A converter and output to a comparator. The comparator carries out feedback control of an actuator made up of a motor and relay, in order to attain the carburetor opening that corresponds to the manipulated variable. Starting and stopping of control are performed by means of the S/S button in Fig. 3.24, and setting of the target speed is done using a simple knob (volume). When a break signal is detected, the measures taken are shutting off the speed control (pause in FLC calculations, closing of carburetor) and resetting the initial value. Resumption of control is carried out when the break signal is not detected or by means of an initialization command from the S/S switch.

The internal construction of the microcomputer is as follows. An Intel 8085 microprocessor is used, and for input and output data an 8-bit A/D, D/A converter is used. The S/S switch signal and brake signal inputs use a parallel I/O chip. Four-kilobit PROMs are used for storage of the

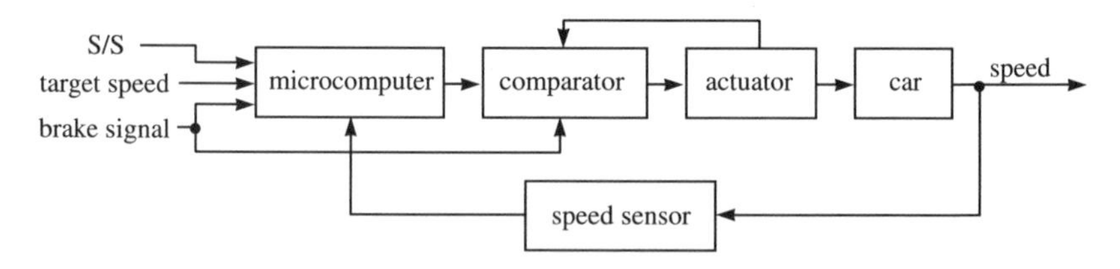

Fig. 3.24 Speed control system.

software for both the FLC and PIDC (PID controller). Besides these, there is a 2-kbit ROM for the monitor and work areas. There is a keyboard for switching between the FLC and PIDC and adjusting the parameters for each of these devices, as well as for hardware and software checks.

3.4.3 Fuzzy Speed Control [9, 10]

Here we will describe fuzzy inference done by means of the indirect method and the linguistic control rules of the FLC, show the algorithms for finding manipulated variables, and finally investigate effectiveness using the results of a simulation and an operational test.

(a) Linguistic Control Rules of the FLC

The FLC is constructed of logic control rules in the following manner:

$$\begin{aligned}
&\text{LCR1:} \quad \text{If } e_k \text{ is } P_1 \text{ then } \Delta u_k \text{ is } P_{u1}.\\
&\text{LCR2:} \quad \text{If } e_k \text{ is } N_1 \text{ then } \Delta u_k \text{ is } N_{u1}.\\
&\text{LCR3:} \quad \text{If } \Delta e_k \text{ is } P_2 \text{ then } \Delta u_k \text{ is } P_{u2}.\\
&\text{LCR4:} \quad \text{If } \Delta e_k \text{ is } N_2 \text{ then } \Delta u_k \text{ is } N_{u2}.\\
&\text{LCR5:} \quad \text{If } \Delta^2 e_k \text{ is } P_3 \text{ then } \Delta u_k \text{ is } P_{u3}.\\
&\text{LCR6:} \quad \text{If } \Delta^2 e_k \text{ is } N_3 \text{ then } \Delta u_k \text{ is } N_{u3}.
\end{aligned} \tag{3.1}$$

Here, for time k, $e_k = r - y_k$ (control deviation), $\Delta e_k = e_k - e_{k-1}$ (first-level difference for control deviation), $\Delta^2 e_k = \Delta e_k - \Delta e_{k-1}$ (second-level difference for control deviation), $\Delta u_k = u_k - u_{k-1}$ (change in the manipulated variable), and r and y_k are the target speed and the car speed, respectively. Furthermore, P and N indicate positive and negative.

LCR1 means "If the speed is below the target speed, press down on the accelerator"; LCR4, "If the speed is increasing, let up on the accelerator"; and LCR5, "If the speed is decreasing rapidly, press down on the accelerator." Here expressions such as "below," "increasing," and "increasing rapidly" are ambiguous. In this case LCR1 and LCR5, "press down" on the accelerator, and LCR4, "let up" on the accelerator, are reciprocal activities. In fuzzy control the manipulated variable is determined by striking a balance among various rules indicating these kinds of reciprocal activities, through fuzzy inference.

Furthermore, if we compare these rules with a PID controller, FLC rules are speed-type rules, so LCR1 and LCR2 correspond to I actions, LCR3 and LCR4 to P actions, and LCR5 and LCR6 to D actions. However, as we will discuss later, each rule is processed using fuzzy inference, so the results are the construction of a nonlinear PID controller.

P_i, P_{ui}, N_i and N_{ui} ($i = 1, 2, 3$) in the rules are fuzzy sets with membership functions for each variable. These membership functions are shown in Fig. 3.25. Here the antecedent membership function is an arctangent function and the conclusion a straight line. The reason we let the antecedent be a saturation curve is because the controller gain becomes large when control deviation, etc., is in the neighborhood of 0, and when control deviation, etc., gets larger, the gain approaches saturation. Because of this, the manipulated variable remains under control, even if there is a large erroneous input into the controller. The determination of these membership functions—that is, the determination of parameters a_i and b_i ($i = 1, 2, 3$)— is important for the construction of the FLC.

(b) Indirect Fuzzy Inference

Many types of fuzzy inference, such as inference using composition rules, have been proposed, but we will discuss Tsukamoto's [11] proposal for fuzzy inference by indirect methods here.

Let us consider the following as a form of fuzzy inference.

$$\begin{array}{ll} \text{Premise 1:} & \text{If } e_k \text{ is } P_1 \text{ then } \Delta u_k \text{ is } P_{u1} \text{ (LCR1)} \\ \text{Premise 2:} & e_k \text{ is } P' \\ \hline \text{Conclusion:} & \Delta u_k \text{ is } C_1 \end{array} . \tag{3.2}$$

Here P' and C_1 are fuzzy variables (linguistic variables). The foregoing means that given the fuzzy conditional proposition of premise 1, we get conclusion (fuzzy output) C_1 when we are given P' as a fuzzy input.

There is a direct method for fuzzy inference in which the value of the membership function of the fuzzy set is interpreted as a truth value and inference is carried out using that value, and an indirect method in which inference is carried out using a fuzzy truth value (linguistic truth value), a fuzzification of this truth value. In the following we will explain converse of

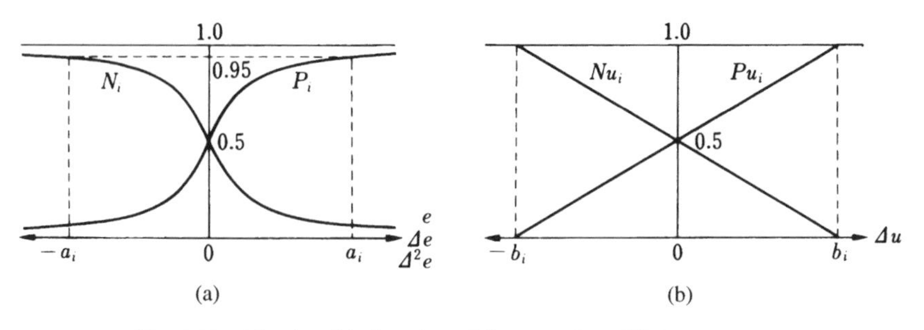

Fig. 3.25 Membership functions. (a) Antecedent. (b) Consequent.

truth qualification, fuzzy modus ponens, and truth qualification, which are used in this indirect fuzzy inference process.

(1) Converse of Truth Qualification. This is finding the fuzzy set for the degree of truthfulness of "e_k is P_1" from the fuzzy proposition in the antecedent of premise 1, "e_k is P_1," and the fuzzy proposition in premise 2, "e_k is P'," in (3.2)—that is, finding fuzzy truth value τ_{P_1}. In other words, it is finding an equivalent fuzzy truth value τ_{P_1} that gives

$$\text{"}e_k \text{ is } P_1\text{" is } \tau_{P_1} \leftrightarrow \text{"}e_k \text{ is } P'.\text{"}$$

(This is called converse of truth qualification.) From this we can find the following equation:

$$\mu_{\tau_{P_1}}(v) = \mu_{P'}\left(\mu_{P_1}^{-1}(v)\right), \qquad v \in [0,1]. \tag{3.3}$$

(2) Fuzzy Modus Ponens. The fuzzy modus ponens is the inferring of the degree of truthfulness of consequent "Δu_k is P_{u1}"—that is, the fuzzy truth value $\tau_{P_{u1}}$—when the degree of truth $\tau_{P_1 \to P_{u1}}$ (fuzzy truth value) for the fuzzy conditional proposition (written $P_1 \to P_{u1}$) and the fuzzy truth value τ_{P_1} (which can be found by (1)) for antecedent "e_k is P_1" are given. In Lukasiewicz logic, the truth value of implication $P_1 \to P_{u1}$ is given by

$$p_1 \to p_{u1} = 1 \wedge (1 - p_1 + p_{u1}) \qquad p_1, p_{u1} \in [0,1]. \tag{3.4}$$

Here p_1 and p_{u1} are numerical truth values. Using this, the fuzzy truth value $\tau_{P_1 \to P_{u1}}$ of fuzzy conditional proposition $P_1 \to P_{u1}$ can be expressed by

$$\tau_{P_1 \to P_{u1}} = 1 \wedge (1 - \tau_{P_1} + \tau_{P_{u1}}). \tag{3.5}$$

Here $+$ and $-$ are fuzzy operations based on the extension principle.

If we use the α-level set that corresponds to the fuzzy truth value in Eq. (3.5), we get

$$\tau^{\alpha}_{P_1 \to P_{u1}} = 1 \wedge \left(1 - \tau^{\alpha}_{P_1} + \tau^{\alpha}_{P_{u1}}\right). \tag{3.6}$$

Here $\tau_{\bullet}^{\alpha} = \{t \mid \mu_{\tau_\bullet}(t) \geqq \alpha\}$, $\alpha \in [0,1]$. Thus, when $\tau_{P_1 \to P_{u1}}$ (usually assumed true) and τ_{P_1} are given, $\tau_{P_{u1}}$ is found by solving Eq. (3.6). Now, if τ_{P_1} is normal (height 1) and convex and $\tau_{P_1 \to P_{u1}}$ is nondecreasing on $[0,1]$, we can write $\tau^{\alpha}_{P_1} = [a_1(\alpha), a_2(\alpha)]$, $\tau^{\alpha}_{P_1 \to P_{u1}} = [r(\alpha), 1]$, so we can find $\tau^{\alpha}_{P_1}$ using the following equation:

$$\tau^{\alpha}_{P_{u1}} = [\{a_1(\alpha) + r(\alpha) - 1\} \vee 0, 1]. \tag{3.7}$$

(3) Truth Qualification. Since we know the fuzzy truth value $\tau_{P_{u1}}$ for consequent "Δu_k is P_{u1}" that we obtained in (2) and fuzzy set P_{u1}, we can find fuzzy set C_1 for the equivalent conclusion "Δu_k is C_1," which gives

$$\text{"}\Delta u_k \text{ is } P_{u1}\text{" is } \tau_{P_{u1}} \leftrightarrow \text{"}\Delta u_k \text{ is } C_1\text{".}$$

(This is called truth qualification.) This can be found with the following equation:

$$\mu_{C_1}(\Delta u_k) = \mu\tau_{P_{u1}}\big(\mu_{P_1}(\Delta u_k)\big). \tag{3.8}$$

Using (1), (2), and (3) above, we find fuzzy set C_1 for manipulated variable Δu_k, when the fuzzy information for control deviation e_k is given. For LCR2–LCR6, we find μ_{C_i} ($i = 2$–6) in the same way using steps (1)–(3). The combined membership function C for each of these rules is found using

$$\mu_C(\Delta u_k) = \min_i \mu_{C_i}(\Delta u_k). \tag{3.9}$$

However, since a nonfuzzy value is required for the controller output, we let manipulated variable Δu_k^* be that value that satisfies the following equation:

$$\mu_C(\Delta u_k^*) = \operatorname*{Sup}_{\Delta u_k \in U} \mu_C(\Delta u_k). \tag{3.10}$$

Here U is the universe of discourse for the manipulated variable.

(c) FLC Design [10]

For the design of an actual automotive speed control system, we use nonfuzzy values for the input information. Next we will find manipulated variable Δu_k^* for the case when input information is nonfuzzy.

Now let $e_1 = e$, $e_2 = \Delta e$ and $e_3 = \Delta^2 e$. For the antecedent membership function from Fig. 3.25, we let

$$\begin{cases} \mu_{P_i}(e_i) = \dfrac{1}{\pi}\tan^{-1}(d_i \cdot e_i) + 0.5 \\ \mu_{N_i}(e_i) = \dfrac{1}{\pi}\tan^{-1}(-d_i \cdot e_i) + 0.5, \qquad i = 1,2,3, \end{cases} \tag{3.11}$$

hold. Here we let

$$d_i = \tan(0.45\pi)/a_i, \qquad i = 1,2,3, \tag{3.12}$$

and the membership function for the consequent be

$$\begin{cases} \mu_{P_{ui}}(\Delta u) = \dfrac{1}{2b_i}\Delta u + 0.5 \\ \mu_{N_{ui}}(\Delta u) = -\dfrac{1}{2b_i}\Delta u + 0.5, \qquad i = 1,2,3. \end{cases} \tag{3.13}$$

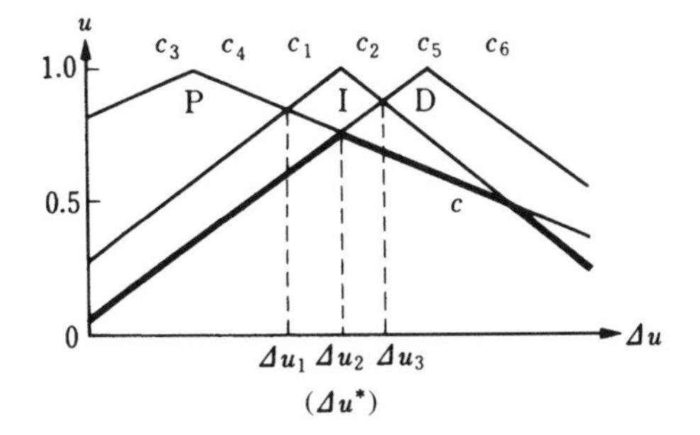

Fig. 3.26 Membership function of manipulated variable.

Using fuzzy inference methods (1)–(3), fuzzy set C for the manipulated variable is found as in Fig. 3.26. Therefore, if we let the intersections of the membership functions in the figure, Δu_1, Δu_2, and Δu_3, be

$$\begin{cases} \Delta u_1 = \dfrac{1}{\pi}\{\tan^{-1}(d_1 e_k) + \tan^{-1}(d_2\,\Delta e_k)\}/(g_1 + g_2) \\ \Delta u_2 = \dfrac{1}{\pi}\{\tan^{-1}(d_2\,\Delta e_k) + \tan^{-1}(d_3\,\Delta^2 e_k)\}/(g_2 + g_3) \\ \Delta u_3 = \dfrac{1}{\pi}\{\tan^{-1}(d_3\,\Delta^2 e_k) + \tan^{-1}(d_1 e_k)\}/(g_3 + g_1) \end{cases} \tag{3.14}$$

where $g_i = 1/(2b_i)$, $i = 1, 2, 3$, the nonfuzzy manipulated variable Δu_k^* is

$$\Delta u_k^* = \text{medium}\{\Delta u_1, \Delta u_2, \Delta u_3\}. \tag{3.15}$$

Here "medium" means taking the middle value. The manipulated variable Δu_k^* determined by Eqs. (3.14) and (3.15) is expressed by two linear combinations out of $\tan^{-1}(d_1 e_k)$, $\tan^{-1}(d_2\,\Delta e_k)$ and $\tan^{-1}(d_3 \Delta^2 e_k)$, and if we look at it from e_k, Δe_k, and $\Delta^2 e_k$, it is a nonlinear combination. Thus, we can say that the FLC is a nonlinear PID control apparatus.

(d) Simulation

The transfer function to the controlled object (operative part + carburetor + automobile) used for the simulation is expressed by the following equation. These are the characteristics based on a 0 km/h → 60 km/h crossover response:

$$G(s) = \frac{35.78}{(1.735s + 1)(16.85s + 1)}[\text{km/h/V}]. \tag{3.16}$$

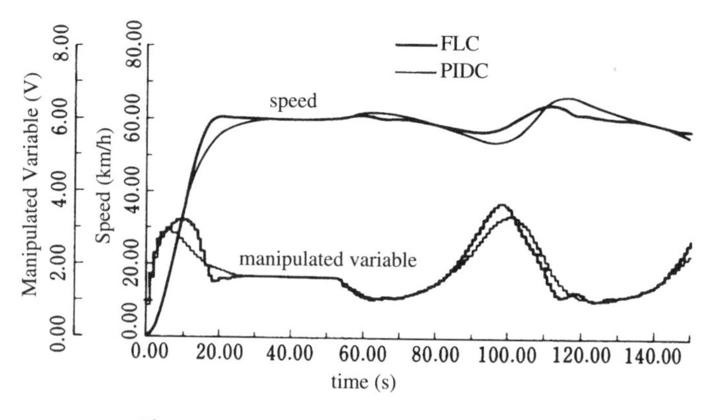

Fig. 3.27 Results of speed control simulation.

The results of simulations for the two controllers, FLC and PIDC, are shown in Fig. 3.27. Here the values of the FLC parameters are as shown in the following:

$$\begin{cases} a_1 = 70, & a_2 = 13, & a_3 = 14 \\ b_1 = 2, & b_2 = 2.13, & b_3 = 2 \end{cases}. \tag{3.17}$$

From the figure, we can see that the FLC overshoots a little more than the PIDC, but in terms of setting time and start-up time it is excellent. In addition, we can see that the FLC is better in terms of changes in motion characteristics after the assumed 50 seconds on inclines and declines. These changes in motion characteristics are cyclic changes of an amplitude of 50% of the gain for the controlled object. From this we can say that the FLC performs control that holds up against disturbance.

(e) Road Test

A microcomputer containing the FLC and PIDC was installed in the car, and a road test was conducted on the Kitakyushu Expressway. The sampling cycle was 1 s. The calculation time per sample was 50 ms for the FLC and 35 ms for the PIDC.

The test results for the PIDC and FLC are shown in Fig. 3.28 and Fig. 3.29, respectively. The results for the PIDC showed that it was affected by the motion characteristics of first and second gears as well as gear changes, and the changes in the manipulated variable were large. This is caused by the fact that the PIDC has a fixed gain, even though the car's characteristics change with first, second, and third gears. In contrast to this, the FLC results show that there were no large changes in the manipulated variable,

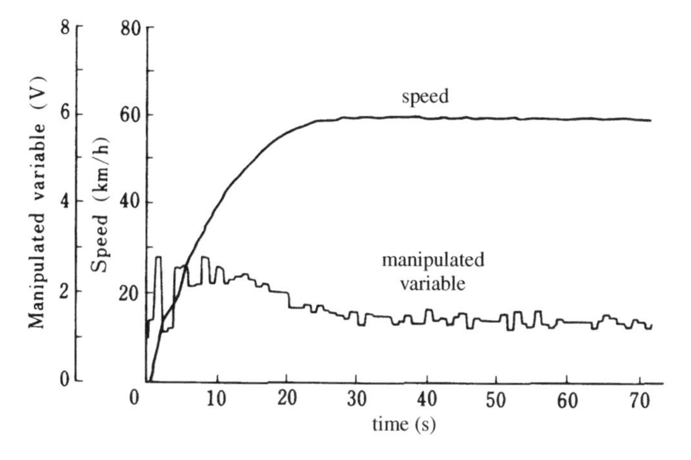

Fig. 3.28 Results of PIDC operation.

and acceleration was smooth, but there was a small amount of overshooting. In addition, it was stable despite the ups and downs in the road and operation over a long period of time.

If we look at it from point of view of the rider's sensation, there is no feeling of insecurity with the FLC, and there is very little difference from the feeling when the driver operates the accelerator. Driving is easy in terms of operability.

3.4.4 Speed Control Using Self-Tuning FLCs [12]

Here we will describe learning rules (tuning rules) that evaluate the control response of the FLC parameters and adjust them and the construction of an FLC with a self-tuning function. Finally we will show the results of a simulation and investigate the effectiveness of this control device.

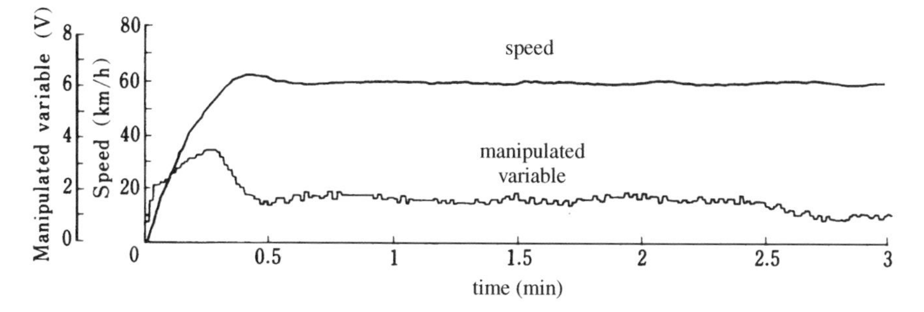

Fig. 3.29 Results of FLC operation over time.

(a) Self-Tuning Function

Under conditions in which there are large changes in load, there are cases in which it is difficult to obtain optimal results when control is performed by an FLC with only one group of parameters. In this type of situation, the parameters must be adjusted. The situation is one in which the parameters can only be adjusted by trial and error or by using experiential knowledge of adjustments.

In order to do this we can consider equipping the FLC with a self-tuning function for adjusting the parameters in a way that resembles the way that specialists would evaluate the control response and adjust them. We can consider two types of learning rules for adjusting FLC parameters, repetitive parameter tuning rules in which the performance of the control results is evaluated and the parameters are adjusted repeatedly, and real-time parameter tuning rules in which the control response is evaluated and the parameters adjusted in real time.

FLC parameters are first determined as well as possible by means of repetitive tuning rules from evaluations of control results. Then these are taken as initial parameters, and during the control operation they are further adjusted based on the control response using real-time tuning rules, at the time of adjustment. In this way the parameters are adjusted to the optimal values through a sequence of repetitive tuning and real-time tuning. Figure 3.30 shows the construction of a self-tuning FLC with these control rules. Here y_k^* is the target response.

(b) Repetitive Parameter Tuning

Repetitive parameter tuning is a method for setting FLC parameters in the best way possible by evaluating the control results, when control is turned off. The amounts of overshooting, time for reaching the target speed, and amplitude shown in Fig. 3.31 can be considered for the evaluation of

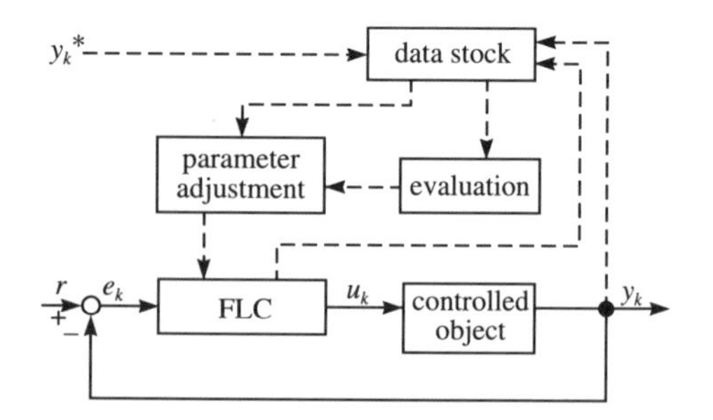

Fig. 3.30 Self-tuning FLC.

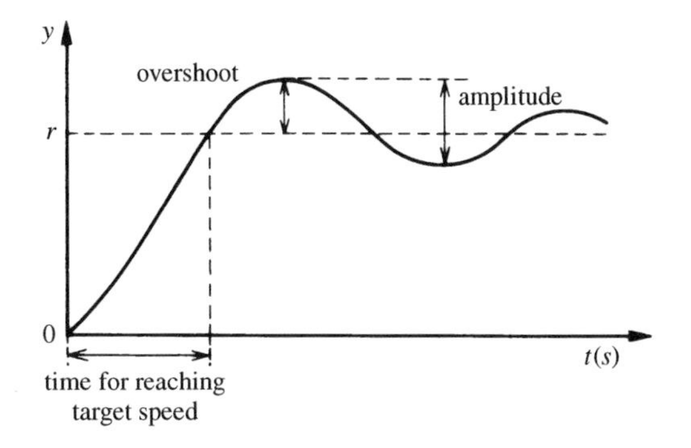

Fig. 3.31 Control performance.

control results. After evaluating the control performance of these and considering the authors' experience and things that resemble FLC and PID controllers, the rules for adjusting the parameters were determined as shown in Table 3.6. The e_{OV}, e_{RT}, and e_{AM} in the table are

$$\begin{cases} e_{OV} = OV - OV^* \\ e_{RT} = RT - RT^* \\ e_{AM} = AM - AM^* \end{cases} . \tag{3.18}$$

OV, OV^* are the actual and target values for overshoot; RT, RT^* are the actual and target values for time to reach target speed; AM, AM^* are the actual and target values for amplitude.

Table 3.6. Repetitive Parameter Tuning Rule

Antecedent		Consequent					
Control Performance		Δa_1	Δa_2	Δa_3	Δb_1	Δb_2	Δb_3
e_{OV}	P_{e1}	P_{a11}	N_{a12}	N_{a13}	P_{b11}	P_{b12}	N_{b13}
	N_{e1}	N_{a11}	P_{a12}	P_{a13}	N_{b11}	N_{b12}	P_{b13}
e_{RT}	P_{e2}	N_{a21}	P_{a22}	P_{a23}	N_{b21}	N_{b22}	P_{b23}
	N_{e2}	P_{a21}	N_{a22}	N_{a23}	P_{b21}	P_{b22}	N_{b23}
e_{AM}	P_{e3}	P_{a31}	N_{a32}	P_{a33}	N_{b31}	N_{b32}	P_{b33}
	N_{e3}	N_{a31}	P_{a32}	N_{a33}	P_{b31}	P_{b32}	N_{b33}

P = Positive, N = Negative

In addition, the membership functions for P and N in the antecedents are arctangent functions, and the P and N in the consequent are bell-shaped.

In order to find parameter corrections Δa_i and Δb_i ($i = 1, 2, 3$) from the rules in Table 3.6, we use the simplified method described in the following as a fuzzy inference method.

Now, let there be n rules given as follows:

$$\begin{cases} R1: & \text{If } x_1 \text{ is } A_1 \text{ and } x_2 \text{ is } B_1 \text{ then } y \text{ is } C_1, \\ R2: & \text{If } x_1 \text{ is } A_2 \text{ and } x_2 \text{ is } B_2 \text{ then } y \text{ is } C_2, \\ \vdots & \\ R3: & \text{If } x_1 \text{ is } A_n \text{ and } x_2 \text{ is } B_n \text{ then } y \text{ is } C_n. \end{cases} \tag{3.19}$$

Here x_1 and x_2 are antecedent variables and y is the consequent variable. A_i, B_i, and C_i ($i = 1, 2, 3, \ldots, n$) are fuzzy sets with membership functions $\mu_{Ai}(x_1)$, $\mu_{Bi}(x_2)$, and $\mu_{Ci}(y)$ ($i = 1, 2, 3, \ldots, n$). When nonfuzzy values ($\bar{x}_1$ and $\bar{x}_2$) are given for x_1 and x_2, the inferred value $\hat{y}$ for y can be found from the following simplified equation:

$$\hat{y} = \frac{\sum_{i=1}^{n} \left(\mu_{A_i}(\bar{x}_1) \wedge \mu_{B_i}(\bar{x}_2) \right) \cdot \bar{y}_i}{\sum_{i=1}^{n} \left(\mu_{A_i}(\bar{x}_1) \wedge \mu_{B_i}(\bar{x}_2) \right)}. \tag{3.20}$$

Here $\bar{y}_i$ ($i = 1, 2, 3, \ldots, n$) is the y for the maximum value of the membership function $\mu_{Ci}(y)$ ($i = 1, 2, 3, \ldots, n$).

Using Eq. (3.20) in the rules in Table 3.6, we can find the parameter corrections Δa_i and Δb_i ($i = 1, 2, 3$). FLC parameters a_i and b_i are corrected using the following equations:

$$\begin{cases} a_i^{\text{NEW}} = a_i^{\text{OLD}} + (1 - FP)\Delta a_i, \\ b_i^{\text{NEW}} = b_i^{\text{OLD}} + (1 - FP)\Delta b_i, \\ \alpha_i \leqq a_i \leqq \beta_i, \quad \gamma_i \leqq b_i \leqq \delta_i, \qquad (i = 1, 2, 3) \end{cases} \tag{3.21}$$

Here FP is a fuzzy performance index that will be discussed later, and α_i, β_i, γ_i, and δ_i ($i = 1, 2, 3$) are constants that determine the possible ranges for the parameters.

It is necessary to end repetitive tuning when optimal control results have been obtained. In order to do this, the fuzzy performance index for overshoot, the time for reaching the target speed, and the amplitude are set using the following equation:

$$FP = \min\{\mu_{OV}(e_{OV}^2), \mu_{RT}(e_{RT}^2), \mu_{AM}(e_{AM}^2)\}. \quad (3.22)$$

Here μ_{OV}, μ_{RT}, and μ_{AM} are membership values that express the quality of overshooting, time for reaching target speed, and amplitude, respectively.

Using this fuzzy performance index, tuning ends when the following equation is satisfied:

$$FP \geqq \theta, \qquad \theta \in [0, 1]. \quad (3.23)$$

Here, θ is the ending standard constant.

(c) Real-Time Parameter Tuning

Real-time parameter tuning is real-time adjustment of FLC parameters such that the control response (y_k) and target response (y_k^*) will agree when control response using parameters corrected by repetitive tuning as initial values is observed. We can consider the four patterns shown in Fig. 3.32a–d as relationships between target response and control response. For example, Fig. 3.32a means "response deviation was positive m samples ago, and the deviation is even more positive at present, so after m more samples the deviation will probably be even larger."

Here the relations in time k of these four patterns and the manipulated variable are as follows:

(1) when $e_k^* \gtrsim 0$ and $\Delta e_k^* \gtrsim 0$, increase Δu_k a large amount;
(2) when $e_k^* \gtrsim 0$ and $\Delta e_k^* \lesssim 0$, decrease Δu_k a small amount;
(3) when $e_k^* \lesssim 0$ and $\Delta e_k^* \gtrsim 0$, increase Δu_k a small amount;
(4) when $e_k^* \lesssim 0$ and $\Delta e_k^* \lesssim 0$, decrease Δu_k a large amount.

Here $\sim$ is a fuzzification symbol, and for time k,

$$e_k^* = y_k^* - y_k, \qquad \Delta e_k^* = e_k^* - e_{k-m}^*. \quad (3.24)$$

e_k^* is the response deviation, and Δe_k^* is the change in response deviation for m samples, where m is the number of samples.

In addition, if we let vertices C_1 and C_2 be the I point, C_3 and C_4 be the P point, and C_5 and C_6 be the D point in Fig. 3.26, the relations between the coordinates of the three vertices I, P, D and positive parame-

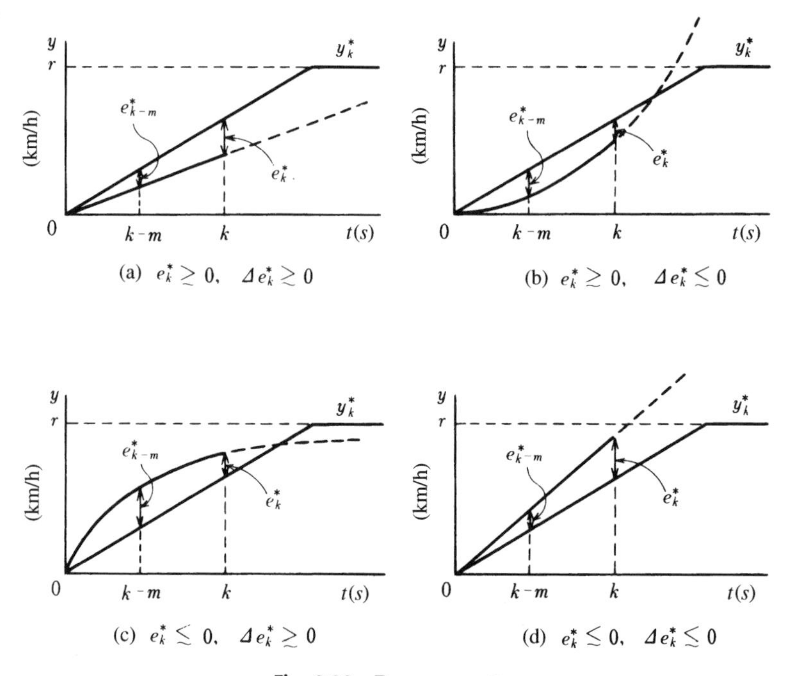

Fig. 3.32 Response pattern.

ters a_i and b_i $(i = 1, 2, 3)$ are

$$\begin{cases} I \text{ point:} & \left(\dfrac{1}{\pi} 2b_1 \tan^{-1}(e_k \cdot \tan(0.45\pi)/a_1), 1\right). \\ P \text{ point:} & \left(\dfrac{1}{\pi} 2b_2 \tan^{-1}(\Delta e_k \cdot \tan(0.45\pi)/a_2), 1\right). \\ D \text{ point:} & \left(\dfrac{1}{\pi} 2b_3 \tan^{-1}(\Delta^2 e_k \cdot \tan(0.45\pi)/a_3), 1\right). \end{cases} \tag{3.25}$$

From the I point of these equations we can see that "when e_k is positive, vertex I will move to the left if a_1 is made larger and b_1 is made smaller." If the vertex moves to the left, it produces a possibility of a reduction in the manipulated variable from that in Fig. 3.26. Thus, in order to vary the manipulated variable the three vertices must be moved simultaneously. Because of this, we can express the relationship between the manipulated variable, vertices, and parameters a_i and b_i in the following way:

(5) When we want to increase Δu_k, if the vertex is in the positive range ($e. \geqq 0$) we reduce $a.$ and increase $b.$. If the vertex is in the negative range ($e. \leqq 0$), we increase $a.$ and decrease $b.$. Here $a.$ and

Table 3.7. Real-Time Parameter Tuning Rule

Antecedent		Consequent	
e_k^*	Δe_k^*	Δa	Δb
P_4	P_5	NB_a	PB_b
P_4	N_5	PS_a	NS_b
N_4	P_5	NS_a	PS_b
N_4	N_5	PB_a	NB_b

P = Positive N = Negative
PB = Positive Big PS = Positive Small
NB = Negative Big NS = Negative Small

$b.$ are parameters related to $e.(e_1 = e_k,\ e_2 = \Delta e_k^*,\ e_3 = \Delta^2 e_k^*)$. (See Eq. (3.25).)

(6) When we want to decrease Δu_k, if the vertex is in the positive range, we increase $a.$ and reduce $b.$. If the vertex is in the negative range, we reduce $a.$ and increase $b.$.

From these relations (1)–(6), we obtain the parameter tuning rules shown in Table 3.7. Here the fuzzy sets for the antecedents are for the case when the vertices in Eq. (3.25) are in the positive range, and when they are in the negative range the symbols are the opposite ($P. \leftrightarrow N.$, for example). The membership functions for the antecedents are arctangent functions, and the conclusion functions are bell-shaped.

Parameters are then corrected by means of the simplified fuzzy inference in Eq. (3.20), following the rules in Table 3.7, but in order to simplify calculations here we will infer the value for a single parameter (a parameter related to the central vertex, Δa_1, for example), and we will use the method of proportional allocations of that value to the other parameters. In this case, after correction, parameter a_i is

$$\begin{cases} a_1^{\text{new}} = a_1^{\text{old}} + \operatorname{sgn}(I) \cdot \lambda_1 \cdot \Delta a_1, \\ a_2^{\text{new}} = a_2^{\text{old}} + \operatorname{sgn}(P) \cdot \lambda_2 \cdot \Delta a_1, \\ a_3^{\text{new}} = a_3^{\text{old}} + \operatorname{sgn}(D) \cdot \lambda_3 \cdot \Delta a_1. \end{cases} \tag{3.26}$$

Here, λ_i $(i = 1, 2, 3)$ is a positive proportional constant, and sgn $(\cdot)$ is the vertex sign (positive or negative). Parameter b_i is found in the same way.

Adjustments of parameters a_i and b_i are not carried out simultaneously; they are carried out alternately. This is because there are cases in which Δu_k will decrease when b_i is moved, even if a_i has been changed and Δu_k increased.

In addition, the parameter tuning timing is determined by a simple rule. This is "When the control deviation is small, the tuning interval is increased, and when the control deviation is large, the tuning interval is decreased."

(d) Simulation

The results of speed control simulations with the self-tuning FLC are shown in Fig. 3.33. The solid line in the figure is the result in the case when the initial parameter is given, which corresponds to an FLC without a parameter tuning function. The x line represents the results of control in which the parameters are corrected using the repetitive method, along with the results shown by the solid line, in which these are taken as initial parameters and adjusted in real time using the real-time parameter tuning method as control is being carried out. As can be seen from the figure, there is less overshooting, smaller amplitudes and more rapid response in the results when parameter tuning is used, when compared to those without it; furthermore, these are excellent results in which the changes in the manipulated variables are smooth. In addition, we can see that there is an improvement in the change in gain 50 seconds after inclines and declines have been registered with the FLC supplied with a learning function.

Finally, a road test of the self-tuning FLC introduced in this chapter was conducted by Takahashi *et al.* [13] at the Electronics Research Institute of Nissan Motor Company. The results are given in Ref. 13. From these we can see that the ride feels good, and there is absolutely no surge going up or down hills, and in addition, highly precise control results were obtained.

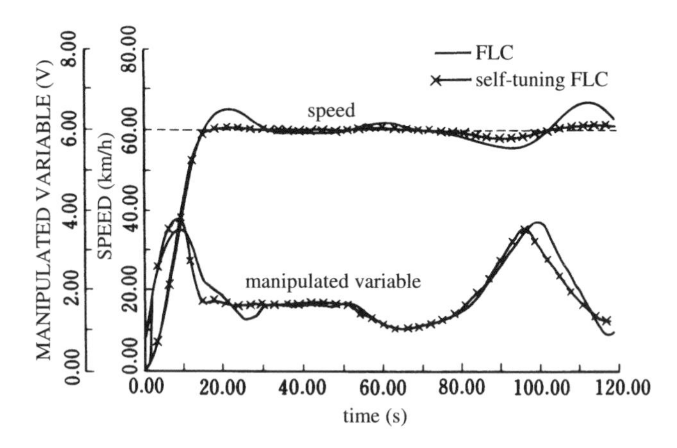

Fig. 3.33 Simulation results for speed control with self-tuning FLC.

3.4.5 Conclusion

In this section we first introduced the design methods for an FLC for automotive speed control and the results of a simulation and road test using it. The results were that speed control by FLC gave smooth acceleration unaffected by gear changes and made possible stable operation over a long period of time even with inclines and declines. In addition, since the feeling of the ride was good and driving easy, we confirmed that FLCs are effective for speed control in cases in which the dynamic characteristics change, as in automobiles.

Next, since learning is required for the establishment of FLC parameters, we designed an FLC with a self-tuning function for parameters in order to support this and in order to obtain good control results in an environment in which there are large changes in dynamic characteristics. From the results of a simulation with this, we saw that the self-tuning FLC certainly warrants application.

3.5 RAINWATER PUMP MANAGEMENT

3.5.1 Outline of Rainwater Pump Operation

We use a large amount of water as we carry out our daily lives and industrial activities. The waste water from everyday life and industry is processed by the sewage system and discharged into rivers and other waterways. The role of the sewage system is, as is shown in Fig. 3.34, not

Fig. 3.34 Rainwater pump facilities.

only to process dirty water and maintain our living environment, but also to quickly remove rainwater to the waterways when rain falls and bear the burden of preventing flood damage within cities. These rainwater pump facilities are located at breaks in the sewer lines and at sewage treatment plants. Automation of their operation is progressing, but the changes in the rainwater outflow process make complete automation difficult.

In recent years expansion of large cities has been advancing, park areas and vacant land decreasing and surfaces being covered with concrete, etc.; therefore, the amount of rainwater that can be absorbed by the earth has decreased. This is accompanied by an increase in the amount of rainwater runoff, and even with a small rainfall, a sudden runoff is produced. For example, flood damage attributable to the concentrated heavy rains of thunderstorms an the end of the rainy season (called *baiu* in Japanese) is a typical example, and this invites city flooding.

As countermeasures, rainwater control using permeable pavements and areas where rainwater can accumulate is being designed, and on the other hand, sewer lines are being expanded, and rainwater pump facilities and their volumes are being increased. However, this expansion of facilities and capacity creates an imbalance between pump output and the normal runoff, which leads to new problems such as increases in the frequency of pumps being activated and shut down and the need for opening and closing operations for inflow gates for times when the runoff exceeds pump capacity. It is difficult to deal with the changes in rainwater runoff and solve these problems by means of the pump well level control methods used up to now. There are many cases in which the ability of experienced people to operate pumps by adapting to circumstances is the only method. Because of this there is a requirement for a control system that eliminates the individual differences in operators and reduces loads.

Thus, in this section, we will describe how fuzzy adaptive control methods are effective for the peculiarities of rainwater pump operation and its automation.

3.5.2 Rainwater Pump Control

When rain falls, the role of rainwater pumps is to remove rainwater that has been collected by sewer lines to rivers and other waterways, and operation must be carried out in accordance with the amount of rainwater. There are examples of rainwater pump systems that employ speed control operation in part, but most pump facilities use control done by means of the number of pumps. Most of these pump quantity control systems use systems in which the relation between pump well level and number of pumps is determined beforehand or systems in which the relation between pump well level and target output for the pumps is determined by a mathematical function beforehand. In the former, one rainwater pump is activated when the well level reaches a certain level, and when it drops below a certain level, one pump is shut down. Therefore, operation management for the rainwater pumps is extremely simple, but since the changes in pump well level and the changes in the amount of rainwater runoff are dynamic, the following problems arise:

(1) Preparation time is needed for activating and deactivating rainwater pumps, and there is little accumulated volume in the wells. The results are that pump operations cannot keep up with the changes in well level resulting from the sudden changes in inflow from typhoons and concentrated heavy rains, and they are delayed.
(2) Information on inflow cannot be reflected in the operation, so we have uniform operation based on well level, not flexible operation. In addition, since operation parameters must be set for some particular circumstances, the system cannot respond to all circumstances that give rise to corresponding changes in rainwater inflow. The result is a high frequency of pump activation and deactivation.
(3) When rainwater inflow exceeds the pump output capacity, operation is carried out using inflow gates to restrain the inflow, but maintaining harmony between the operation of the gates and pumps is difficult.

Because of this, predictive operation of rainwater pumps using physical models and regression models to predict the amount of rainwater inflow has been tested. However,

(1) Reliable predictions are limited to water level and amount of rainfall, but there is a limit to locations for measurement.
(2) It is difficult to predict natural phenomena such as rainfall.
(3) The rainwater runoff mechanism is complex, changes in relation to things like the accumulation in the lines and the descent conditions, and has high nonlinearity.
(4) It is impossible to measure the amount of rainwater inflow, which is necessary for identification of parameters for a process model, and since use of assumed values cannot be avoided, there is a drop in model reliability.

And because of this, there is a great deal of work and many unsolved problems remaining in the way of actual application.

For these reasons, pump facilities, which cannot be fully automated, have used experienced operators who carry out safe pump operation management that keeps pump well levels low using a large amount of information not only about pump well levels, but also about rainfall conditions, rainwater inflow conditions and pump operation conditions, as well as the peculiarities of the rainwater inflow of their particular sewer system and their past operational experience. However, water accumulates in the sewer lines in relation to rainfall conditions; large changes are allowed in water level to hold down the frequency of pump switching as much as possible to extend the life of the pump facilities, and operation that makes possible a reduction in workload for operators is becoming necessary. In other words, rainwater pump operation management that allows for harmony between two opposing operational goals, depending on the rainwater inflow conditions, is required.

It is difficult to carry out this kind of operation using older methods, and in order to realize operation management that adapts to the circumstances, it is necessary to use the knowledge of experienced operators who work at pump facilities.

3.5.3 Characteristics of Fuzzy Adaptive Control

Fuzzy control is the modeling of the operational methods of experienced operators using control rules, and it attempts to realize control of the same quality as that of experienced operators using computers. Looked at another way, control rules are a tying together of circumstantial judgments of the control object and operational methods using an if/then form; therefore, we can interpret it as a fuzzy partition of the space determined by input variables and a description of the operational methods for each of these subspaces. Because of this, the control rules naturally increase if the number of input variables increases, and the design of control rules becomes difficult. Therefore, control rules are structured, based on the characteristics of the plant. An example of the structuring of control rules is shown in Fig. 3.35.

Figure 3.35a is an example of correcting the values from a mathematical model using fuzzy inference, and it is used for controlling the addition of coagulants [14] at water purification plants. Figure 3.35b is an example of finding correction values for the basic values for control outputs using fuzzy inference for each, and it is used for prechlorination addition control [15] at water purification plants. Figure 3.35c is an example of evaluating plant conditions using the fuzzy inference of the previous steps and using fuzzy inference to find control output values by means of those values and measured values. These structuring methods have control output values as their objects, but there are also fuzzy adaptive control methods in the methods for structuring the control rules themselves.

Fuzzy adaptive control is a control method that makes adaptations to the changes in plant parameters and changes the control algorithm. This

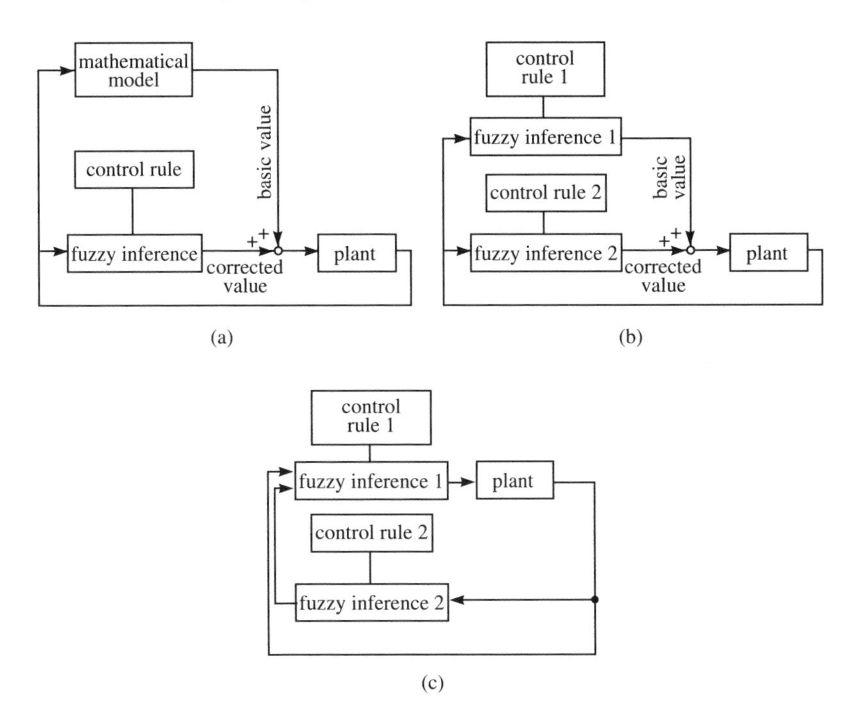

Fig. 3.35 Examples of fuzzy inference structuring.

method has a large effect in cases where plant characteristics such as process gain and dead time change with operational conditions and where a number of control objectives must be harmonized according to the circumstances. First let us consider two conditions for parameter a, which expresses the plant condition (we will let these be a_1 and a_2, and call them condition a_1 and condition a_2 hereafter), and design control rules parameters c_1 and c_2 for these conditions. Next we infer parameter a^*, which expresses the current plant condition, and from that value find the degree of closeness ω of it to conditions a_1 and a_2. Let it be 0 for condition a_1 and 1 for condition a_2 and define it as having a value between 0 and 1 between the two conditions. Control rule parameter c^* for condition a^* is determined from control rule c_1 for condition a_1 and control rule c_2 for condition a_2 using the weighted mean, with $1/\omega$ and $1/(1-\omega)$ for weighting:

$$c^* = c_1 \cdot (1 - \omega) + c_2 \cdot \omega. \tag{3.27}$$

As is shown in Fig. 3.36, the control rule parameter c^* reacts to changes in plant parameter a from a_1 to a_2 and changes continuously from c_1 to c_2. In other words, fuzzy adaptive control is a method for reacting to plant

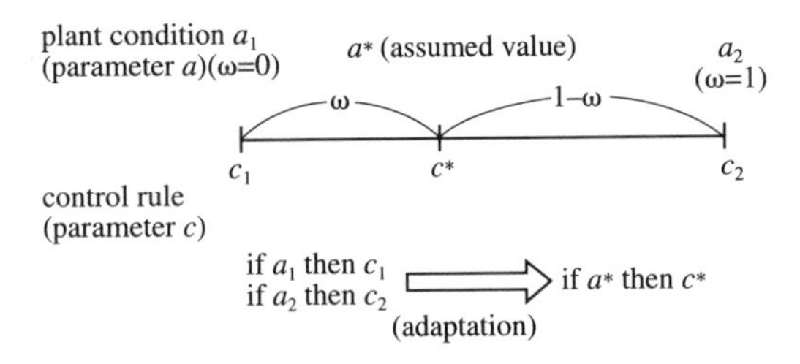

Fig. 3.36 Fuzzy adaptation operations.

conditions and adapting the control rule parameter using Eq. (3.27). In general, membership function parameters are used for control rule parameters. Therefore, one group of control rules is set up, but to respond to plant conditions a_1 and a_2, membership functions for two groups are defined for the simultaneously occurring fuzzy variables.

The preceding was for the case of a single plant parameter, but let us give a little more concrete explanation. Plant characteristics often change according to operational conditions. For example, we see examples of process gain and dead time changing with loads of 50% and 100% again and again. In cases like these we use load condition as a plant parameter. In addition, when the plant conditions change according to a number of factors, parameters that show the characteristics of plant conditions are inferred using fuzzy inference with these factors as input values. These plant parameters are given one-to-one correspondences with ω, and the membership function for any load condition can be found from Eq. (3.27) using the control rule membership functions for loads of 50% and 100%. Whether the antecedent membership function or the consequent membership function is made adaptive depends on plant characteristics. For example, when time constants or delay time changes according to load, the antecedent membership functions are made adaptive, and when process gain changes, the consequent membership functions are adapted. Even in cases of multiple objective control in which the control objectives change with loads of 50% and 100%, parameter ω is used to show which condition is the closest, using this same process, and harmonious control is possible through the adaptation of the membership functions for each of the objectives.

3.5.4 Control Rule Design

The following is an example of judgmental standards for the operation of rainwater pumps by experienced operators:

(1) When the present pump well level is high and the forecast is for further increase, one more pump should be activated.

(2) Conversely, when the pump well level is low and the forecast is for its lowering further, one pump should be deactivated.
(3) When there is a forecast of a sudden increase in rainwater inflow, pumps should be activated in advance, and pump well level maintained at a low level.
(4) If there is a forecast for a rise in pump well level that will present no danger, the pumps should be left as they are, even if the well level rises.
(5) When there is a large volume of rainwater inflow, a low well level should be maintained for safety.

The main characteristic of these judgments is that they are two-stage judgments, as is shown in the following.

Step 1: Are the rainwater inflow conditions safe, or do they require caution?
Is present inflow large or small?
Will inflow increase or not?

Step 2: Judging from the rainwater inflow conditions, is the present pump well level appropriate?
Are additional pumps necessary?
Should one pump be deactivated?

Therefore, if we let the judgment of the safety of rainwater inflow conditions be the plant parameter, let the standard conditions be the safe condition of a small rise in pump well level that comes with weak rains, and the dangerous condition be the large inflow that comes with typhoons and other storms, and let the index used for evaluating the present inflow conditions be ω (defined as $\omega = 0$ for fine weather and weak rains and $\omega = 1$ for typhoons), we can employ fuzzy adaptive control. In addition, since the judgment of pump well level and its change varies with the inflow conditions at the time, it is appropriate to adapt the antecedent membership function using ω. In other words, we set up two membership functions, one with the object of reducing the frequency of pump activation and deactivation as much as possible during weak rains and the other with a safety-first objective for typhoons, and if we adapt them using ω, we can obtain harmonious pump operation [16] that responds to rainwater inflow conditions.

Figure 3.37 is a block diagram of these control calculations. First of all, the rainwater inflow condition is inferred from the present inflow and amount of rainfall, which gives us a prediction of the inflow to come. Since rainwater inflow cannot be measured directly here, pump output is used instead. Using this inferred value ω, the antecedent membership functions for control rule 1, which is used for inferring pump output, are adapted

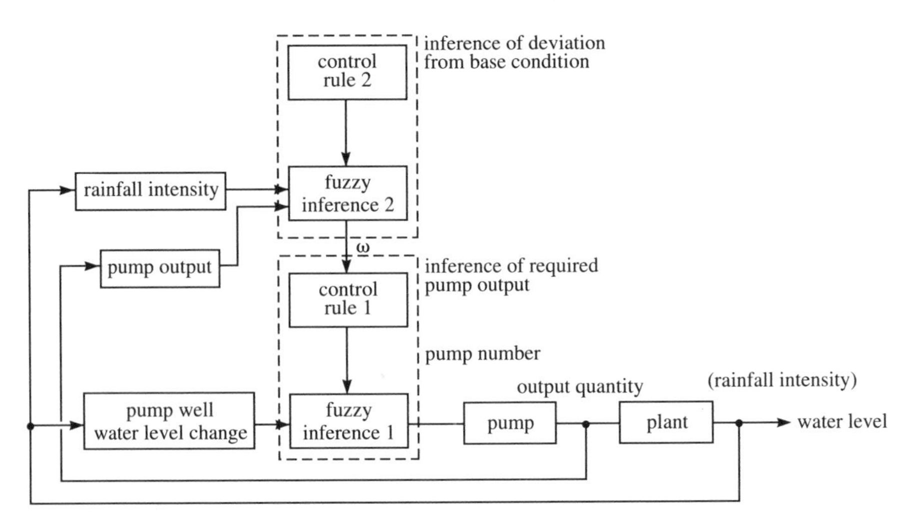

Fig. 3.37 Block diagram of fuzzy adaptive control.

from pump well level and its change. At this time the number of operating pumps is determined from the pump output that is obtained through fuzzy inference.

The method for fuzzy inference is the algebraic product of the consequent membership functions of the control rules at the minimum grade for each antecedent proposition and finding the center of gravity [17] of the membership function, which is obtained through a MAX operation with each control rule. In addition the antecedent and consequent membership functions are triangular membership functions. Therefore, the adaptation operations for the antecedent membership functions come out as follows, using parameter ω:

$$\begin{cases} p_1 = p_{10} \cdot (1 - \omega) + p_{11} \cdot \omega, \\ p_2 = p_{20} \cdot (1 - \omega) + p_{21} \cdot \omega, \\ p_3 = p_{30} \cdot (1 - \omega) + p_{31} \cdot \omega. \end{cases} \tag{3.28}$$

For example, if we let the target water level during weak rains and the target water level during typhoons be the solid lines in Fig. 3.38, the target water level (MM for water level H) when ω is 0.5 will be the dashed line. In the same way all of the membership functions that express the pump well level and the water level change are calculated using the value for ω and Eq. (3.28). If fuzzy inference with control rule 1 is then carried out using the membership functions obtained, fuzzy adaptive control with ω as the adaptive index can be performed.

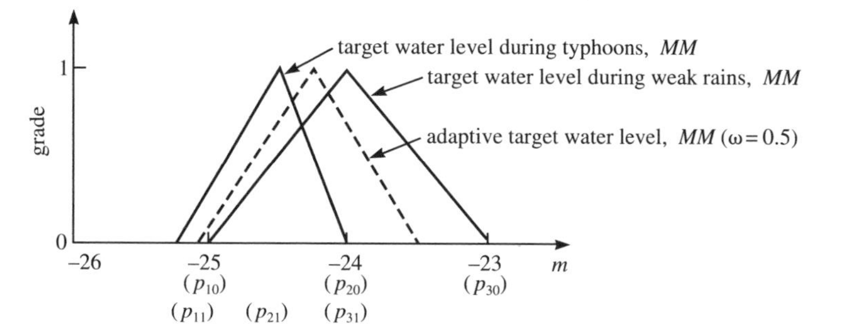

Fig. 3.38 Adaption of membership functions for water level (*MM*).

Examples of control rules for fuzzy adaptive control of rainwater pumps are shown in Table 3.8 and Table 3.9, and the meanings of the fuzzy variables are shown in Table 3.10 and Table 3.11. All of the fuzzy variables for the antecedents have five levels (SA, SM, MM, ML, LA), and the fuzzy variables for the consequents have six or seven levels. Therefore, if we express the meaning of the control rules in words, we get, "When the rainfall intensity is normal (MM) and pump output is somewhat large (ML), the inflow is somewhat large (ML)" for the case indicated by parentheses in Table 3.9. Furthermore, an example of membership functions that show the content of the fuzzy variables is given in Fig. 3.39.

The number of control rules shown here is 50, but there are usually four input items with fuzzy inference, so we get 625 (5^4), and we can see that there has been a large reduction. This means that control rule design is easy and there is an improvement in the reliability of the control.

3.5.5 Effects of Fuzzy Adaptive Control

We will show the effects of using fuzzy adaptive control for the operation of rainwater pumps using the results of a computer simulation. The

Table 3.8. Examples of Control Rule 1

		H Water Level				
		SA	*SM*	*MM*	*ML*	*LA*
DH Water Level Change	*SA*	*NB*	*NM*	*ZE*	*ZE*	*ZE*
	SM	*NM*	*NS*	*ZE*	*ZE*	*PS*
	MM	*ZE*	*ZE*	*PS*	*PS*	*PM*
	ML	*ZE*	*PS*	*PM*	*PM*	*PM*
	LA	*PS*	*PM*	*PM*	*PM*	*PB*

Table 3.9. Example of Control Rule 2

		QP Pump Output				
		SA	*SM*	*MM*	(*ML*)	*LA*
R Rainfall Intensity	*SA*	*ZE*	*ZE*	*SM*	*MM*	*ML*
	SM	*SM*	*SM*	*MM*	*ML*	*ML*
	(*MM*)	*MM*	*MM*	*MM*	(*ML*)	*LL*
	ML	*MM*	*MM*	*ML*	*LL*	*LL*
	LA	*MM*	*ML*	*ML*	*LL*	*LL*

process to be controlled is conceptualized in Fig. 3.40. The length of the sewer pipe is 2.05 km; it is a round pipe with a diameter of 5.5 m and a slope of 1/1000, and the area of the pump well, including sand settling basins, at the pumping station is 450 m^2. There are nine rainwater pumps, each with a rated output of 5.5 m^3/s.

When it rains, the rainwater flows through the sewer line, and the rainwater pumps remove it from the pumping station. The main characteristics of this process are the flow delay within the sewer line and the accumulation of rainwater. Since the latter is larger than the capacity of the pump well, these effects cannot be ignored. Because of this, a detailed mathematical model is necessary, and a continuous equation and a one-dimensional motion equation for unsteady flow are used. The computer simulation imitates a real plant by solving the mathematical model for the process, making for changes in the flow of rainwater over time, and simultaneously carrying out control calculations that use pump well level, its change, rainfall intensity, and total pump output for each timing as inputs. Fuzzy adaptive control of rainwater pumps and the pump well level-based pump number control used up to now were carried out for a variety of rainfall patterns, and the differences in pump well level and pump operational conditions for each were compared.

Table 3.10. Meaning of Fuzzy Variables in Antecedent

Fuzzy Variable	Water Level	Water Level Change	Pump Output	Rainfall Intensity
SA	Low	Falling	Low	Weak
SM	Somewhat low	Somewhat falling	Somewhat low	Somewhat weak
MM	Close to objective	Almost steady	Normal	Normal
ML	Somewhat high	Somewhat rising	Somewhat high	Somewhat heavy
LA	High	Rising	High	Heavy

Table 3.11. Meaning of Fuzzy Variables in Consequent

Fuzzy Variable	Pump Output Change	Fuzzy Variable	Rainwater Inflow Condition
NB	Large reduction	*ZE*	Low
NM	Reduction	*SM*	Somewhat low
NS	Small reduction	*MM*	Normal
ZE	As is	*ML*	Somewhat high
PS	Small increase	*LA*	High
PM	Increase	*LL*	Very high
PB	Large increase	—	—

Figure 3.41 shows the results of a simulation of fuzzy adaptive control when the maximum rainfall is 80 mm/h. In this instance the control rules and membership functions are those discussed in the previous section. For comparison, the changes in pump output and well level in the case of pump number control are shown with dashed lines. With fuzzy adaptive control we know that the inflow of rainwater is not so large 30 min after the beginning of the rain from the valve for ω, which is the value of the judgment of flow conditions, and from the rainfall conditions we also determine that there is no danger, so the pump well level is allowed to rise and it is handled by accumulation in the line. After this the rainwater inflow is judged to be getting closer to the danger point with the increase in rainfall, so one rainwater pump after another is activated, and the pump well level goes down. After 60 minutes all pumps are activated and output cannot be increased, so the water level can be seen to rise, but we can see that overall there has been pump operation management that corresponds to the inflow conditions. From this we can confirm that in the case of the pump number control indicated by the dashed line, there is a high level of flood danger, with the maximum water level reached being -22.33 m and a variation in water level of 3.86 m, when compared with the low values of

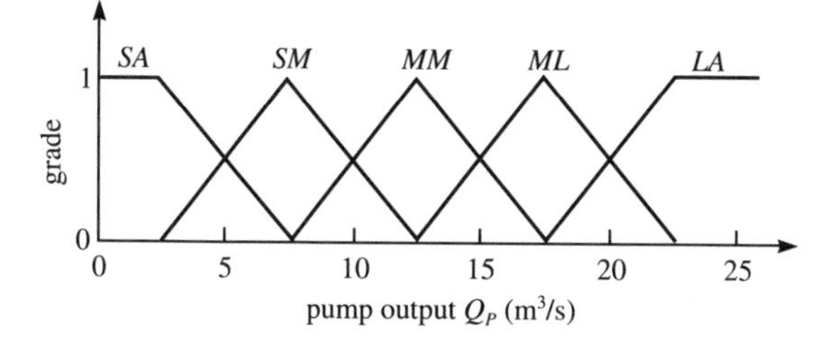

Fig. 3.39 Example of membership functions for pump output.

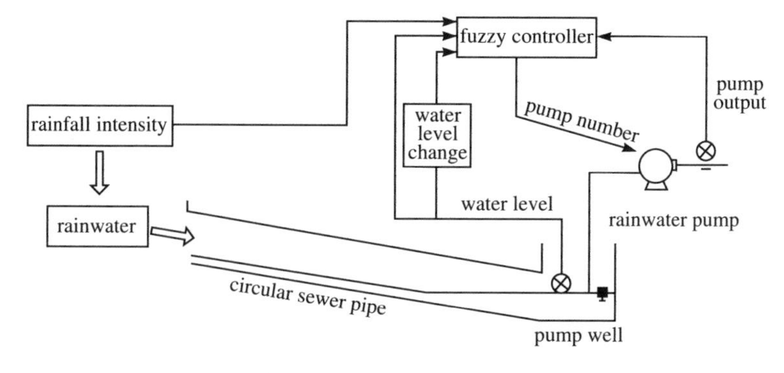

Fig. 3.40 Process concepts.

−23.48 m and 2.71 m, respectively, for the solid lines showing fuzzy control.

On the other hand, the results of a simulation in which fuzzy adaptive control is not carried out, where the judgment is normally $\omega = 0$ or $\omega = 1$, is shown in Fig. 3.42. The solid line shows the changes in pump output and pump well level for when $\omega = 0$ is the normal judgment, and the dashed

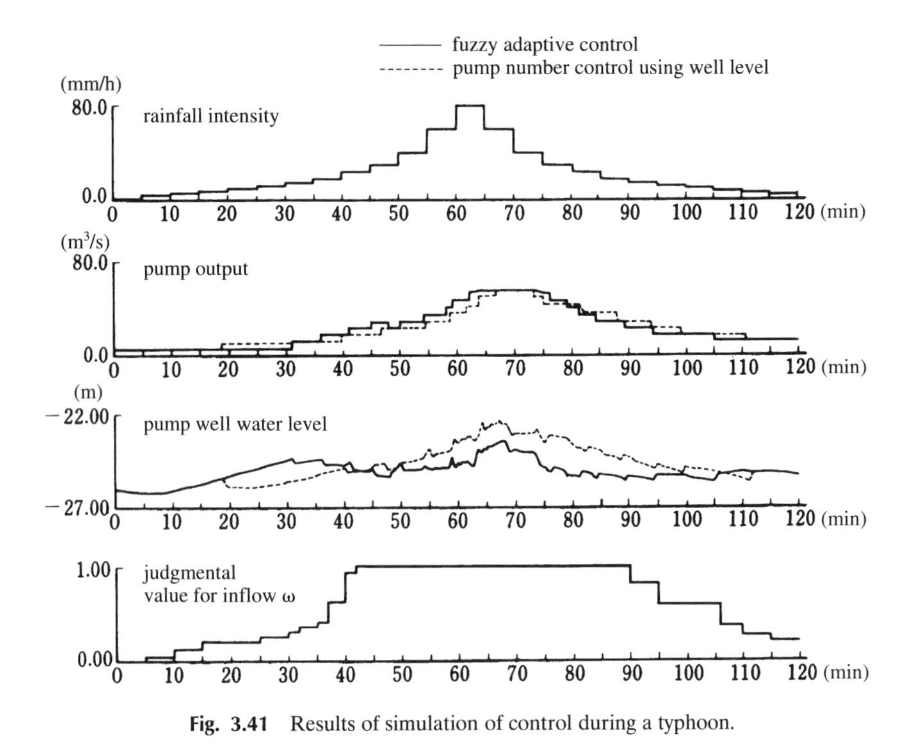

Fig. 3.41 Results of simulation of control during a typhoon.

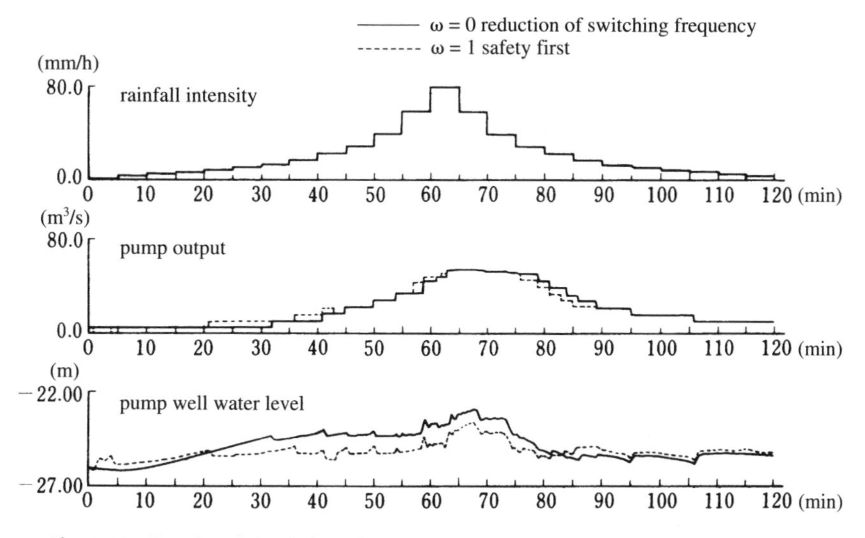

Fig. 3.42 Results of simulation of control when harmonious control is not carried out.

line shows them when $\omega = 1$ is the normal judgment. With the solid line, pump operation follows the pump well level with a delay in order to make use of accumulation in the line, and with the dashed line pumps turn on and off with even a small change in water level, so the well level is kept at a fixed low level.

From the results of these simulations we can see that fuzzy adaptive control makes for harmonious operation with two methods of operation: that which minimizes the activation and deactivation of pumps by making use of accumulation in the line, and that which puts safety first and maintains a low pump well level. We can presume that this kind of fuzzy adaptive control will also be effective when there are changes in the strength of rainfall. Figures 3.43 and 3.44 show the results of simulations with a maximum rainfall of 7 mm/h. The maximum increases for pump well level are -25.33 m and -23.55 m, respectively. With pump well level-based pump number control there was irregular operation with pumps turning on and off frequently, but with fuzzy adaptive control rainwater inflow is judged to be low, so we see that the scheme of operation that makes use of accumulation in the line functions well.

We have used fuzzy adaptive control for the operation of rainwater pumps, and shown the effects using simulations. At the same time we have confirmed that we get results such as

(1) Control rule design is easy, and we can get by with a small number of control rules.
(2) There is an increase in the robustness of control.

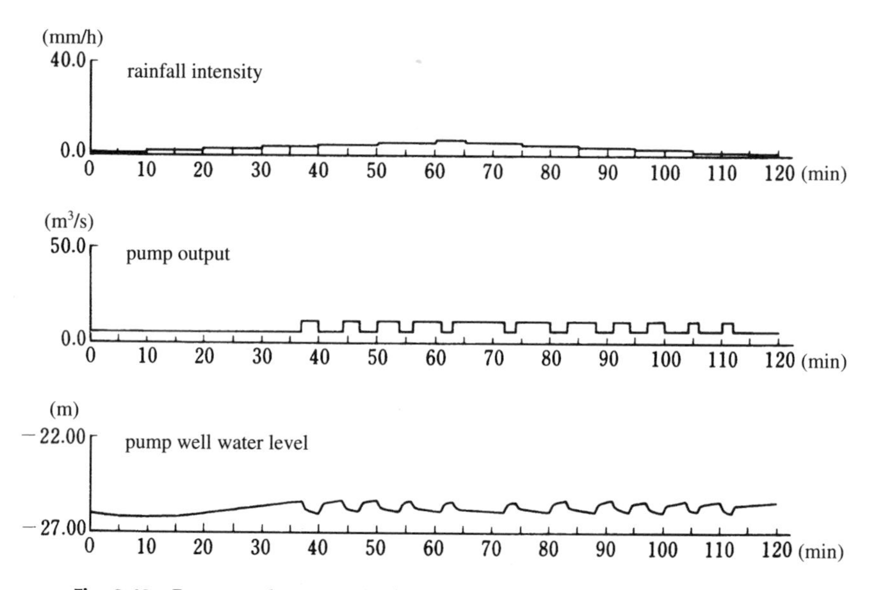

Fig. 3.43 Pump number control using pump well water level during weak rainfall.

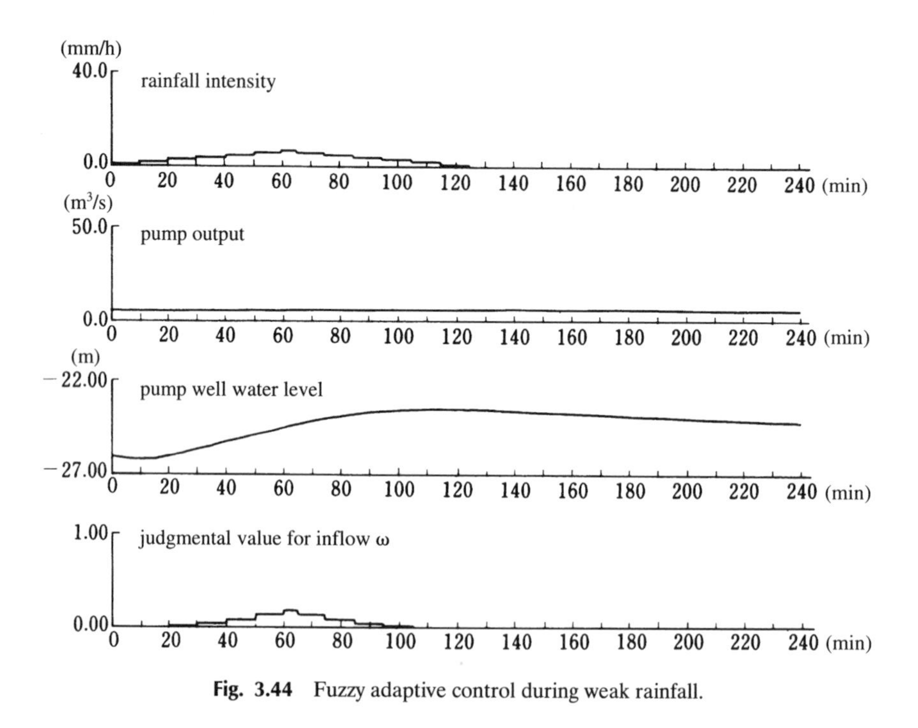

Fig. 3.44 Fuzzy adaptive control during weak rainfall.

These effects play a very important role in rainwater pump management, which is a process in which the conditions change drastically and for which operations must vary according to those changes. This is because there is a great need to improve the understandability of control calculations, the ease of design and maintenance and the reliability of pump operation, in which a judgmental mistake in operations raises the fear of inviting disaster.

We have discussed the application of fuzzy adaptive control to rainwater pump management, but in terms of applications in process control, there is a need for designing control systems that are shaped to process characteristics. In order to do this, past control methods must be coordinated, and control system design technology that brings out the virtues of all of them must be developed.

3.6 AUTOMATIC TRAIN OPERATION

Fuzzy control is one experiment in the attempt to automate operations (control) that have been performed by people up to now by making use of microcomputers. In other words, it is an attempt to make up algorithms for the intelligent activities of experienced operators involved in operation and to perform these intelligent activities regularly in the same way experienced operators would. In recent years applications in various actual systems have been carried out with great enthusiasm, as will be shown in the various sections of this chapter.

Because of this, the authors proposed a predictive control system that can determine control commands in real time and in which the balance among the various control objectives peculiar to the system is thoroughly evaluated, and at that time, when a control order is output, what will happen to the system is predicted by simulation in order to make for an ideal form of operation. Applications in actual systems are moving forward [18–24]. Here we will use this system for automatic train operation and introduce the application of the predictive fuzzy controller for automatic train operation in the Sendai subway system (Fig. 3.45) that started operations in July 1987.

3.6.1 Train Operation and Conditions for Automation

Train operation involves control that begins acceleration at a departure signal, regulates train speed so that it does not exceed the control speed, and stops the train at a fixed position at the next station that requires a stop. The control objectives (evaluation indices) that experienced opera-

Fig. 3.45 Cars of the Sendai subway system operated using fuzzy control.

tors must consider during the overall operation of trains are things such as (1) guaranteeing the safety of passengers, (2) maintenance of a comfortable ride, (3) conformation of the control speed, (4) stopping accuracy at stations, (5) shortening of running time between stations, and (6) reduction of power consumption. Operators evaluate these things in an appropriate way, specifically and generally, according to circumstances, and produce good operation that satisfies people (Fig. 3.46).

The attempt to use a computer instead of operators to operate trains is automatic train operation. This type of system has been put into operation in various places in Japan using conventional control methods that generate target speed patterns from control speed signals and surface devices (ground point signals) and coordinate train speed with these by calculating control commands for applying power (acceleration) and braking (deceleration).

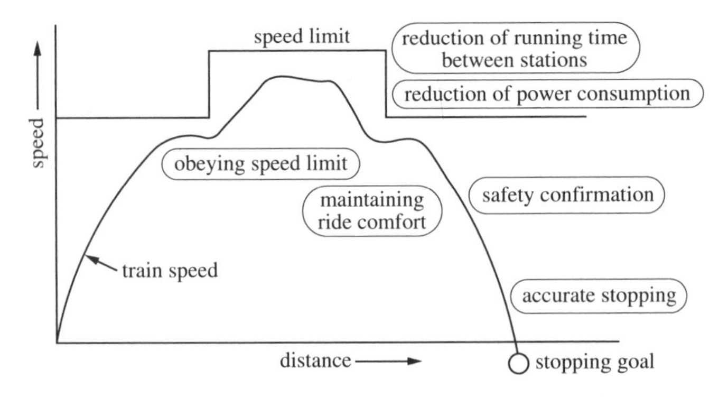

Fig. 3.46 Objectives of train operation by experienced operators.

However, with these systems, the actual controllers that are installed on trains only perform minute-by-minute evaluations of speed deviation, which tells how far the train's actual speed is from the target speed pattern. Therefore, the ride is not good, and the basic control objective of operating trains for accurate stopping means indirect consideration using small variations in control constants for the speed control and the shape of the target speed pattern that corresponds to speed and distance.

3.6.2 Experiential Rules for Train Operation

Here we will consider train operation by experienced operators. When people operate trains they do not regulate train speed by accurately following a predetermined target speed pattern as computers do. For example, for stopping control, they carry out quality operation by considering the number of passengers, the strength of the brakes, and whether they can stop and maintain passenger comfort at the time of acceleration.

For example, when an operator is trying to stop the train and is 20–30 m before the target he thinks something like "It looks like I'll be able to stop just fine as is," or "If I let up on the brakes a little, I'll be able to make an accurate stop and there will be no problem with passenger comfort," as he operates the brakes. In other words, operators consider control objectives like "stop just fine," "accurate stop," and "passenger comfort" as they drive trains. In addition, the operation characteristics of the train, such as stopping ability if the brakes are maintained or pressure lessened, are also included in the judgment.

Therefore, the experiential rules considered by the operator can be expressed in the form of "If, when braking is maintained, stopping can be carried out properly, maintain braking," or "If, when braking is lessened, the ride will be good and stopping can be carried out accurately, lessen braking." As in the preceding, train operation can be taken as a decision-making problem for determining optimal control commands based on these experiential rules.

Therefore, the authors assembled the control command operations of operators in the form of experiential rules as follows, based on our experience in the automation of train operation using conventional control methods. Furthermore, in train operation, control commands are dealt with dispersedly and are known as notches, so we will call the following control commands notches. In addition, train operation can be divided into two types of operation: between-station running, which mainly involves minding the speed limits, and fixed-position stopping control, in which the train must be stopped at a prescribed target position. We will describe the experiential rules for these.

(a) Station-to-Station Running Control

On departure, operators accelerate using the highest notch. When they reach the speed limit, they maintain the train at a slightly lower speed. If there is a section with a speed limitation ahead, the brakes are applied so as to reduce the speed by the beginning of it. In addition, in cases where time allows before the next station, they drive with the thought of reducing power consumption by coasting (running freely without a control notch output).

Concretely, control notches are selected through a continuous parallel consideration of the following experiential rules:

(C-1) When it seems that the speed limit will be exceeded, the output is positioned somewhere between the current notch and the emergency brake. By this means, the emergency brake shocks caused by ATC (Automatic Train Control/Protection) systems when the speed limit is exceeded are eased, and safety and riding comfort are maintained.

(C-2) Even when coasting, if the train is in a position to be on time, there should be no braking or power notch outputs. In this way power consumption is reduced.

(C-3) Even if a maximum acceleration command is being output, if the speed is far lower than the speed limit, a maximum acceleration command should be output. This decreases running time.

(C-4) If it seems that an appropriate speed just below the speed limit can be maintained with the current notch, the current notch should be maintained. This eliminates wasteful notch changes and improves the ride.

(C-5) If an appropriate speed cannot be maintained with the current notch, the notch within the range of $\pm n$ notches that will most accurately adjust the train speed to the appropriate speed should be chosen. In order to maintain a comfortable ride at this time, the size of the change from the previous notch and the elapsed time should be considered, the notch changed and the running speed and ride maintained appropriately.

(b) Fixed-Position Stopping Control

When the train passes the marker (showing the point at which fixed-position stopping control should begin, according to approach speed and the characteristics of that station, hereafter called point B) situated before the station, the operator begins braking. After this, operation is carried out with consideration for the comfort of passengers and stopping accuracy.

Concretely, control notches are selected through a continuous parallel consideration of the following experiential rules:

(T-1) Braking should not be applied before point B. In this way unnecessary deceleration before the station is avoided, and running time is shortened.

(T-2) After point B has been passed, a little braking should be applied in order to stop at the fixed point. This makes for an improvement in riding comfort when fixed-position stopping control begins in earnest.

(T-3) After passing point B, if it seems that stopping can be accomplished well without changing the current braking notch, that notch should be maintained. This way wasteful notch changing is eliminated and riding comfort is improved.

(T-4) After passing point B, if it does not seem that stopping can be accomplished with the current braking notch, then the notch, within the range of $\pm n$ notches, that will most accurately stop the train should be selected. In order to maintain a comfortable ride at this time, the size of the change from the previous notch and the elapsed time should be considered, the notch changed and the ride and stopping precision improved.

3.6.3 Predictive Fuzzy Control Method

This is a computerization method that works through an organic combination of three techniques (Fig. 3.47): (1) fuzzy theory that evaluates the multiple objectives of the system to be controlled in a non-crisp way, (2) artificial intelligence that is a formulation of the experiential rules for control knowledge in the form of if/then rules, and (3) real-time simulations that make real-time predictions of control results.

As is shown in Fig. 3.48, the construction of this system is one in which simulations of control results for all control command value candidates are conducted based on observed values for system conditions, evaluation quantities that generate concern as control objectives are predicted, multiple objective evaluation is carried out along with these predictions using fuzzy functions, these values and control knowledge are processed by means of fuzzy inference, and optimal control commands are determined. The following is a simple introduction to what this entails [18].

(a) Control Rule Formulation

Control rules for predictive fuzzy control can be shown generally as

$$\text{If } (u \text{ is } C_i \rightarrow x \text{ is } A_i \text{ and } y \text{ is } B_i) \text{ then } u \text{ is } C_i \qquad (i = 1, \ldots, n). \quad (3.29)$$

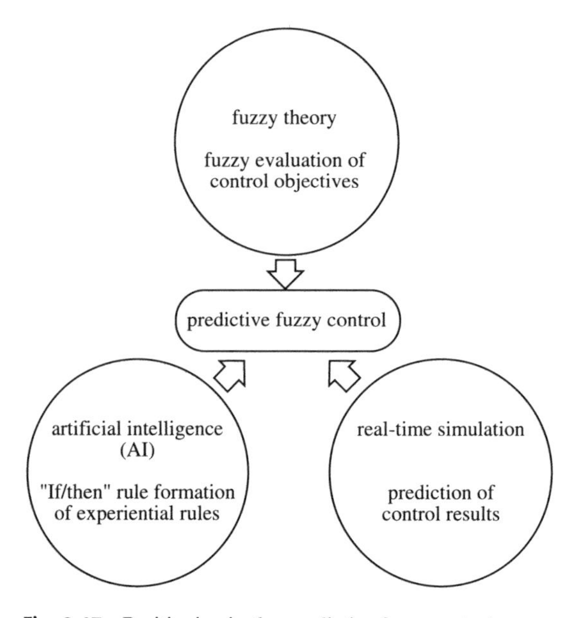

Fig. 3.47 Positioning in the predictive fuzzy control system.

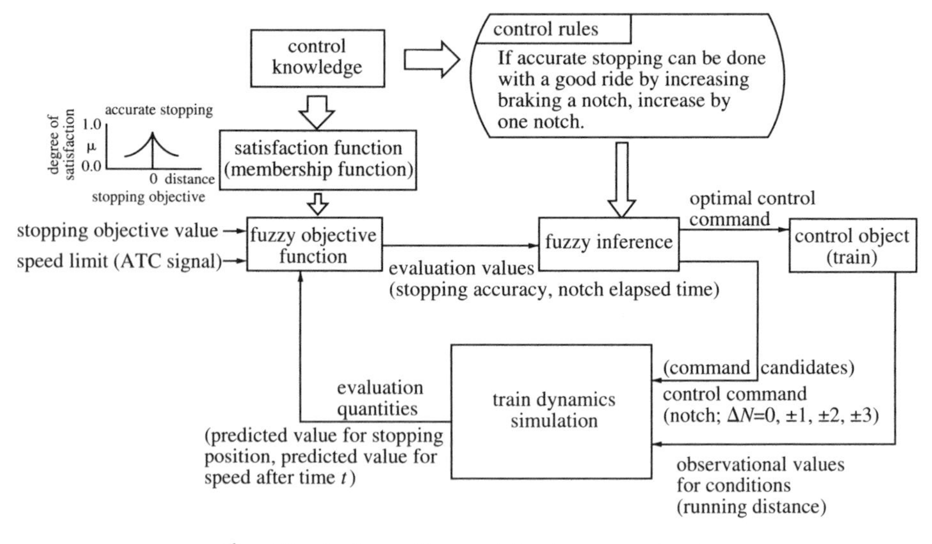

Fig. 3.48 Outline of the predictive fuzzy control system.

The point that shows a large difference from the fuzzy control used by Mamdani *et al.* [3] in the past is the fact that the x and y values are the evaluation indices for predicting what will happen to the control objectives when the control command is put into effect as C_i. Of course there are cases in which the evaluation indices do not depend on control command C_i, the attempted output, and if not all of them exist, evaluation is carried out with the current evaluation indices; this is consistent with older fuzzy control, which just evaluated the conditions.

If we rewrite Eq. (3.29) in words, we can express it as, "If at this time we let control command u be C_i, if evaluation index x is A_i and evaluation index y is B_i, we use this control rule R_i and output C_i as the control command."

If we let the premise of control rule R_i be P_i and its membership function be $\mu_{Pi}(C_i : x, y)$, the fuzzy set can be expressed as

$$P_i = \int \mu_{p_i}(C_i : x, y)/(x, y) = A_i \times B_i : u = C. \tag{3.30}$$

This is the Cartesian product of membership functions μ_{Ai} and μ_{Bi} for A_i and B_i.

(b) Method for Determining Control Commands

Here we will let the time that control is being performed be t and consider control rule R_i. If at this time the control command "u is C_i" is performed, we predict what will happen to evaluation quantities x and y as fuzzy sets, based on partial knowledge (an ambiguous model) of the controlled system, and say that fuzzy sets $x(C_i : t)$ and $y(C_i : t)$ are found as those control results. The control rule premise at this time, $P_i|t$, is found by means of

$$P_i|t = A_i \cap x(C_i : t) \times B_i \cap y(C_i : t). \tag{3.31}$$

The degree of satisfaction of control command C_i can be found from the maximum values for the membership functions for each control rule R_i at time t.

The optimal control command C_0 at current time t can be found from the degrees of satisfaction for each control command found in this way. The methods determining the output control command are the method of using control command C_j, which has the control rule R_j with the highest degree of satisfaction, just as it is, or the method of using the center of gravity for each control command, and the appropriate method is used for the object being controlled.

As was mentioned earlier, this type of control makes use of a system model along with a definition of each control rule by means of fuzzy sets, so operation that takes care of all strategic points can be carried out by means of control rules, the most important characteristic being their small number.

3.6.4 Automatic Train Operation System with Predictive Fuzzy Control

The problem in computerization of the experiential rules for operators that were discussed in Section 3.6.2 is how to define and quantify the evaluations of the control objectives related to safety, comfort, energy savings, speed, elapsed time, and stopping precision by means of fuzzy sets. In order to do this we first quantify the linguistic meanings of the evaluation indices for each control objective by means of fuzzy sets as shown in Fig. 3.49, and make the experiential rules over into fuzzy control rules.

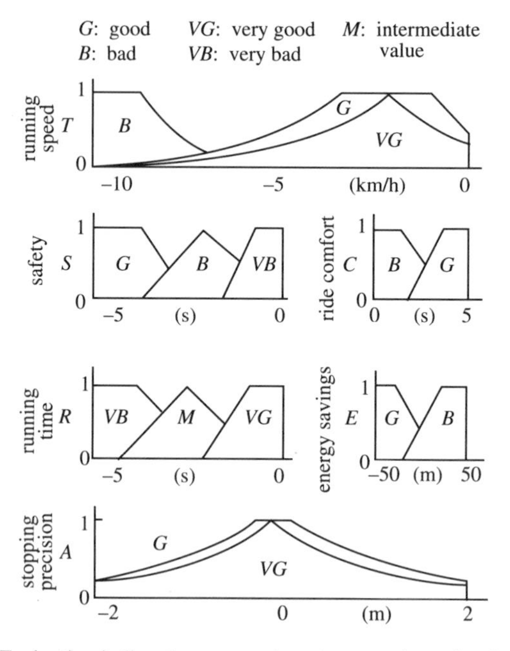

Fig. 3.49 Evaluation indices for automatic train operation using fuzzy control.

(a) Fuzzy Sets for Evaluation Indices

(1) Definition of the Evaluation Index for Safety (S). The basic safety measure of preventing collisions with the train ahead is insured by the ATC system, which takes the top position in the automatic train operation system. With the ATC system, the emergency brakes are applied if the speed limit is exceeded, and the train speed is reduced to a speed below that speed. At this point, if there is a section ahead for which the speed limit is lower than the current speed of the train, danger is evaluated in terms of the time required to arrive at the point of speed reduction, in order to reduce the braking shock upon entering that section. Here the evaluation index is defined as "Safety is good: S is G" when the required time is large and "Safety is bad: S is B" when the required time is small.

(2) Definition of the Evaluation Index for Comfort (C). The problem here is how to control forward and backward rocking of passengers during automatic train control and improve the comfort of the ride. The quantified evaluation index for this comfort is jerking motion of time-based changes in acceleration. However, measuring this value and getting feedback on it is difficult. Therefore, comfort is evaluated in terms of the elapsed time in a fixed notch position after the last notch change, since passengers receive large shocking vibrations when control notches are changed frequently. Here, the evaluation index is defined as "Comfort is good: C is G" when the time after the notch change is long and "Comfort is bad: C is B" when it is short.

(3) Definition of the Evaluation Index for Energy Savings (E). Energy is saved when there is no notch output and the train travels forward on inertia (coasting). This coasting naturally increases the running time. Therefore, in cases when it is possible to save energy with the knowledge that coasting from one specific place to another between stations increases the running time, the train is put under coasting control from its current location to that location. The evaluation index is defined as "Energy savings is good: E is G" when the train is operating within one of these low-energy sections.

(4) Definition of the Evaluation Index for Speed (T). In terms of speed, the train speed must be in an appropriate range a little below the speed limit. This speed is not evaluated in terms of the current speed, but in terms of what will happen if the current control notch is maintained and what will happen if the control notch is changed. This is evaluated by presuming these control notch changes and estimating the train speed 2–3 seconds later. Here the evaluation index is defined as "Speed is good: T is G" if the predicted speed falls a little below the speed limit and is near the

appropriate speed, "Speed is very good: T is VG" if it equals it, and "Speed is bad: T is B" if it is very low.

(5) Definition of the Evaluation Index for Running Time (R). Running time is controlled to a great extent by braking deceleration for stopping at stations. Normally, the point for engaging the brakes (point B) is determined considering the braking capacity of the train. If fixed-position stopping control commences before this point B, there is more braking deceleration and running time is lengthened. Here the evaluation index is defined as "Running time is very bad: R is VB" far before point B, "Running time is medium: R is M" a little before, and "Running time is very good: R is VG" after it.

(6) Definition of the Index of Stopping Accuracy (A). For braking precision, where the train will stop is predicted from the current speed and position under the conditions in which the current braking notch is maintained, braking increased and braking decreased, and this is evaluated in terms of the predicted stopping point versus target point. Here the evaluation index is defined as "Stopping precision is good: A is G" if this predicted stopping point is near the target and "Stopping precision is very good: A is VG" if they are the same.

(b) Determination of Fuzzy Control Rules

In the foregoing we formulated fuzzy control rules from the experiential rules for train operation control that were qualitatively described in Section 3.6.2, using the evaluation indices we defined. Here DN denotes the amount of change in the current control notch value; P_n, power notch n; and B_n, braking notch n. In addition, we denote the emergency brake as $B_{\max}$ and maximum acceleration as P7.

For example, let us rewrite the experiential rule for improving ride comfort and stopping precision, which are described under experiential rule (T-4) for fixed position stopping, in the form for predictive fuzzy control. For this experiential rule we have "If, when braking is increased n notches, running speed is good, comfort is good, and stopping precision is very good, braking should be increased n notches." The various parts of this control rule have the following correspondences:

When braking is increased n notches	$\rightarrow DN$ is n,
Running time is good	$\rightarrow R$ is VG,
Comfort is good	$\rightarrow C$ is G,
Stopping precision is very good	$\rightarrow A$ is VG.

In other words we can express the predictive fuzzy control rule as

(T-4) If (DN is $n \rightarrow R$ is VG and C is G and A is VG) then DN is n
$(n = \pm 1, \pm 2, \pm 3)$.

In the same way we get

(T-1) If (N is $0 \rightarrow R$ is B and C is G) then N is 0,

(T-2) If (N is B2 $\rightarrow R$ is M and C is G) then N is B2,

(T-3) If (DN is $0 \rightarrow R$ is VG and A is G) then DN is 0,

for the other experiential rules, (T-1) to (T-3), and the control rules for running between stations are

(C-1) If (N is $(N(t) + B_{max})/2 \rightarrow S$ is B) then N is $(N(t) + B_{max})/2$.

(C-2) If (N is $0 \rightarrow S$ is G and$C$ is G and E is G) then N is 0.

(C-3) If (N is P7 $\rightarrow S$ is G and C is G and T is B) then N is P7.

(C-4) If (DN is $0 \rightarrow S$ is G and T is G) then DN is 0.

(C-5) If (DN is $n \rightarrow S$ is G and C is G and T is VG) then DN is n.
$(n = \pm 1, \pm 2, \pm 3)$.

(c) Implementation of Fuzzy Control Rules

As far as train motion characteristics are concerned, the time constant for the brake system is 0.6 seconds, and in addition, the capacity undergoes large changes according to the changes in the weight of the cars, which varies according to the number of passengers. Because of this, the fuzzy inference described above occurs every 0.1 seconds, and when there has been no change in the control notch for 2 seconds or more, the control performance of the drive and brake apparatus at that control notch are inferred. In addition, the control commands that are obtained moment by moment from the each of the sets of control rules, for station-to-station running and for fixed-position stopping, can make a flexible response to changes in conditions by giving preference to braking and reducing speed during fixed-position stopping because of the preceding train, etc.

3.6.5 Evaluation of the Predictive Fuzzy Control System for Automatic Train Operation

The program for predictive fuzzy control of automatic train operation was installed in an actual automatic train operator that currently makes use of a conventional PID control method, and the performance of the two

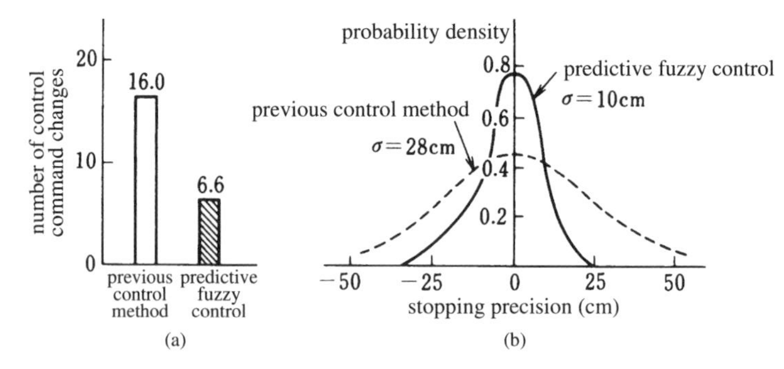

Fig. 3.50 Comparison with previous control method using simulation. (a) Evaluation of number of control command changes. (b) Evaluation of stopping precision.

systems, predictive fuzzy control and conventional control, was compared. The comparison was carried out by giving conditions that conformed with actual automatic train operation to each automatic operation device and through simulations.

Figure 3.50 shows the results in the case of 30% changes in the rated value for braking capacity and 5‰ changes in slope. Compared to the older form of control, predictive fuzzy control had a standard deviation of stopping errors of about 10 cm and one-third the number of notch changes, and the realization of a comfortable ride along with accurate control was confirmed.

Furthermore, predictive fuzzy control also emulated the actions of experienced operators in terms of the robustness of control toward changes in environment, such as changes in outside force due to slope along the train line and sudden changes in braking strength that accompany switching from regenerative brakes to air brakes, and made responsive control possible.

In addition, in terms of electricity consumed, which is an important index for quantitatively evaluating the economics of train operation, the ability to realize low-energy operation by effectively bringing coasting know-how into automatic operation and making use of margins in running time was confirmed.

3.6.6 Usage in the Sendai Subway System

The fuzzy automatic train operation system that was developed as described earlier and the performance of which was confirmed by computer simulation was used in actual trains in the Sendai subway system. Modeling of the characteristics of the trains was installed and adjusted, and it

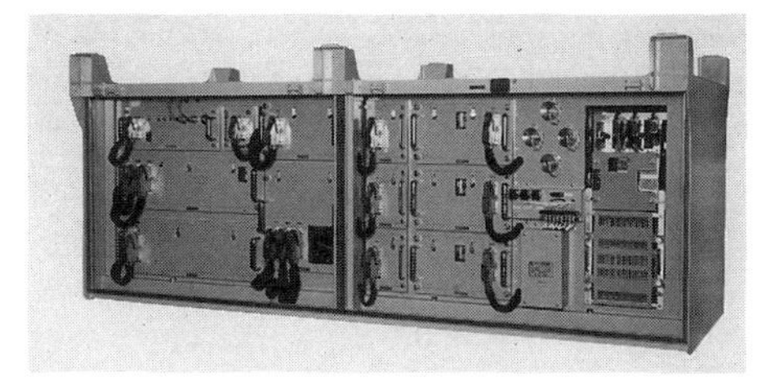

Fig. 3.51 Fuzzy automatic train operation controller.

began operation on July 15, 1987. The automatic train operating device (Fig. 3.51) with the microcomputer in which we developed the program was installed under the floor of the train. Figure 3.52 gives an example of the results of operational tests that were performed before the opening of the system. Even though the speed limit varied between 45 km/h and 75 km/h during runs between stations, the appropriate speed was maintained smoothly, and when the speed started to drop on inclines, it was easily regained with an output of power notch 5 ($P5B$). In addition, braking for stations moved smoothly into fixed-position stopping control with an initial braking output of notch 2 ($B2N$). Brakes were then adjusted slowly on the order of 4 to 5 seconds after that, and operation with a comfortable ride was obtained. As far as stopping accuracy is concerned, the results of 20 repetitions with each of the 19 configurations of trains that make up this subway system (more than 10,000 actual tests) are shown in Fig. 3.53. The results are that the occurrence of stopping errors of more than 30 cm was

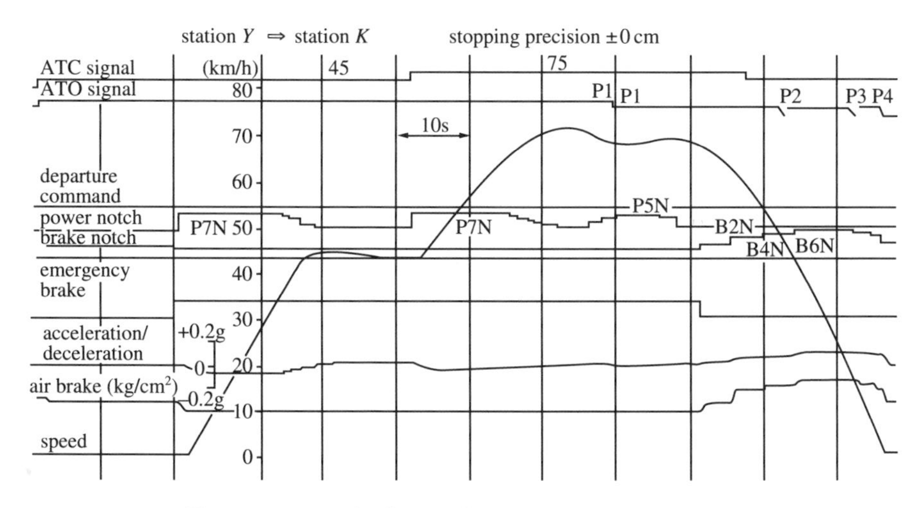

Fig. 3.52 Example of results from actual train operation tests.

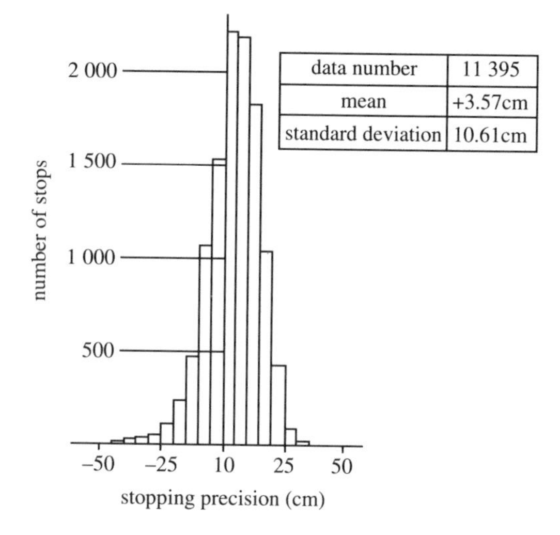

Fig. 3.53 Results of stopping precision measurements.

less than 1%, with a standard deviation for stopping precision of 10.61 cm, which confirms that excellent stopping is possible with changes in train and station conditions. This system has been working on all configurations without any problems since the beginning of operations.

Finally, we would like to express our thanks to all of those who gave us assistance, including those at the Sendai Transportation Bureau, those at the Mito Works of Hitachi Ltd., and the Systems Development Laboratory, Hitachi, Ltd.

3.7 SPEECH RECOGNITION

"Man/machine interface" is a term often heard in reference to the point of contact between people and machines. If we look at it from the human side, it would be ideal if we could converse with machines in order to convey our aims. People have several ways, other than speech, to convey their intentions, but from ancient times they have used speech for things near them, and more than anything else it required no new training or rules. In these days when it has become necessary to think about convers-ing with computers and robots, it is natural to think about using language to make for a friendly exchange. However, there are numerous problems that remain to be solved in order to do this, but even though conversation is a condition that is still a long way off and despite several limitations, practical application has started. Since the examples we will introduce here are ones in which the concept of fuzzy sets has been introduced into speech recognition, these are ones that can handle differences in speakers and temporal variations.

3.7.1 Problems in Speech Recognition

Let us think about our speech generation mechanism. The source of vowel sounds is the vocal chords. The shape of the vocal tract is changed to change the resonance, and speech information is added to sound waves. In the case of consonants, the source is one part of the vocal tract, rather than the vocal chords. Speech recognition is the extraction of linguistic information from the speech output that is formed in this way.

There are discrimination methods for speech recognition, but in recent years, the main current has been pattern matching. The improvements in electronic devices such as CPU capacity and the expansion of memory are largely responsible for this. Pattern matching means selecting the pattern with the greatest similarity as a recognition result by converting speech into characteristic patterns, matching them with previously stored standard patterns, and calculating the degree of similarity.

There are several problems when speech is recognized by means of pattern matching, but the most representative of these are

(1) Time-based fluctuation of the characteristic speech patterns. Speech lengths are not uniform, the cause being that even for the same speech the speed of utterance varies. Naturally, the differences in length are large with different people, even if the words are the same.

(2) The effects of differences in the size of the organs of speech. As was mentioned earlier, the size of the organs of speech varies with the person. Because of this, there is a difference in resonant frequency, even if sound is generated with the speech organs taking the same shape. This shows up as the individual nature of patterns.

In addition to this, sounds are influenced in different ways according to the types of sounds that come before and after them; in other words, there are problems with so-called co-articulations and problems generated by differences in the living environment of the speaker such as dialects, but here we will hold our discussion to two points, (1) and (2).

First, for problem (1), we must make a correspondence between time lengths when patterns are matched (time normalization). Linear matching is known to be a simple method for this purpose. However, the time variations of speech do not always fluctuate linearly, so DTW (dynamic time warping), in which dynamic programming is introduced into pattern matching, has been proposed as an answer to this [26]. DTW, which is a typical form of nonlinear matching, sets up time warping in which the difference between a pair of characteristic patterns is minimized or the similarity of the pair is maximized when they are matched, so a dynamic programming technique is used to determine this correspondence. The amount of calculations for implementation of this method is large, but it improves match accuracy and contributes to advances in speech recognition.

On the other hand, the differences due to speakers in problem (2) might be considered easy, since all of us human beings can recognize anyone's voice, but it is actually an extremely difficult problem. At present the differences in speakers are coped with using statistical processing and multi-template methods. In the multi-template method, various patterns can be recognized by storing a number of templates that have typical fluctuations in one category. Just how many patterns must be prepared for one category cannot be determined categorically, but there are reports [27] of 20 to 60. But it cannot be denied that the more there are, the greater the calculation time during matching.

Besides this, a larger utterance unit, such as the word, is often adopted in order to get around the problem of co-articulations. If the group of users can be kept to a single person, no special care needs to be taken, even for the frequency problems in (2). Methods in which users are limited in this way and the system is used after training with that person's voice are known as speaker-dependent methods. The fact that most of the speech recognition devices that have been popularized are speaker-dependent word-recognition systems is because of this background. On the other hand, those in which anyone's voice can be recognized without training are called speaker-independent systems, and recognition devices of the speaker-independent type have been created. However, the number of words recognized runs from several words to several dozen words.

The example we will introduce here is an actual speaker-independent recognition device in which the concept of fuzzy sets [28] has been introduced into word recognition, making for simple calculations for more than 100 words.

3.7.2 Fuzzy Pattern Matching

We will first explain the characteristic patterns used for speech recognition.

The well-known characteristic features that are extracted from speech are LPC (linear prediction coefficients), cepstrum, and spectra. Of these, spectra, for which physical meanings are understandable, are employed.

Speech is divided into short intervals, and the TSP (time spectrum pattern) —the plotted pattern of frequency and time obtained by analyzing the utterance within that interval—expresses the speech characteristics quite well; people can also read the spoken sounds from TSP by means of spectrum reading.

Since, as was mentioned before, people produce words by varying the resonant frequencies of the speech organs, what is especially important with TSP is resonant frequency, that is, the pattern peaks. The resonant frequencies of vowels are known as *formants*, and *local peak* [29] is the name used for the concept when it is also expanded to consonants. The method that will be described here is one that attempts to discriminate spoken words using the nature of the local peaks and how they change over time. In this case, the two problems that were described in Section 3.7.1 show up as fluctuations in pattern time lengths and local peak frequency variations caused by different speakers.

Since we are primarily concerned with the position in which the local peak exists, it is convenient to designate peak positions with "1" and other positions with "0" and use binalized processing, since it makes positions clear and allows for compression of data. This is called BTSP (binary time spectrum pattern), and it is employed as a characteristic feature. Words are expressed by means of BTSP, and it is used for pattern matching, taking into account how the frequency variations of different people appear in BTSP and what happens with time fluctuations. This is called fuzzy pattern matching [30].

An example of an actual pattern for a spoken word is shown in Fig. 3.54. Figure 3.54a is the TSP for the word "end" pronounced by a certain man, and Fig. 3.54b is the BTSP obtained by binary conversion of the same TSP. The horizontal axis in the figure is frequency, the vertical time, and this is the taking of 15 samples on the frequency axis and lining up 15 samples every 10 ms. In Fig. 3.54a there are eight bits for each element, but in the BTSP of 3.54b, the 15 pieces of data can be expressed with two bytes and it is easily handled by a computer.

The number of words is represented by n, the set for those word names by $I = \{i_1, i_2, i_3, \ldots, i_n\}$, and the set of patterns for the words by $X = \{x_1, x_2, x_3, \ldots, x_j\}$. I is the general set for element number n, but for X, x_j can be thought of as a fuzzy set that expresses the various patterns for word i_j. Therefore, we can define the set of membership functions $M = \{m_1, m_2, m_3, \ldots, m_n\}$, which in turn defines the set for word pattern x_{ij}, for word i_j. The fuzzy pattern matching described here uses membership function M, calculates degree of similarity S_j, and gives word name j from

$$j = \max_{j \in I} \{S_j\} \tag{3.32}$$

as the recognition result.

NO. 113end

51 0

0	0	0	0	0	0	0	0	1	0	0	0	0	4	2
0	0	0	0	0	0	0	0	4	2	1	4	5	14	9
0	0	0	1	0	1	2	7	24	22	14	11	13	28	14
0	2	2	5	5	2	3	9	26	25	19	16	21	40	22
2	10	2	4	4	3	3	7	25	24	19	19	25	67	34
28	23	16	29	27	25	12	14	27	28	30	29	37	95	56
35	32	50	122	84	51	37	52	126	82	61	112	72	70	52
24	29	62	138	84	49	43	56	131	96	77	110	70	72	48
26	28	54	141	68	45	37	53	119	106	73	102	62	66	50
29	32	66	166	70	45	38	53	139	109	63	64	52	66	49
26	34	70	178	66	43	32	45	103	91	58	77	61	72	53
33	34	69	146	59	46	29	42	91	86	61	69	58	69	50
26	33	60	124	50	39	26	36	84	78	83	87	67	79	49
25	35	48	101	38	33	23	32	81	64	65	69	55	59	36
26	42	45	76	31	26	19	27	69	55	70	58	46	45	29
[illegible]	48	46	60	27	21	15	23	64	43	69	46	32	31	20
27	52	49	51	25	18	12	19	57	36	66	39	25	28	15
29	54	51	42	21	15	10	15	46	28	53	29	18	28	14
33	41	41	23	10	8	6	7	24	16	29	13	9	23	9
44	22	22	10	6	8	6	6	9	11	26	11	7	18	4
31	29	25	9	5	7	6	7	10	13	35	12	7	19	4
31	27	26	9	5	7	6	8	10	15	42	14	8	20	5
29	26	26	9	5	6	6	8	10	13	36	12	7	19	4
30	25	24	8	5	6	6	8	10	11	31	11	7	16	3
30	24	22	7	4	5	5	7	9	11	31	10	5	13	2
33	24	19	6	4	5	6	8	8	10	31	9	4	11	1
35	24	16	6	3	4	5	8	6	7	23	7	2	7	0
38	25	14	5	3	5	5	8	6	7	25	7	2	8	0
48	27	14	6	3	5	5	7	6	7	22	6	2	10	1
51	29	15	7	3	6	5	7	6	6	20	6	1	6	0
55	29	16	7	4	6	4	7	6	8	26	7	2	8	0
58	26	15	7	4	7	4	6	4	7	20	5	1	7	0
59	23	13	7	4	7	3	4	3	3	12	2	0	1	0
61	21	11	5	3	6	1	1	1	1	7	0	0	0	0
65	22	10	4	1	4	0	0	0	0	4	0	0	0	0
68	24	10	3	1	5	0	0	0	0	1	0	0	0	0
70	25	10	4	2	4	0	0	0	0	0	0	0	0	0
66	21	10	4	2	2	0	0	0	0	0	0	0	0	0
46	10	2	0	0	0	0	0	0	0	0	0	0	0	0
16	2	0	0	0	0	0	0	0	0	0	0	0	1	0
9	0	0	0	2	3	3	6	11	7	11	24	11	6	2
3	4	5	6	9	11	10	13	23	22	37	91	58	38	23
3	3	4	5	6	6	4	6	13	18	38	97	59	37	21
1	0	5	3	2	1	2	4	9	13	28	81	58	32	20
0	0	0	0	0	0	0	2	7	9	20	59	44	21	12
0	0	0	0	0	0	0	1	6	7	16	47	36	16	8
0	0	0	0	0	0	0	0	5	4	10	37	19	9	6
0	0	0	0	0	0	0	0	4	3	10	21	13	7	2
0	0	0	0	0	0	0	0	4	1	5	12	12	5	0
0	0	0	0	0	0	0	0	3	0	1	4	4	1	0
0	0	1	0	0	0	0	0	1	0	0	0	1	0	0

(a)

NO. 113end

51

0	0	0	0	0	0	0	0	0	0	0	0	0	1	0	0
0	0	0	0	0	0	0	0	1	0	0	0	0	1	1	0
0	0	0	0	0	0	0	0	1	1	0	0	0	1	0	0
0	0	0	1	1	0	0	0	1	1	0	0	0	1	0	0
0	1	0	0	0	0	0	0	1	1	0	0	0	1	0	0
1	0	0	1	1	1	0	0	1	0	0	0	0	1	0	0
0	0	0	1	0	0	0	0	1	0	0	1	0	0	0	0
0	0	0	1	0	0	0	0	1	0	0	1	0	0	0	0
0	0	0	1	0	0	0	0	1	1	0	1	0	0	0	0
0	0	0	1	0	0	0	0	1	1	0	0	0	1	0	0
0	0	0	1	0	0	0	0	1	1	0	1	0	1	0	0
0	0	0	1	0	0	0	0	1	1	0	1	0	1	0	0
0	0	0	1	0	0	0	0	1	1	1	1	0	1	0	0
0	0	0	1	0	0	0	0	1	0	0	1	0	0	0	0
0	0	0	1	0	0	0	0	1	1	1	1	0	0	0	0
0	1	1	1	0	0	0	0	1	0	1	0	0	0	0	1
0	1	1	1	0	0	0	0	1	0	1	0	0	0	0	1
0	1	1	1	0	0	0	0	1	0	1	0	0	1	0	1
1	1	1	0	0	0	0	0	1	0	1	0	0	1	0	1
1	0	0	0	0	0	(	0	0	0	1	0	0	1	0	1
1	1	1	0	0	0	0	0	0	0	1	0	0	1	0	1
1	1	1	0	0	0	0	0	0	0	1	0	0	1	0	1
1	1	1	0	0	0	0	0	0	0	1	0	0	1	0	1
1	1	1	0	0	0	0	0	0	0	1	0	0	1	0	1
1	1	1	0	0	0	0	0	0	0	1	0	0	1	0	1
1	1	1	0	0	0	0	0	0	0	1	0	0	1	0	1
1	1	0	0	0	0	0	1	0	0	1	0	0	1	0	1
1	1	0	0	0	0	0	1	0	0	1	0	0	1	0	1
1	1	0	0	0	0	0	1	0	0	1	0	0	1	0	1
1	1	0	0	0	0	0	0	0	0	1	0	0	1	0	1
1	1	0	0	0	0	0	0	0	0	1	0	0	1	0	1
1	0	0	0	0	0	0	0	0	0	1	0	0	1	0	1
1	0	0	0	0	1	0	0	0	0	1	0	0	0	0	1
1	0	0	0	0	1	0	0	0	0	1	0	0	0	0	1
1	1	0	0	0	1	0	0	0	0	1	0	0	0	0	1
1	1	0	0	0	1	0	0	0	0	0	0	0	0	0	1
1	1	0	0	0	1	0	0	0	0	0	0	0	0	0	1
1	0	0	0	0	0	0	0	0	0	0	0	0	0	0	1
1	0	0	0	0	0	0	0	0	0	0	0	0	0	0	1
1	0	0	0	0	0	0	0	0	0	0	0	0	0	0	1
1	0	0	0	0	0	0	0	1	0	0	1	0	0	0	0
0	0	0	0	1	1	0	0	1	0	0	1	0	0	0	0
0	0	0	1	1	0	0	0	0	0	0	1	1	0	0	0
0	0	1	0	0	0	0	0	0	0	0	1	1	0	0	0
0	0	0	0	0	0	0	0	0	0	0	1	1	0	0	0
0	0	0	0	0	0	0	0	0	0	0	1	1	0	0	0
0	0	0	0	0	0	0	0	1	0	0	1	0	0	0	0
0	0	0	0	0	0	0	0	1	0	1	1	1	0	0	0
0	0	0	0	0	0	0	0	1	0	0	1	1	0	0	0
0	0	0	0	0	0	0	0	1	0	0	1	1	0	0	0
0	0	0	0	0	0	0	0	0	0	0	0	0	0	0	0

(b)

Fig. 3.54 Example of feature pattern (word: "END"). (a) TSP. (b) BTSP.

In applications of fuzzy theory, the problem is often how to define membership function. Here too, one point is how we define word pattern x_j-ness. There are membership functions that are constructed in relation to people, but for the reasons we will discuss in Section 3.7.5 it is convenient for these membership functions to be built into the recognition device. Because of this, and bringing it together with the primary objective of expressing x_j-ness, we set up the construction in the following order.

Many utterances, by different speakers, of the words to be stored were collected and converted to BTSP. Aggregates of the patterns for each word were made, and two-dimensional membership functions that expressed the variations within identical words were constructed. Concretely, they were constructed using the additive mean of the patterns chosen from the BTSPs for the same word, following a certain standard. There is one problem at the time of this addition. In terms of the two variations mentioned earlier, with frequency variations, "1," which is the element that expresses the resonant frequencies within the pattern, only varies along the frequency axis, while with time variations the pattern length changes, making addition difficult.

In order to deal with this, pattern lengths are unified using linear time warping before addition. Since linear time warping is a simple process of eliminating the differences and making the lengths of the patterns to be compared the same through uniform insertions and deletions, the amount of calculations is much less than for nonlinear time warping. Inclusive of this construction process, an example of a membership function is shown in Fig. 3.55. The word name is "start," and the time lengths of the BTSP on the left and the BTSP next to it are made equal and one is superimposed on the other pattern. Then the corresponding elements are added. The next BTSP is added to the pattern, and by repeating this process the pattern on the right is obtained. This is used as the membership function. Normally, membership functions are defined between values from 0 to 1, but in this case, these are defined by integer values 0–15; that is, one element is expressed by four bits.

Next we define the degree of similarity. Let the unknown speech input be $y(y \in X)$. If we can find the degree of attribution to each fuzzy set using membership function m_j, we can find which word it is similar to. However, if we use the previously defined membership function, we have the following problems. We will explain using Fig. 3.56.

Figure 3.56a shows the local peaks for a certain vocalization, and if we say there are peaks at f_1 and f_2, we can write

$$y = (f) = \mathbf{1}(f - f_1) + \mathbf{1}(f - f_2). \tag{3.33}$$

1 in Eq. (3.33) is a function like the following:

$$\mathbf{1}(f) = \begin{cases} 1 & \text{for } f = 0 \\ 0 & \text{for } f \neq 0 \end{cases}.$$

Figures 3.56b and 3.56c are membership functions that define the sets of patterns of word j and word k, 3.56b having two local peaks and 3.56c having one. When the two local peaks of Fig 3.56b are in complete agreement with the maximum points in the membership function, the degree of attribution is 1, so the maximum values of the membership

```
000000000000111 0
000000000000111 0
000001000000111 0
000000001001111 0
000010000000111 0
000000001000111 0
001000000000111 0
000000101000111 0
000000100000111 0
000100100000111 0
001001000000111 0
000011000000110 0
001001000000110 0
001000000000111 0
000000001000111 0
000000000000111 0
000000000000000 0
000000000000000 0
000000000000000 0
000000000000000 0
000000000000000 0
000000000000000 0
010100001000100 0
011100001001100 0
001100000011100 1
000100000011000 1
000110010001000 1
000011010001000 1
000011110001000 1
000011110001000 1
000011110001000 1
000011110001000 1
000011100001000 1
000011010001000 1
000010110001000 1
000010110001000 1
000010011101000 1
000010011000000 1
000010011000000 1
000100011000000 1
001100011100000 1
001100011000000 1
001100011000000 1
001100011000000 1
001000001000000 1
011100001000000 1
011100001000000 1
000000000000000 0
000000000000000 0
000000000000000 0
000000000000000 0
000000000000000 0
000000000000000 0
000000000011100 0
001000100001100 0
011000001001100 0
```

```
000000000001110 0
000000000000100 0
000000000000110 0
000000001000110 0
000000001001111 0
000010001001111 0
000000001001110 0
000000100000110 0
000000000000100 0
000000000000110 0
000000000001110 0
000000001001111 0
000000000000110 0
000000000000100 0
000000000000000 0
000000000000000 0
000000000001100 0
011100000000100 0
001110001010100 1
000010001001100 1
000010001001000 1
000011000000000 1
000011010101000 1
000001010001000 1
001011110001000 1
001011010001000 1
001011010101000 1
001011011101000 1
000011011001000 1
000011011001000 1
000010011001000 1
000010011000000 1
000110011000000 1
001110011000000 1
001100001000000 1
001110001000000 1
001000001100000 1
011000001000000 1
000000000000000 0
000000000000000 0
000000000000000 0
000000000000000 0
000000000000000 0
000000000000000 0
000000000000000 0
000000000000000 0
000000000000000 0
000000000000000 0
001000100001100 0
011100000001100 0
001000000101100 0
001100000000100 0
```

```
000000001001110 0
000000000001110 0
000000000001100 0
000001001001100 0
000000001000100 0
010000000001110 0
010000000000110 0
000101000000100 0
000101000000110 0
000000000000110 0
000001100000110 0
000000001000110 0
000000001000110 0
000000001001110 0
000000000001110 0
000000001001110 0
000000001001111 0
000000000011100 0
000000000001000 0
000000001001000 0
000000000001000 0
000000000000000 0
000000000000000 0
000000000000000 0
000000000000000 0
000001001111000 0
001001001001000 0
001000001001100 1
011100010001100 1
101100010001100 1
000110010001100 1
000110010000100 1
000110100000100 1
000110100000100 1
000010000000000 1
000010100100000 1
000110100000100 1
000010000001100 1
000110000100000 1
000110110000000 1
000110010000000 1
000110011000000 1
000100011000000 1
000100011000000 1
001100011000000 1
001100011000000 1
001000001000000 1
101010001000000 1
000000000000000 0
000000000000000 0
000000000000000 0
000000000000000 0
000000000000000 0
000000000000000 0
000000000000000 0
000000000000000 0
000000000000000 0
000000000000000 0
001011101001000 0
001010001001100 0
001000001001100 0
001000001100100 0
001010001101100 1
```

Fig. 3.55 Example of membership function (word: "START").

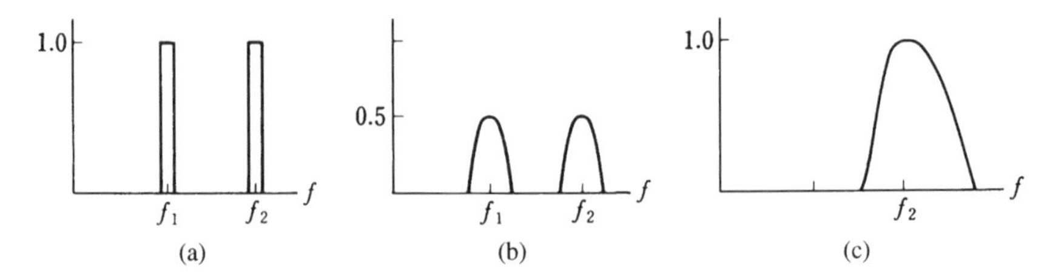

Fig. 3.56 Unknown pattern and membership functions of phonemes.

function are normalized at 0.5, and we let

$$D_j = \int m_j(f) \wedge y(f)\, df \tag{3.34}$$

be the degree of attribution D_j of y in x_j. In the figure, (a) is just about 1, so the word has j-ness.

On the other hand, Fig. 3.56c is the type with one local peak, which is often seen with consonants, etc., and has one maximum value. By the way, if we find the degree to which Fig. 3.56a is attributable to x_k, which is defined by the membership function in 3.56c, using Eq. (3.34), the answer is just about 1. At this point we have a contradiction in that the pattern in Fig. 3.56a seems to resemble x_j and x_k to the same degree. Therefore, we define the degree of non-attribution, $\overline{D}_j$:

$$\overline{D}_j = \int \overline{m_j(f)} \wedge y(f)\, df. \tag{3.35}$$

Here $\overline{m}$ is the well-known membership function that describes the complement of the pattern set for j. Introducing Eq. (3.35), the degree of non-attribution between Figs. 3.56a and 3.56c is large, so we can make the distinction that 3.56a more closely resembles 3.56b.

For this discussion, we reduced the dimensions, and the preceding discussion was carried out with frequency only, but we actually have a two-dimensional distribution. Normalizing the peak values for the membership functions while keeping the number of local peaks that occur in this kind of two-dimensional distribution in consideration is not only difficult, but also loses the meaning that each element is allotted four bits. Therefore, the ratio of Eq. (3.34) and Eq. (3.35) is used to define the degree of similarity, in order to calculate it without this kind of normalization:

$$S_{jy} = D_j | \overline{D}_j. \tag{3.36}$$

When there is no normalization, D_j, the numerator for the preceding equation, has a large value when there are a large number of local peaks in y, but in Eq. (3.36), normalization is made unnecessary by letting the denominator be $\overline{m}_j$, which can easily have large values when the number of local peaks is large.

Since digital values are dealt with in actual recognition devices, the equation for similarity is modified in the following way:

$$S'_{jy} = \frac{P_{j0}}{P_{jd} - P_{j0}} \cdot \frac{P'_{j0}}{P_y - P'_{j0}}. \tag{3.37}$$

P_{jd}, P_y, P_{j0} and P'_{j0} are each given as follows:

$$P_{jd} = \sum_f \sum_t m_j(f,t),$$

$$P_y = \sum_f \sum_t y(f,t),$$

$$P_{j0} = \sum_f \sum_t m_j(f,t) \cdot y(f,t),$$

$$P'_{j0} = \sum_f \sum_t m_j(f,t) * y(f,t).$$

Here, $\cdot$ means the product of the various elements of m_j and y, and $*$ means the α-level logical product of y and m_j, which is

$$m_j(f,t) * y(f,t) = \begin{cases} 1 & \text{if } m_j \geq \alpha,\ y = 1 \\ 0 & \text{otherwise} \end{cases}.$$

When each element of the membership function is expressed in four bits, the values are usually $0 \leqq \alpha \leqq 3$.

Equation (3.37) is made up of two terms, and both of them take the form of Eq. (3.36). The numerator and denominator of the term on the left are $y \wedge m_j$ and $\bar{y} \underline{\wedge} m_j$, respectively, and those for the term on the right are $y * m_j$ and $y * m_j$.

3.7.3 System Construction

An actual recognition system was constructed using the recognition method described. A block diagram of it is shown in Fig. 3.57. As is shown in the diagram, speech is converted into an electric signal by a microphone, and after frequency analysis using a filter bank, it undergoes binary conversion to create the BTSP. The parts with dashed lines concern the training that will be discussed later, and they are not used in the standard speaker-independent system. The following is an explanation of each of the sections.

(1) Filter Bank. A filter bank is a collection of a number of band pass filters, and in this system it is made up of 15. The center frequencies of these are 250–6,300 Hz, they are arranged by 1/3 oct and the resonance bandwidth of the filter is $Q = 6$. After this output has been smoothed out, it is quantized with a sampling period of 10 ms and a 15-dimensional vector is created. Each value is expressed in eight bits.

(2) Conversion to BTSP. The peak value from the 15 pieces of data from the filter bank output is found, and binalization is carried out with that

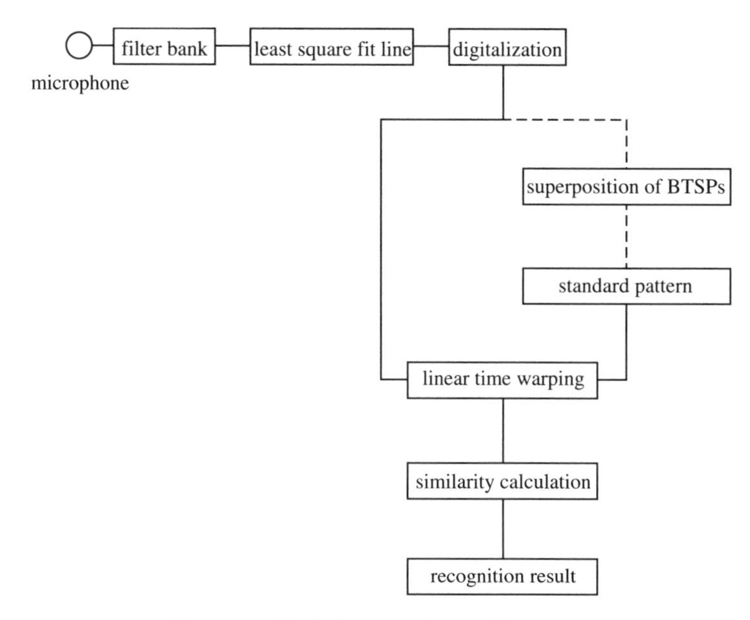

Fig. 3.57 Block diagram of recognition system.

neighborhood, including the peak, being labeled "1" and the rest being labeled "0." The characteristics of the sound source are normalized before binalization. The characteristics of the vocal chords, which decrease with higher frequency, appear in speech. Therefore, the least-square fit line is found for the 15 pieces of data before binary conversion processing, and by using the line, the peaks are corrected [31]. Next, the peaks above this least-square fit line are found and digitalized. The least-square fit line is not calculated in the actual device, but it is found using a filter [32] from which a similar result is obtained. In addition, since the type of source can be determined from the slope of the line obtained, it is considered to be a voiced sound when the slope is negative and unvoiced when it is positive, and these are delineated by "1" and "0" and added as a sixteenth piece of data in the BTSP.

(3) Matching Section. The degree of similarity between the unknown input pattern and the stored membership functions is calculated using Eq. (3.37). At this time, the time lengths of the two patterns being compared are unified using linear time warping in the same way as when the membership functions were constructed. The time fluctuations are stored in the membership functions in the form of variations from the time warping standards. In other words, nonlinear time warping appears as intermediate values other than "0" and "15" in the standard patterns.

Table 3.12. Recognition Rate Using the Speaker-Independent System

Language	Japanese	English	German
Recognition rate	93.2%	92.8%	93.7%

3.7.4 Recognition Test

This system has undergone speaker-independent recognition testing, and we will give the results here. Tests were conducted in Japanese, English, and German. The Japanese was 110 pieces of OA machine command terminology that included words and numbers [33], and the English and German had the same content as the Japanese with the addition of animal names and colors to make 120 words [34]. No special care was taken in the selection of words. Because of this, a number of groups with confusing lineups of sounds were created. For example, in Japanese there were "roku," "koku," and "oku," in English "quick" and "quit," and in German "nein" and "neun," and there were several others besides these.

For these utterances, A/D conversion was performed on the voices of speakers that have been recorded previously on audio tape in soundproof rooms or nearly soundproof rooms; they were input into a microcomputer and standard patterns generated. For reference we note that the English was collected on the American West Coast, and the German from the northern part of Germany.

The tests were carried out under the preceding conditions, and the average recognition rates that were obtained are shown in Table 3.12. We can see that a recognition rate of around 93% was obtained for each of the languages.

3.7.5 Group Use

The speaker-independent system that has been discussed so far is one in which standard patterns are constructed by processing the speech of a large number of speakers, and ideally it should be able to recognize the speech of anyone without any training. However, depending on usage, it is not always necessary to recognize everyone's voice—for example, in the case of an office, it would be sufficient to recognize a limited number of people's voices. However, if the recognition rate is improved even a little by limiting the users, it would be more convenient. As was touched upon earlier, the reason for having a function for constructing membership functions in the device that uses this system is so that users can create membership functions for their own group through training. This is called group use.

In the recognition testing for group use, four types of separate men's and women's groups of 10 people each were created and recorded, and recognition was carried out. The data used were the same as for the tests of the speaker-independent method, but after training with the voices of the 10 people in the group, speech by the same people uttered on another occasion was input. The results were that an average of 95.9% was obtained for the four groups; compared to a completely speaker-independent system, there was an improvement in the recognition ratio of several percent.

Naturally, if one person were to do the training, this system could be used as a speaker-dependent system. In this case, the frequency and time fluctuations would be fewer than when the objective was for speaker independence; therefore, the values for each element of the membership functions would be expressed in two bits, and it would be sufficient to store three utterances of each word. In this case, average recognition rates of 98–99% have been obtained for all three of the previously mentioned languages. This is the average for 10 men and women for each language, with 120 words for Japanese and the other languages.

Here and above, we have introduced the use of fuzzy theory in speech recognition as an example of its application. One can find other examples of tests of the introduction of fuzzy theory into speech recognition [35, 36], but the one shown here takes a different approach and uses the concept for ambiguous pattern matching. The authors call this fuzzy pattern matching. Fuzzy pattern matching makes for a high recognition rate with simple calculations, and because of this more than 100 words can be recognized in a speaker-independent system. With the system of membership function training, it can work for speaker-dependent usage, speaker-independent usage, and group usage. A photograph of the actual device is shown in Fig. 3.58. This one used a standard 8086 CPU, but within it are the recognition and training functions. An RS232C interface is provided

Fig. 3.58 Recognition device.

for communication with other devices, and if it is connected to a personal computer, it can be used for applications immediately.

3.8 IMAGE RECOGNITION

3.8.1 Outline

Since the history of image recognition as a subcategory of research on artificial intelligence is long, and there is a practical need for it, there has been a great deal of research done on it up to this point. The oldest example of an application is mechanical reading of hand-written numbers, but now its use is spreading to every field, beginning with inspecting, sorting assembly, and measurements for production processes all the way to remote sensor image analysis, the reading of hand-written characters, image-based medical diagnosis, quantification of experimental data, identification of people, automatic design, and understanding of images for robot eyes. However, the human processes for image recognition are complex, and they are not just the processing of simple image information; psychological factors play a large role. Semantic inference is especially indispensable for the understanding of images, but since the collection of a huge amount of data and, over and above logic, intuitive judgments are necessary for carrying it out, it is extremely difficult for computers to simulate it.

If we take a look at the flow of image recognition technology up to now, the methods differ according to whether the objects to be recognized are artificial or natural. With the former, we are dealing with objects with a clear uniform shape, and most research involves performing pattern matching after detecting outlines and edges and using geometrical rules to infer three-dimensional shapes. On the other hand, since most natural objects' light and dark colorations are not governed by rules, they are normally broken down into homogeneous areas through cluster analysis, and the content is determined from the characteristics of the shapes of these. Furthermore, in robotics there is a necessity for processing dynamic images in real time, and recognition speed is an important problem.

In general, the processes for computerized image recognition are as follows:

(1) Image data is input by means of a camera and converted into digital information. In essence, this divides the picture into many smaller elements, and each of these is apportioned a color and brightness.

(2) Preprocessing: removal of noise, normalization for comparison with standard patterns, segmentation (extraction of the part of the information necessary for recognition), etc.
(3) Extraction of characteristics: There are various levels of image characteristics. Strictly speaking, segmentation is also a part of this extraction of characteristics. Methods for extraction of characteristics can be broadly broken into the local and general. An example of the former is boundary detection, and examples of the latter are clustering and domain extension. In contrast to boundary detection, which makes use of the lack of uniformity between areas, clustering segments things by finding areas of uniformity. In either case, the image information includes noise that is not removed during preprocessing, so ambiguous information must be processed for segmentation. Next, the extraction of general characteristics is performed for the shape, qualities, and relative position of the areas that have been obtained. This is very important for discrimination, the next stage.
(4) Understanding and discrimination: The clusters obtained are compared with known models, classified, and identified; a three-dimensional construction done by means of inference is what is known as image understanding. Outputting this is the final goal of image recognition.

Up to this point a huge amount of research has been done on each of these stages. Despite this, we cannot avoid saying that the level that has been attained at present is very low. For example, problems such as understanding of complicated images, general conversion of linguistic and image information, recognition of curved surfaces and/or objects without shape rules, recognition of out-of-focus images, high-level extraction of characteristics, and semantic inference and imagining have hardly been touched. These image recognition problems alone are so difficult that it is important to bring in fuzzy theory and approach them with various methods.

The methodologies used up to now have centered upon statistical theory, cluster analysis, and two-valued logical inference, but all of them are different from the recognition process that humans use. The most important thing in image recognition is extraction of characteristics, but this is extremely difficult. Properly speaking, what are the characteristics of images? Why do cartoon portraits look more like the actual people than photographs? Information that would only be seen as noise by a computer probably plays an important role in human recognition processes, and this is expressed in exaggeration. It is human sensitivity that grasps this kind of characteristic, not logic. In addition, when ambiguous images are recognized, the ability to make overall judgments is at work more than analytical capability, and this is something intuitive. In order to imitate this, research into methods for processing subjective and emotional information and the handling of macro information are necessary. Research into fuzzy image recognition is just starting, but there are expectations for the development of new methodologies in response to this type of requirement.

Next we will give a simple explanation of the current state of fuzzy image recognition. Since image information is disturbed by a great deal of noise, even if it is clear, there are many examples of its being used in the detection of boundary lines. Class division of picture elements by means of fuzzy clustering is representative of these. Since there are very few picture elements that are exactly the same, ambiguous clustering is necessary. In addition, the same method is used for recognition of hand-written characters or a voice whose patterns are distributed around the standard patterns.

Noise is the problem when boundary lines are detected directly, but this cannot be completely removed using filters, and inference is also necessary for supplying dropout portions. Heuristic rules are used for this, but this is qualitative ambiguity. Furthermore, when we progress to the stage of image understanding, higher-level ambiguous pattern-matching problems occur. This is because we not only need shape matching, but also meaning matching. Especially when we come to problems such as diagnosis using x-ray photographs, we have a situation of not being able to create rules *per se*.

In the following section we will introduce several representative studies on image recognition that make use of fuzzy theory.

3.8.2 Image Recognition Using Fuzzy c-Means Clustering [37]

Clustering is a method for dividing scattered groups of data into several groups. In this instance, the data within a group have similar characteristics, and they are divided so as to make the average difference in characteristics between groups as large as possible. Here we let n pieces of data be $x_1, x_2, \ldots, x_n$ and the set of these data be X. Each x_j is a vector of d dimensions:

$$X = \{x_1, x_2, \ldots, x_n\}, \qquad x_j \in R^d. \tag{3.38}$$

We can consider dividing this into c $(2 \leqq c < n)$ clusters. Here we let u_{kj} be the attributability of x_j to cluster k. With hard clustering, u_{kj} is one of two values, 0 or 1, but with fuzzy clustering it can be any value from 0 to 1:

$$u_{kj} \in \{0, 1\} \quad \text{Hard clustering}$$

$$u_{kj} \in [0, 1] \quad \text{Fuzzy clustering} \tag{3.39}$$

However, for each case we have

$$\sum_j u_{kj} > 0, \qquad \sum_k u_{kj} = 1.$$

Fuzzy clustering allows one piece of data to belong to two or more clusters. However, the total of the degrees of attribution is 1, and u is the weight of attachment to each cluster. In this case there are many algorithms for finding u, but since the fuzzy c-means (FCM) method that was proposed by Dunn and generalized by J. C. Bezdek is a way of finding u using fuzzy sets, it is frequently used in pattern recognition [38].

Now we label the set of $c \times n$ matrices U, the elements of which are u_{kj}, which satisfy Eq. (3.39) M_{fc}. On the other hand, R^d in Eq. (3.38) is called the characteristic space, and $x_k = (x_{k1}, x_{k2}, \ldots, x_{kd})$ is the characteristic vector. Clustering is no more than coupling data set X and division matrix U. The result of this coupling is denoted U_X. In order to find the optimal U_X with the FCM algorithm, we let the objective function be the following generalized sum of square error within groups:

$$J_m(U, v) = \sum_{j=1}^{n} \sum_{k=1}^{c} (u_{kj})^m \|x_j - v_k\|^2, \qquad 1 \leqq m < \infty. \tag{3.40}$$

Here, x_j is d-dimensional measured data, and v_k is the center of cluster k and is d dimensions. $\|*\|$ is any norm, and it expresses the similarity between any measured data and the center. If $m = 1$ and $u_{kj} = \{0, 1\}$ the minimalization of Eq. (3.40) turns out to be what is called the ordinary k-means, but the points m and of weighting $u_{kj} = [0, 1]$ are different in FCM. To the extent that this m is greater than 1, we have fuzzy clustering. In this way, the characteristic of FCM is the ability to independently adjust ambiguity.

The u_{kj} and v_k (these are designated by $\hat{u}_{kj}$ and $\hat{v}_k$) that minimize Eq. (3.40) satisfy the following equation for $m > 1$:

$$\text{Condition 1:} \quad \hat{u}_{kj} = \left(\sum_{g=1}^{c} \left(\frac{\|x_j - \hat{v}_k\|}{\|x_j - \hat{v}_g\|} \right)^{2/(m-1)} \right)^{-1}, \qquad \forall j, k \tag{3.41}$$

$$\text{Condition 2:} \quad \hat{v}_k = \frac{\sum_{j=1}^{n} (\hat{u}_{kj})^m x_j}{\sum_{j=1}^{n} (\hat{u}_{kj})^m}, \qquad \forall k \tag{3.42}$$

The $\hat{u}_{kj}$ that minimizes Eq. (3.40) can be found by repeating the following process:

(Step 1) m and cluster number c are assumed, and a norm in Eq. (3.41) is defined appropriately. In addition, an appropriate initial value, $U^{(0)} \in M_{fc}$, is given for U. ($U^{(0)}$ can be chosen at random without any relation to u_{kj}.)

(Step 2) The cluster center $\{v_k^{(0)}\}$ is calculated using $U^{(0)}$ and Eq. (3.42).

(Step 3) $U^{(1)}$ is calculated using $\{v_k^{(0)}\}$ and Eq. (3.41).

(Step 4) Defining an appropriate norm and threshold value ϵ, the preceding steps are repeated until $\|U^{(p)} - U^{(p-1)}\| \leqq \epsilon$.

Since element $\hat{u}_{kj}$ of U obtained here expresses the degree to which x_j belongs to cluster k, the boundary between k and g is the place where $\hat{u}_{kj}$ and $\hat{u}_{gj}$ are about the same when we compare them, for example.

In some problems, the value of c is unknown. In such instances, various values must be used for c and the ones that express the physical processes best determined. In addition, since Eqs. (3.41) and (3.42) give extreme values for J_m, there is a possibility that the final value may change when the initial setting for U is changed. The exactness of the final value depends on the selection of $U^{(0)}$. In addition, small, isolated areas are difficult to recognize, no matter what.

There is a great variety of examples of applications of FCM image recognition [39–42], but we will explain an example of remote sensing image analysis studied by M. M. Trivedi [43].

In his study, the levels of image recognition were varied and recognition carried out. In other words, rough clustering is carried out at low levels. For example, in an aerial photograph of a city, commercial, residential, and agricultural areas are recognized for the most part. Uniformity tests are then performed, and if an area is not uniform, higher-level recognition is carried out, and things such as buildings, houses, and cars are distinguished. If we show this in a diagram, fuzzy clustering gives rise to a gray area when the image elements are large (Fig. 3.59). If this kind of element is divided into smaller elements at a higher level, we get something like that in Fig. 3.60. In this way recognition can be carried out with a gradually increasing precision. In this study, FCM was performed using a remote sensing image like that in Fig. 3.61 as an example. The same scene was photographed using three different spectra. One image is divided into 32×32 elements. Each element is made up of 10×10 pixels. Here four clusters were separated out: rivers, reclaimed land, agricultural land, and forests. The results of carrying out FCM and those of having an experienced person draw the boundaries were compared. Here we let $m = 2$.

To begin with, if we look into each of the average gray levels M_R (rivers), M_M (reclaimed land), M_A (agricultural land), and M_F (forest) for

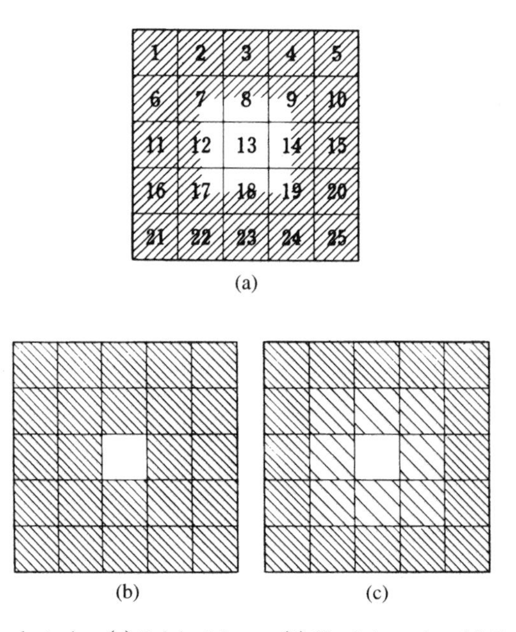

Fig. 3.59 Fuzzy clustering. (a) Original figure. (b) Hard clustering. (c) Fuzzy clustering.

representative points in the clusters as divided by humans, we come up with the following as to three photographs:

$$\left.\begin{pmatrix} M_{\mathrm{R}} = (35.9 \quad 27.1 \quad 4.6) \\ M_{\mathrm{M}} = (46.6 \quad 57.5 \quad 24.0) \\ M_{\mathrm{A}} = (26.0 \quad 44.4 \quad 21.0) \\ M_{\mathrm{F}} = (13.5 \quad 54.3 \quad 29.6) \end{pmatrix}\right\} \quad \text{Clustering done by humans.}$$

On the other hand the convergence values for the cluster centers from

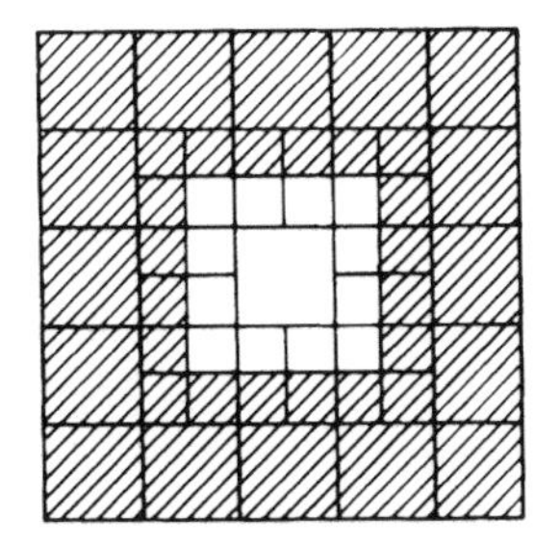

Fig. 3.60 Higher-level clustering.

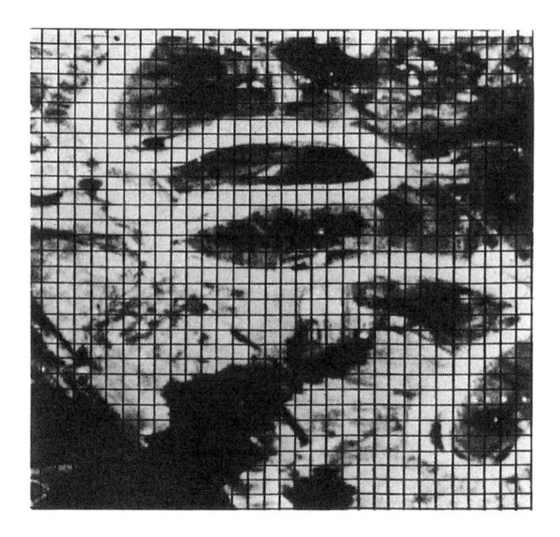

Fig. 3.61 Analysis example. (M. M. Trivedi.)

the results of $c = 4$ calculations using FCM come out as follows:

$$\left.\begin{pmatrix} v_1 = (34.1 & 28.9 & 6.4) \\ v_2 = (22.5 & 52.7 & 26.6) \\ v_3 = (27.1 & 41.1 & 18.3) \\ v_4 = (16.5 & 58.6 & 31.9) \end{pmatrix}\right\} \text{Initial FCM.}$$

Both are about the same except for reclaimed land. When viewed through FCM, v_2 can be seen as a subcluster of v_3 or v_4. If we look carefully, the area covered by reclaimed land within the image is extremely small. Because of this, we can see that it is difficult to find the center of the v_2 cluster. Therefore, FCM was carried out again with $c = 3$ and $m = 2$ for image elements with a maximum membership value of 0.5 or less. The results of this are shown below:

$$\left.\begin{pmatrix} v_5 = (24.4 & 45.5 & 21.4) \\ v_6 = (18.3 & 47.4 & 23.9) \\ v_7 = (41.0 & 58.6 & 25.9) \end{pmatrix}\right\} \text{Second FCM.}$$

Here we can see that v_7 is close to M_{M}, and v_5 and v_6 are close to M_{A} and M_{F}, respectively. Furthermore, two levels of FCM were performed in this study, and the first one was 16×16 element (20×20 pixels for an element), $c = 3$, $m = 2$ clustering. The results were that 175 elements were seen as nonuniform, and these were further split into four divisions.

The second level of FCM ($c = 4$, $m = 2$) was carried out with these 700 new elements (10×10 pixels), and results roughly the same as those just given were published.

3.8.3 Detection of Ambiguous Boundary Lines

Image recognition is even difficult for humans with images in which the boundary lines are blurred, as in medical x-ray images. It is even more difficult a job to make computers understand them. This type of research has been increasing in the field of artificial intelligence, but the difficulty is that there is no single method for solving the problems, so unless the approach is one in which every possible method is available, the solutions are dubious. Naturally, using fuzzy sets is one such approach. As an example of this we will describe the research of S. K. Pal here [44].

This research uses x-ray photographs of the skeleton as an example, and proposes a method for transforming the blurred outlines into clear lines. In other words, an algorithm [45] for detecting the curves and loops proposed by J. T. Tou is used, and for this the basics are a joint use of fuzzy sets for smoothing and segmentation.

First of all, membership functions for the three basic types of lines (vertical, horizontal, and diagonal) in the blurred outlines are defined as follows:

$$\text{Vertical:} \quad \mu_V(x) = \begin{cases} 1 - \left| \dfrac{1}{m_x} \right|^{Fe}, & |m_x| > 1 \\ 0, & \text{otherwise} \end{cases} \tag{3.43}$$

$$\text{Horizontal:} \quad \mu_H(x) = \begin{cases} 1 - |m_x|^{Fe}, & |m_x| < 1 \\ 0, & \text{otherwise} \end{cases} \tag{3.44}$$

$$\text{Diagonal:} \quad \mu_{Ob}(x) = \begin{cases} 1 - \left| \dfrac{\theta - 45}{45} \right|^{Fe}, & 0 < |m_x| < \infty \\ 0, & \text{otherwise} \end{cases} \tag{3.45}$$

Here m_x ($= \tan \theta$) is the slope of line x (Fig. 3.62a). In addition, F_e is a positive parameter for adjusting the degree of fuzziness.

In the same way the "curvedness" of line segment x is expressed by the following membership function:

$$\mu_{\text{arc}}(x) = \left(1 - \frac{a}{l}\right)^{Fe}. \tag{3.46}$$

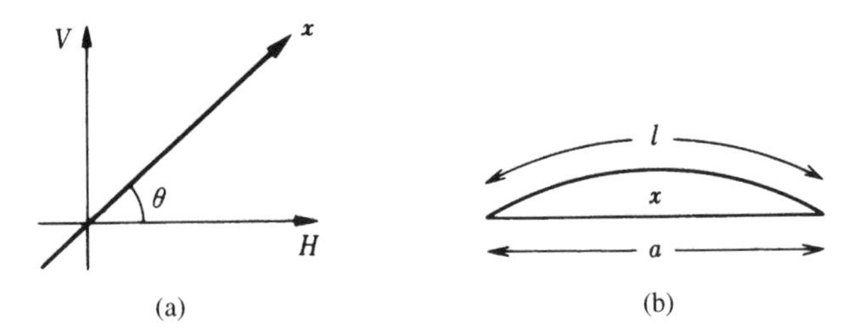

Fig. 3.62 Membership function. (a) Line. (b) Curve.

Here a is the length of the straight line from one end of the line segment to the other, and l is the length of the segment itself. The smaller a/l is, the greater the degree of curvature (Fig. 3.62b).

The outline is detected as an $M \times N$ dimension gray image. In order to transform this into one-dimensional code series, the direction of the line segment made up of w number of elements is expressed by one number from the eight-direction code in Fig. 3.63. In order to search out the outline, elements in which the darkness is not zero are located through scanning, and the direction with the strongest directionality is followed. However, rather than determining directionality for one element, if we locate the maximum membership value for several elements together as the direction of the line segment, the amount of calculations and memory can be reduced.

When there are two or more directions with the same strength, the preceding information is used and the direction with the greatest continuity is chosen. Moreover, this information is used later for junctions. In addition, the start point of a line segment is changed after the outline has some length. In this way recognition of whether the outline is forming a loop is possible, and when part of the outline is lost it can be retraced.

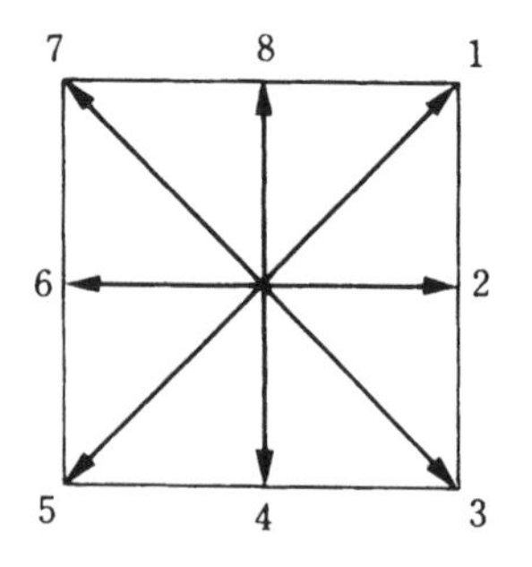

Fig. 3.63 Directional codes.

Fuzziness is not used for smoothing out the outline, but the following four processes are used: (1) When there are four or more of the same code in succession and one different code (or two in succession) is put between them, the codes are changed to those of before/after, or they are removed. (2) When four or more of the same code precede two codes that differ, or when the values are 6 and 3, the same measures are taken. (3) When two adjacent elements have opposite directions, they are removed. (4) A rule is made for converting the vector sum of a pair of codes into an intermediate direction, and minute changes in direction are removed.

Segmentation of the outline is dividing it up where the characteristics of the line segments change. In other words, places where the directional changes cease to be uniform are partitions. When segmentation is performed, the curvature is determined by Eq. (3.46).

Figure 3.64 is an x-ray of a wrist joint, and it is an example of processing the outline of the bones with this method using a boundary line with a 145 × 128 dimensional gray gradient. Here the segment unit was 2 and $F_e = 0.5$.

If we do the calculations for the membership values for the linearity and curvature (concave and convex) for each segment, we can choose the characteristic set for the bones from them. This is also helpful for creating rules for recognition of undetected sections of the outline.

In order to show the nature of the codification and smoothing, a portion of Fig. 3.64 is shown in Fig. 3.65. Part (a) in this figure is prior to smoothing and (b) is after smoothing. This is neater and makes it easy to extract the characteristics.

Fig. 3.64 Example of x-ray processing.

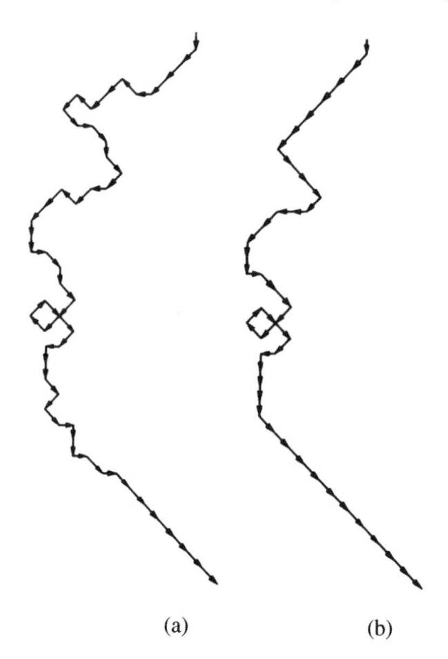

Fig. 3.65 Effects of smoothing. (a) Before smoothing. (b) After smoothing.

3.8.4 Clustering by Means of Possibility Theory

Clustering is not just used for image recognition; it is also used widely in general data classification. There are an infinite variety of methods, but most of them are statistical. However, in order to do statistical calculations, the amount of data must be much greater than the number of clusters. In the fields of medicine and astronomy, there is a great deal of data that does not satisfy this condition. Here we will introduce research on extracting characteristics from comparatively small amounts of data by means of the joint use of statistical methods and possibility theory studied by V. Di Gesu *et al.* [46].

Now let g_X be the membership function of fuzzy set A ($A \subseteq X$). Then we define the following set as the Ψ-image of g_X, for a certain real number Ψ:

$$X_g^{\phi} = \{x \in X: g_x(x) \geqq \psi\}. \tag{3.47}$$

For the membership function f_X of another fuzzy set, we can define the following function, with the addition of restriction by the Ψ-image of g_X:

$$f_{X_g^{\phi}} = (x) = f_x(x), \qquad \forall x \in X_g^{\phi}. \tag{3.48}$$

This restricts f_x to A, so we write f_A, and call it the possibility function. The possibility function has the following properties when we let x be an element of A:

(1) The possibility function increases monotonically as the attributability of x to A increases.

(2) $f_{A \cup B}(x) = \max\{f_A(x), f_B(x)\} \quad \forall A, B \subseteq X. \quad (3.49)$

(3) $f_{A \cap B}(x) = \min\{f_A(x), f_B(x)\} \quad \forall A, B \subseteq X. \quad (3.50)$

The extraction of characteristics is as follows. Let there be N pieces of data on a characteristic space of n dimensions, and let this be set X. First we choose the K_1 number of one-dimensional subspaces using a statistical method (uniformity test), and next we use the same method to choose the K_2 number of two-dimensional subspaces from the $(n - K_1)/2$ number of the combination. Finally, we use an appropriate possibility function to find the characteristics in the remaining groupings.

The statistical method divides data when the relative distance between pieces of data exceeds a certain threshold value, but this threshold value is chosen so that the probability Q of a difference between the number of clusters at that time and the theoretical number of clusters when the data are assumed to have a Poisson distribution falls below a certain value.

Possibility functions are used to find the shape of the clusters and characteristics. For example, in a case in which $n = 2$, the cluster is uniform if it is a circle. If it is an ellipse the characteristic lies in the direction of elongation. The shape of a cluster can be characterized by a second moment on r axes. If we let this be m_i $(i = 1, 2, \ldots, r)$,

$$m_i = \sum_j (x_j - y_j \tan \alpha_i)^2 / (1 - \tan 2\alpha_i), \tag{3.51}$$

$$\alpha_i = i \cdot (360/r). \tag{3.52}$$

Here, α_i is the direction of the axis. If we express two standard clusters (circle and line) with fuzzy sets C and T, respectively, we can determine whether a certain cluster A belongs to C or T through a comparison of $f_C(A)$ and $f_T(A)$.

In this study, fuzzy entropy f^E and Minkowski's distance are chosen as the two possibility functions, and it is reported that under simulation tests separation was better with the former. The form of the possibility functions is shown in the following:

$$f^{E_i} = 1 - \frac{1}{r} \sum_{j=1}^{r} \left(-\xi_{ij} \ln \xi_{ij} - (1 - \xi_{ij}) \ln(1 - \xi_{ij}) \right) \quad \in [0, 1] \tag{3.53}$$

$$f^{M_i} = 1 - \frac{1}{r} \left(\sum_{j=1}^{r} (m_{i,j} - m_j) h \right)^{1/h} \quad \in [0, 1] \tag{3.54}$$

where

$$\xi_{ij} = |m_{i,j} - m_j| < 1$$

$m_{i,j}$ = moment of standard shape i for axis j
i = number of standard clusters, h = dimensions ($h = 2$)
r = number of axes of rotation
m_j = moment for the j axis of the data

This method was used in the clustering of data on patients with brain contusions. The number of samples N was 46, and the number of characteristic parameters was 17; however, some of the parameters for the data were missing. When one-dimensional subspaces were chosen by means of a $Q = 10^{-2}$ uniformity test, $K_1 = 8$ was obtained. Of the two-dimensional subspaces (((17 − 8)/2) = 36 groups) for what remained, $K_2 = 9$ satisfied the uniformity conditions for the same Q value. The final remaining portion was divided using possibility functions. Both f_M and f_E were tested as possibility functions. Partial results are shown in Fig. 3.66. These are pretty much in agreement with human judgments.

3.8.5 Recognition of Vegetables Using Linguistic Expressions [47]

The pattern recognition research that we have discussed so far has inherited the statistical thinking used up to now, just substituting fuzzy theory into a portion of the statistical processing. Research into expressing shapes linguistically and processing them by means of fuzzy theory is completely different from this. This differs from detection of outlines and recognition by means of simple pattern matching; it is an attempt to give consideration to the meaning of the patterns in the recognition of objects.

First, the characteristics of the outlines of standard shapes are analyzed, and the overall and local characteristics are expressed in natural language. These are saved in a frame-type knowledge base using fuzzy sets. On the other hand, observational results are processed numerically, and each characteristic is output in terms of fuzzy numbers. Then recognition is carried out by means of ambiguous pattern matching of the two fuzzy sets. Not only is this method convenient for recognition of objects for which the shape and dimensions are not clearly delineated, but also, since various meanings for the pattern can be expressed, high-level information processing is possible through the use of inference. For example, if there is a figure in which there are many children and one adult in a large room, it could be inferred to be a scene in a school classroom. To put it another way, this method is an intermediate between the linguistic information

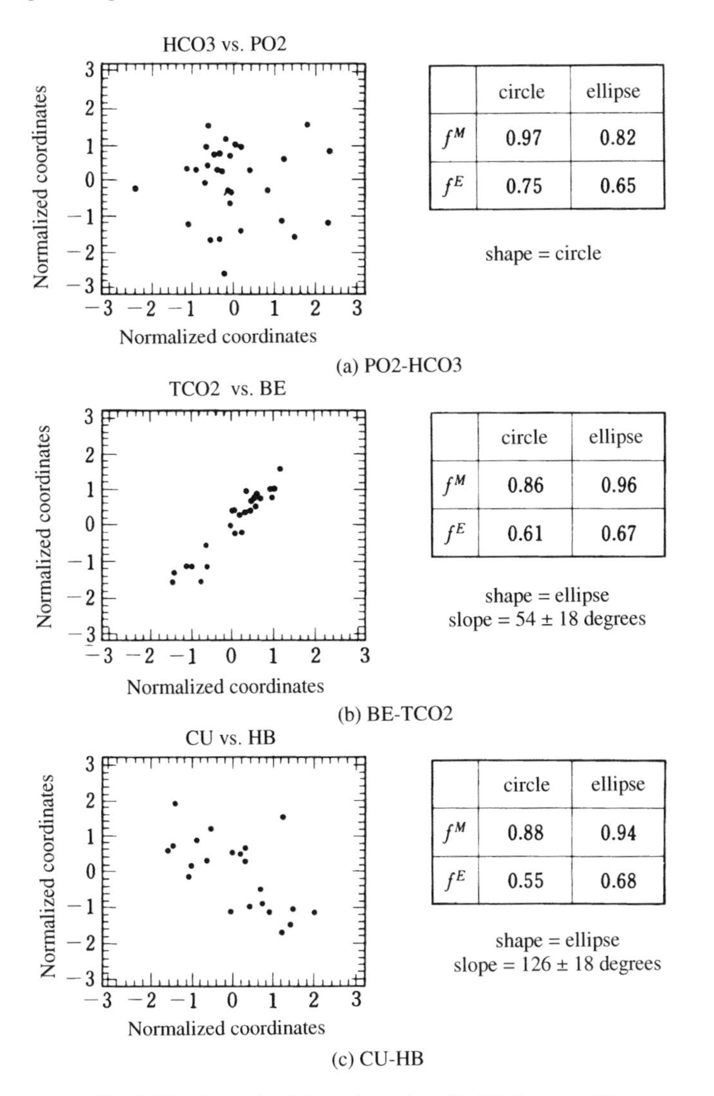

	circle	ellipse
f^M	0.97	0.82
f^E	0.75	0.65

	circle	ellipse
f^M	0.86	0.96
f^E	0.61	0.67

	circle	ellipse
f^M	0.88	0.94
f^E	0.55	0.68

Fig. 3.66 Example of data clustering. (V. Di Gesu, *et al.*)

processing of artificial intelligence and pattern recognition. However, the problem is how the characteristics are to be expressed linguistically. This job is entrusted to humans at present and cannot be automated. This means that this system is like expert systems in that it is a man/machine system.

Here several kinds of vegetables were chosen as recognition objects. First, the objects were filmed from directly above in black and white, using

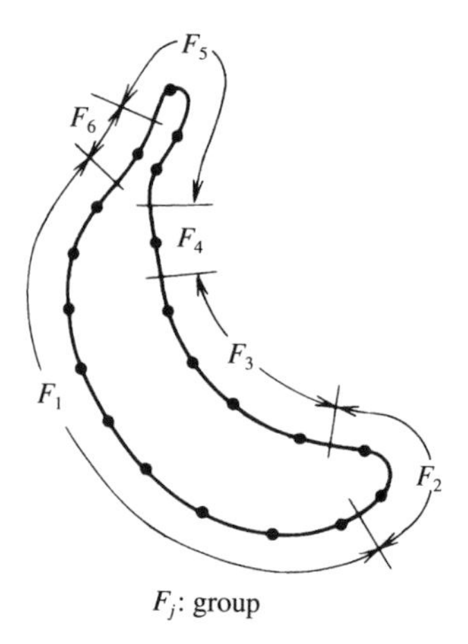

Fig. 3.67 Outline grouping.

a CCD camera. After this image had undergone smoothing in order to eliminate noise, density was digitalized and expansion and contraction processes for interpolation of dropout carried out. A normal Laplacian method was used for outline detection. Once the outline was obtained, the length, width, and area of the object were calculated instantaneously.

In order to learn the local characteristics of the outline, it was divided into short segments of a fixed length, and the angles of adjoining segments investigated. Segments for which the change in angular deviation was less than 10 degrees were grouped as in Fig. 3.67. Each group was classified as one of three standard patterns—straight line, curve, or angle—and fuzzy pattern matching was used to do this. In other words, the standard patterns just given are a fuzzy set, the elements of which are angular deviations, and its membership functions were determined as B_1, B_2, and B_3 and are in Fig. 3.68. The characteristics are given by equilateral triangular function F, the vertex of which is the average angular deviation

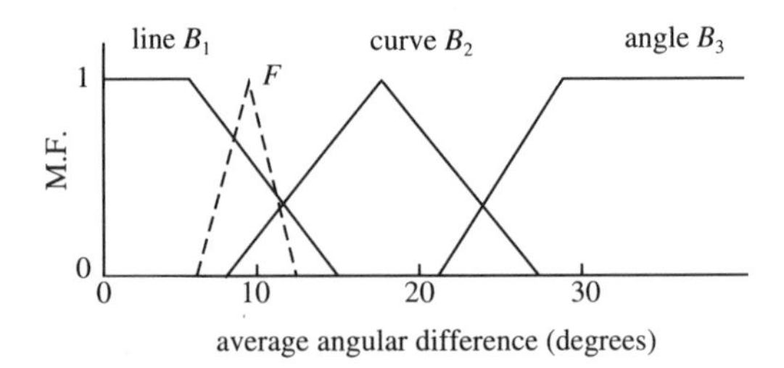

Fig. 3.68 Standard pattern.

Table 3.13. Local Characteristics of Outline

Group	Group Length	Line	Curve	Angle
F_1	9	0	0.17	0
F_2	2	0	0	1.0
F_3	4	0	−0.79	0
F_4	1	0.43	0.02	0
F_5	3	0	0	0.78
F_6	1	0	−0.55	0

and the ambiguous width of which is the standard deviation. Converse truth qualification of numerical truth values for the following equation, which was proposed by Y. Tsukamoto, is used for determining the similarity of F and B:

$$N(F/B) = \{\text{Sup}(\mu_F \wedge \mu_B) + \inf(\mu_F \vee \neg \mu_B)\}/2. \tag{3.55}$$

By this means, the local characteristics of what was investigated in Fig. 3.67 turn out as in Table 3.13.

When the local characteristics have been obtained, the macro characteristics of the entire shape can be judged. Six shapes—circle, ellipse, triangle, rectangle, rod, and bow—are chosen as standard macro patterns, and each of them is expressed linguistically. For example, for circle characteristics, "The aspect ratio is close to one; the curved section takes up the largest part; the angles are either near zero or more than 6; of these almost 100% are convex and 0% are concave." This is given by the fuzzy sets in Fig. 3.69. The entire shape is expressed in the same way as ambiguous sets of sectional characteristics in the five remaining standard shapes. If we investigate the macro characteristics of Fig. 3.67 by the same method as before, we see that the rodlike character is 0.65 and the bowlike character is 0.78.

The aspect ratio and the size that were first investigated are expressed as an ambiguous measurement of small, medium, or large, and it is entered in the knowledge base as a fuzzy number. Besides this, we let it have unevenness if that condition for the periodically changing code for the curve or angle of the sectional characteristics continues three times, in order to express the roughness of the outline.

As a concrete example, recognition was carried out with the 12 types of vegetables (artificial objects intentionally mixed in) in Fig. 3.70. In terms of standard knowledge, a banana, for example, would be given linguistic expression as in Table 3.14 and entered into the knowledge base in the form of a fuzzy frame. The efficiency of the recognition of unknown objects depends on which characteristics are searched for first, but as a

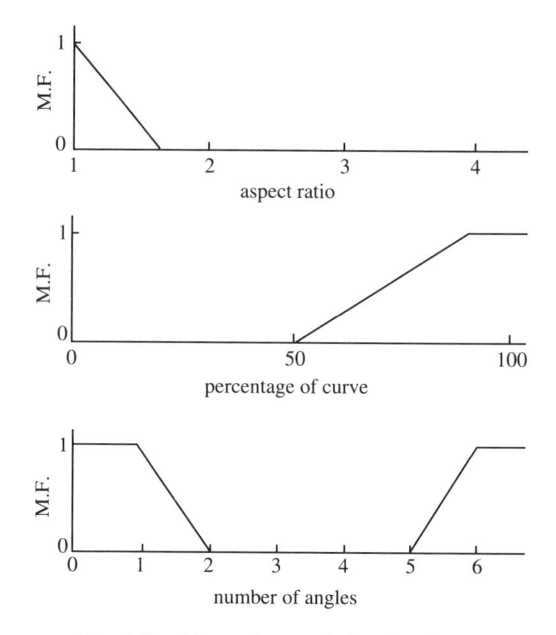

Fig. 3.69 Macro characteristics of a circle.

result of investigations into the recognition activities of humans, it was determined that for this example the degree of importance was in the following order: size, aspect ratio, number of angles, curvature, degree of arcs, unevenness. So a search tree was constructed on this order.

Figure 3.71 is the result of the recognition of the green pepper from Fig. 3.70, using this method.

3.8.6 Recognition of Moving Objects [48]

In the recognition of moving objects, not only accuracy but also speed presents a problem. Here we will describe research of K. Hirota in which fuzzy sets are used in order to gain a substantial characterization of the shape of the object, and adjustability and importance of characteristic quantities are considered in order to improve the effectiveness of discrimination.

The objects used were the 12 two-dimensional patterns in Fig. 3.72. The major characteristic was size, but in order to match this to human senses, the area of a rectangle circumscribing the object was used rather than the area of the object itself. This was expressed as an ambiguous measurement of small, medium, or large, and the membership functions of these are shown in Fig. 3.73. Letting the area of the rectangle circumscribed around

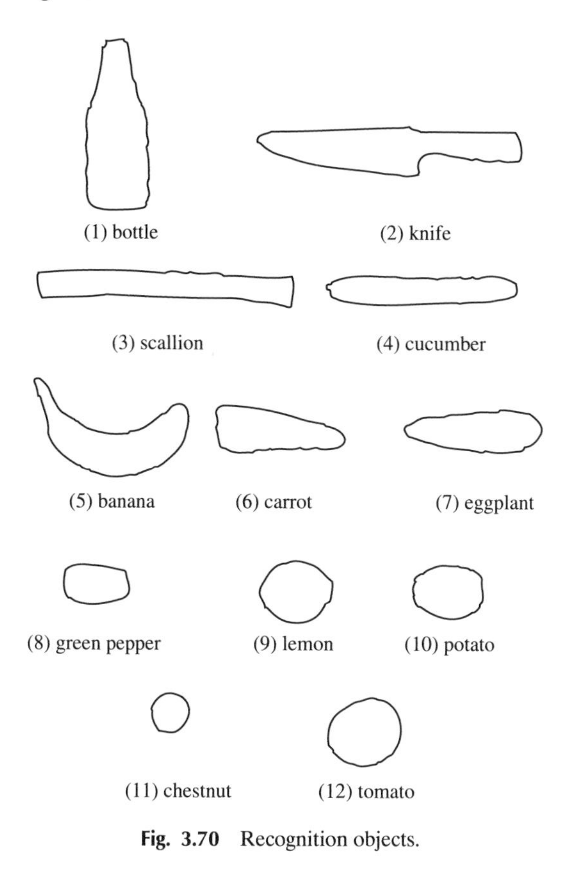

Fig. 3.70 Recognition objects.

Table 3.14. Characteristics of a Banana

Banana		
Size:	Medium	
	Area:	Medium
	Length:	Medium
Macro shape:	Rod or bow	
	Aspect ratio:	Large
	Angularity:	2 angles
	Roundness:	About 80%
	Concavity:	About 30%
	Convexity:	About 70%
Roughness:	Even	

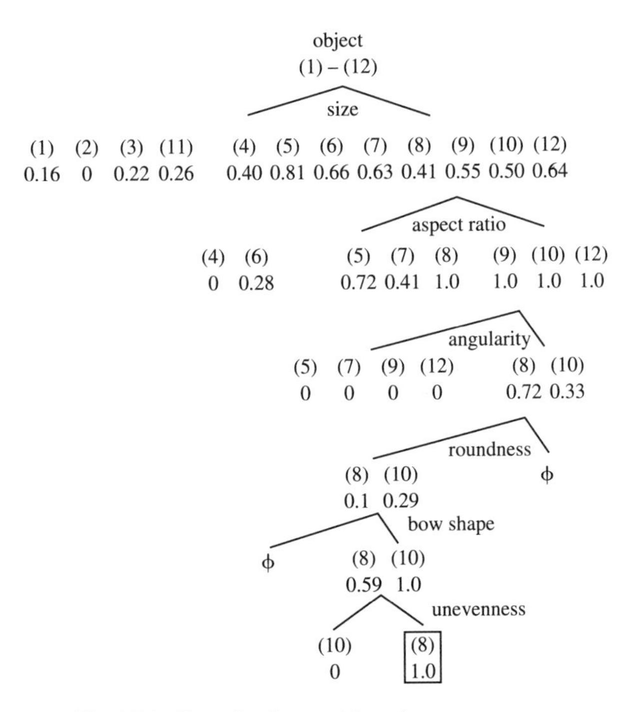

Fig. 3.71 Example of recognition of a green pepper.

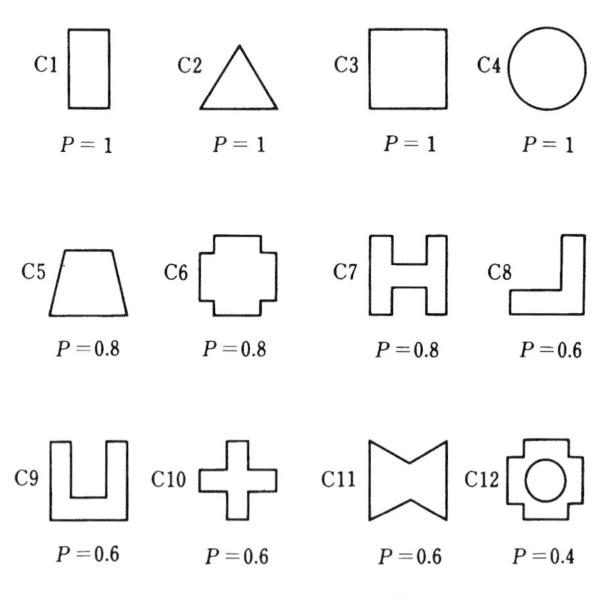

Fig. 3.72 Recognition objects.

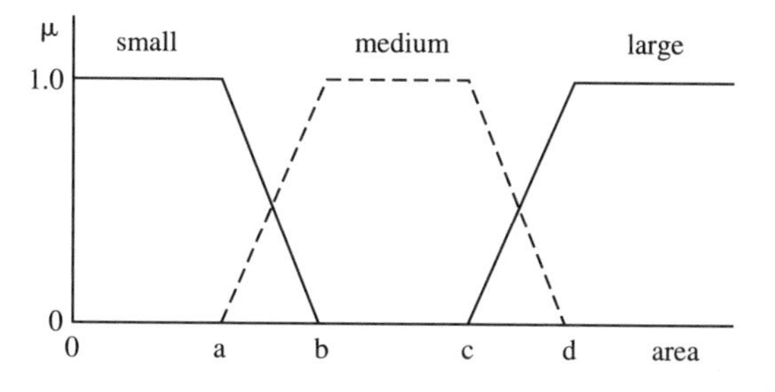

Fig. 3.73 Membership function for area.

the object be A, the classifications were: $A < a$, small; $A > d$, large; $b < A < C$, medium. However, if $a < A < b$ or $c < A < d$, the surface density (the ratio of A to the actual area) was found, and classification was done by comparing this ratio with the membership functions.

The other characteristic quantities were aspect ratio; marginal distribution and the ratio of its average and the maximum value; actual area; surface density; center of gravity and its inclination; circumference; ratios of area and circumference; and variance of marginal distribution. Definite values were obtained for all of these.

Next comes the investigation of how effective the preceding characteristic quantities are for discrimination. We let five types of objects be $C_1, C_2, \ldots, C_5$. To what degree characteristic quantity F_i can clearly distinguish C_j can be expressed ambiguously by membership functions like those in Fig. 3.74. The overall distance between patterns when F_i is used was determined by these membership functions. One example of this is shown in Table 3.15. We can say that the sum total of the figures in the matrix expresses the discrimination effectiveness of F_i.

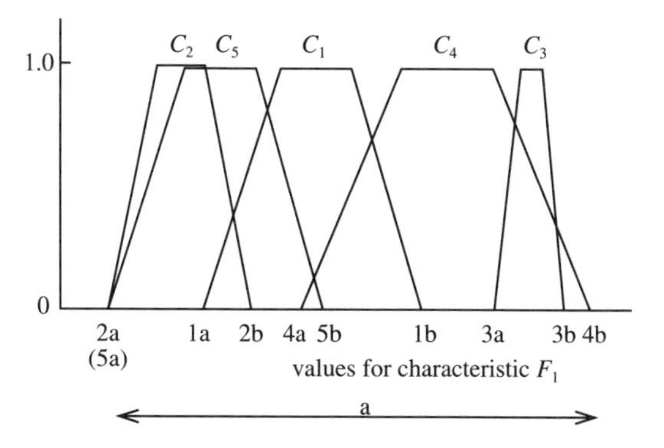

Fig. 3.74 Characteristic quantities for C_i (for E_1).

Table 3.15. Effectiveness of F_1

	C_5	C_4	C_3	C_2	C_1
C_1	0.005	0.0145	0.25	0.045	
C_2	0	0.25	0.575		
C_3	0.45	0.005	e′		
C_4	0.129				
C_5					

$E_1 \Sigma e' = 1.7235$

If in addition, we multiply the frequency of appearance of each pattern by the elements in the preceding matrix, we get Table 3.16. The value obtained by dividing the sum of these figures by the processing time is termed the "degree of importance" of characteristic F_i. For the discrimination of a given pattern, it is most effective to begin by investigating the characteristic quantities with the greatest degrees of importance.

In an actual test, the 12 patterns shown in Fig. 3.72 were placed on a conveyer belt, moved at a speed of 3 cm/s and detected using a CCD camera. The P in Fig. 3.72 shows the probability of appearance. The entire system was controlled by a PC-9800 personal computer. When an object with an indicated shape came along, the system recognized it, and it was picked up by a robot arm and placed on a work table. Nine characteristic quantities were used here, and the processing times for them are shown in Table 3.17.

This experiment shows that real-time pattern recognition of moving objects can be carried out with a comparatively simple apparatus.

Table 3.16. Importance of F_1

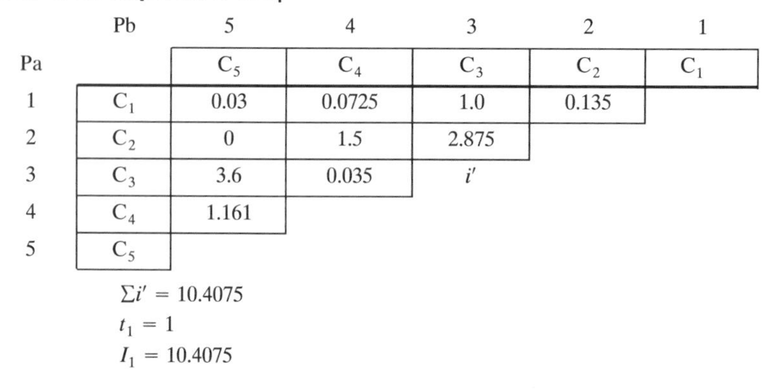

Pa \ Pb		5	4	3	2	1
		C_5	C_4	C_3	C_2	C_1
1	C_1	0.03	0.0725	1.0	0.135	
2	C_2	0	1.5	2.875		
3	C_3	3.6	0.035	i'		
4	C_4	1.161				
5	C_5					

$\Sigma i' = 10.4075$

$t_1 = 1$

$I_1 = 10.4075$

Table 3.17. Processing Time for Characteristic Quantities

F_1	Aspect ratio	0.25 s
F_2	Marginal distribution (x axis)	0.808 s
F_3	Marginal distribution (y axis)	0.807 s
F_4	Average and max (x axis) length	0.778 s
F_5	Average and max (y axis) length	0.60 s
F_6	Area density	0.59 s
F_7	$(\text{Circumference})^2/\text{area}$	1.04 s
F_8	Center of gravity offset (x axis)	0.59 s
F_9	Center of gravity offset (y axis)	0.59 s

REFERENCES

1. Tsunozaki, Y., *et al.* Nippon Kokan Technical Report, *Overseas* **51**, 1–8 (1987) (in Japanese).
2. Iwamoto, M., *et al.* Nippon Kokan Technical Report, *Overseas* **50**, 1–8 (1987) (in Japanese).
3. Mamdani, E. H., "Application of Fuzzy Algorithms for Control of a Simple Dynamic Plant," *Proc. IEEE* **121**(12), 1584–1588 (1974).
4. Suzuki, H., "New Method for Calculating Poss Schedule of Tanden Mill," *Journal of the Japan Society for Technology of Plasticity* **18**(80), 460–466 (1967).
5. *Theory and Practice of Plate Rolling*, Japan Iron and Steel Association, pp. 283–294 (1983) (in Japanese).
6. Zadeh, L. A., "Outline of a New Approach to the Analysis of Complex Systems and Decision Process," *IEEE Trans.* **SMC-3-1**, 28–44 (1973).
7. Sugeno, M., "Application of Fuzzy Sets and Fuzzy Logic in Control," *Measurement and Control* **18**(2), 150–160 (1979) (in Japanese).
8. Tong, R. M., "Fuzzy Control of the Activated Sludge Wastewater Treatment Process," *Automatica* **16**, 695–701 (1980).
9. Maeda M., and Murakami, S., "Design of Fuzzy Logic Controllers," *Kyushu Institute of Technology Research Report No. 49*, pp. 43–51 (1984) (in Japanese).
10. Maeda M., and Murakami, S., "Speed Control of Automobile by Fuzzy Logic Controller," *Transactions of the Society of Instrument and Control Engineers* **22**(9), 984–989 (1985) (in Japanese).
11. Tsukamoto, Y., "Fuzzy Logic Based on Lukasiewicz Logic and Its Application to Diagnosis and Control," Doctoral Dissertation of T. I. T. (1979).
12. Maeda M., and Murakami, S., "Self-tuning Fuzzy Controller," *Transactions of the Society of Instrument and Control Engineers* **24**(2), 191–197 (1988) (in Japanese).
13. Takahashi, H., *et al.*, "Application of Self-tuning Fuzzy Logic System to Automatic Speed Control Devices," *Proceedings of 26th SICE Annual Conference*, II, pp. 1241–1244 (1987).
14. Yagishita, O., Itoh, O., and Sugeno, M., "Application of Fuzzy Theory to Control of Chemical Addition for Water Purification Plant," *Systems and Control* **28**(10), 597–604 (1984) (in Japanese).
15. Itoh, O., "Application of Fuzzy Theory to Control," *'87 Conference for Instrumental Control Technology Preprints*, pp. 2-6-1–2-6-7 (1987) (in Japanese).
16. Itoh, O., Migita, H., Yagishita, O., and Aoki, T., "Fuzzy Adaptive Control for Rainwater Pump Operation," *Preprints for the Third Symposium on Fuzzy Systems*, pp. 121–126 (1987) (in Japanese).

17. Itoh, O., "Fuzzy Controller," *Mathematical Science* **284**, 55–62 (1987).
18. Yasunobu, S., Miyamoto, S., and Ihara, K., "A Fuzzy Control for Train Automatic Stop Control," *Transactions of the Society of Instrument and Control Engineers* **19**(11), 873–880 (1983) (in Japanese).
19. Yasunobu, S., Miyamoto, S., and Ihara, K., "A Predictive Fuzzy Control for Automatic Train Operations," *Systems and Control* **28**(10), 605–613 (1984) (in Japanese).
20. Yasunobu, S., "Automatic Operation of Container Cranes Based on Predictive Fuzzy Control," *Transactions of the Society of Instrument and Control Engineers* **22**(10), 1066–1073 (1986) (in Japanese).
21. Yasunobu, S., and Hasegawa, T., "Evaluation of Automatic Container Crane Operation System Based on Predictive Fuzzy Control," *Control-Theory and Advanced Technology* **2-3**, 419–432 (1986).
22. Yasunobu, S., and Hasegawa, T., "Predictive Fuzzy Control and Its Application for Automatic Container Crane Operation System" *Preprints of Second IFSA Congress*, pp. 349–352 (1987).
23. Yasunobu, S., "A Proposal of Fuzzy Logic Control Method Based on Human Control Strategies," *Transactions of the Society of Instrument and Control Engineers* **23**(9), 969–976 (1987) (in Japanese).
24. Yasunobu, S., "Proposal of Elevator Supervisory Control Method by Predictive Fuzzy Control," *Preprints of the 26th SICE Annual Conference*, pp. 443–444 (1987) (in Japanese).
25. Syoji, T., Tashiro, R., Sekino, S., Sugata, N., Yasunobu, S., and Arai, K., "Automatic Train Operation Devices for the North–South Line of the Sendai Subway System." *Proceedings of the 27th Japanese Symposium on the Use of Cybernetics for Railroads*, pp. 243–246 (1988) (in Japanese).
26. Sakoe, H., and Chiba, S., "Continuous Word Recognition Based on Time Normalization of Speech Using Dynamic Programming," *J. Acoust. Soc. Jap.* **27**(9), 483 (1971) (in Japanese).
27. Sugamura, N., Shikano, K., Aikawa, K., and Koda, M., "Speaker Independent Word Recognition Using the SPLIT Multitemplate Method," *Research Report from the NTT Communications Research Institute* **34**(12), 1687 (1985) (in Japanese).
28. Zadeh, L. A., "Fuzzy Sets," *Inform. Contr.* **8**, 338 (1965).
29. Matsuoka, T., and Kido, K., "Investigation of Phonemic Information with Vocalization Spectra with Static Local Peaks," *J. Acoust. Soc. Jap.* **32**(1), 12 (1976) (in Japanese).
30. Fujimoto, J., Nakatani, T., and Yoneyama, M., "Speaker-Independent Word Recognition Using Fuzzy Pattern Matching," *Fuzzy Sets and Systems* **32**, 181 (1989).
31. Miwa, J., and Kido, K., "Investigation of Speaker Normalization for Speech Recognition," *Acoust. Soc. Jap.*, Onsei Kenkyu-kai, s79–24 (1979) (in Japanese).
32. Nakamura, S., *et al.*, "Simulation of BTSP Method Using Digital Filters," *Proc. Acoust. Soc. Jap.* 1-2-8 (March 1983) (in Japanese).
33. Muroi, T., Nakagawa, M., Fujimoto, J., and Yoneyama, S., "Speaker-Independent Word Recognition—A Method Using Partial Linear Expansion and Weighted Average Templates," *Trans. IEICE* **J69-A**(1), 150 (1986) (in Japanese).
34. Fujimoto, J., *et al.*, "Speaker-Independent Word Recognition Using Fuzzy Theory," *International Fuzzy System Association Congress* (*IFSA '87*), p. 819 (1987).
35. De Mori, R., and Laface, P., "Using Fuzzy Algorithms for Phonetic and Phonemic Labeling of Continuous Speech," *IEEE Trans. PAMI* **PAMI-2**(2) 136 (1980).
36. Morishita, S., and Harashima, S., "A Merger of Acoustic and Symbolic Processing for Construction of a Word Recognition System," *Trans. IEICE* **J70-D**(10), 1890 (1987) (in Japanese).
37. Bezdek, J. C., "Partition Structures: A Tutorial," in *Analysis of Fuzzy Information*, Vol. III (J. C. Bezdek, ed.), pp. 81–107, CDC Press, Boca Raton, Florida (1987).

38. Osumi, N., "Fuzzy Clustering," *Mathematical Science* **191**, 34–41 (May 1979) (in Japanese).
39. Bezdek, J. C., "Some Recent Applications of Fuzzy C-Means in Pattern Recognition and Image Processing," *IEEE Workshop of Lang. Autom.*, pp. 247–252 (1983).
40. Hirota, K., and Iwama, K., "Use of an Additive Data Variation of F-ISODATA in Image Area Partitioning," *2nd Fuzzy Systems Symposium of the Japan Chanter of IFSA*, pp. 94–99 (1986) (in Japanese).
41. Huntsburger, T. L., Jacobs, C. T., and Cannon, R. C., "Iterative Fuzzy Image Segmentation," *Pattern Recognition* **18-2**, 131–138 (1985).
42. Huntsburger, T. L., Rangarajan, C. and Jayaramamurthy, S., "Representation of Uncertainty in Computer Vision Using Fuzzy Sets," *IEEE Trans. Computer*, **C-35-2**, 145–155 (1986).
43. Trivedi, M. M., "Analysis of Aerial Images Using Fuzzy Clustering," in *Analysis of Fuzzy Information*, Vol. III (J. C. Bezdek, ed.), pp. 133–151, CDC Press, Boca Raton, Florida (1987).
44. Pal, S. K., King, R. A., and Hashim, A. A., "Image Description and Primitive Extraction Using Fuzzy Sets," *IEEE Trans* **SMC-13-1**, 94–100 (1983).
45. Tou, J. T., "An Approach to Understanding Geometrical Configurations by Computer," *Int. J. Comput. Inform. Sci.* **9**, 1–13 (Feb. 1980).
46. Di Gesu, V. and Maccarone, M. C., "Feature Selection and Possibility Theory," *Pattern Recognition* **19-1**, 63–72 (1986).
47. Terano, T., Masui, S., Kono, S., and Yamamoto, K., "Recognition of Crops by Fuzzy Logic," *2nd IFSA congress*, Tokyo, pp. 474–377 (July 1987).
48. Hirota, K., Arai, Y., and Hachisu, F., "Robot for Recognition of Moving Objects and Replacement of Moving Objects Using Fuzzy Logic," *2nd Fuzzy Systems Symposium of the Japan Chapter of IFSA*, pp. 15–22 (1986) (in Japanese).
49. Hirota, K., and Arai, Y., "Robot for Grasping Moving Objects Using Membership and Vagueness," *1st Fuzzy Systems Symposium*, pp. 139–144 (1985) (in Japanese).

Chapter 4

Expert Systems

It is indispensable for expert systems to be able to handle fuzziness. The capabilities of experts, who can be thought of as workers in specialized technologies, include processing of ambiguous information and the ability to reason. Fuzzy control is actually a successful example of a very early expert system, and we are now in an era in which many of the expert systems that are heading toward practical use are constructed as fuzzy expert systems.

In this chapter we will discuss bus scheduling, evaluation of structural reliability, applications of schema systems for decision-making, and processing of natural-language information and systems for medical diagnosis as examples of fuzzy expert systems.

4.1 OUTLINE

Expert systems can be thought of as one of the fields of application of artificial intelligence. Even so, research into artificial intelligence is not aimed at the construction of expert systems. Research on artificial intelligence, which began in the 1950s and came into the limelight in the 1960s, suggested that machines with intelligence like that of human beings could be realized in the near future. However, as this research progressed, we learned that this was just a dream, and research slowed down for a period of time. Most research into artificial intelligence at the time either used formal logic under unrealistic assumptions or solved puzzles and board games by means of tree searches with computers. The program for solving a certain puzzle depends on a construction specific to that puzzle and cannot solve a different puzzle. Things such as the GPS (general problem solver), which solved problems not in the way people do, but by means of mathematical methods matched to computers, were constructed, but the solution of real problems was still far away. Naturally, the puzzle researchers had the image of high-level artificial intelligence before them, and they thought that the generalization of puzzle research would be the key to solving the problems of artificial intelligence—but puzzles are puzzles, and the fact is that it did not go beyond this.

The blockage in artificial intelligence research was probably the fact that recognition of human intelligence and, in addition, the essential differences between people and machines was not thorough enough; what helped research into artificial intelligence was the appearance of expert systems and the thinking behind knowledge engineering that began in the 1970s. DENDRAL was created at the end of the 1960s and MYCIN in the first half of the 1970s. The success of expert systems led to a return of prosperity to research into artificial intelligence, but whether expert systems helped artificial intelligence research or had a negative effect on it is a difficult question. At any rate, there is no question that it lowered the level of the goals of research into artificial intelligence. One of the larger problems of artificial intelligence is the realization of inductive reasoning with machines, but Satoshi Watanabe (Professor Emeritus, University of Hawaii), an excellent researcher in the field of the theory of pattern recognition and thinking, indicates the following. Induction is finding the rules managing the movement from a finite number of pieces of data to an infinite number. In principle, however, this kind of thing cannot be done. Therefore, the results of human inductive reasoning are individual and ambiguous rather than uniform. What helped research into artificial intelligence the most was throwing away the dream of making machines carry out inductive reasoning, and the creating of expert systems is what made for this change in direction.

It seems that expert systems will do well, but now that we are in the 1990s, the research into artificial intelligence that is still behind them seems to be beginning to stagnate. Under these circumstances, fuzzy theory, which is expected to be a methodology that will help research into artificial intelligence, naturally has an important relationship with expert systems. Applications of fuzzy theory began in the first half of the 1970s with fuzzy control. Since we can say that the idea of fuzzy control is an expert system for control, it is not too much to say that the very first example of a successful expert system was actually fuzzy control. The method of predicating algorithms in an if/then form was described by

Zadeh in 1968 in his paper entitled "Fuzzy Algorithms." Standard expert systems do not handle fuzziness, but since we can find the essence of expert systems in Zadeh's idea of expressing ambiguous algorithms linguistically, we can say that it is a very advanced idea. What this says is that the focusing of attention on the problem of uncertainty in expert systems is a late development. Naturally, uncertainty of facts or rules can be expressed in expert systems by degrees of confidence, as in the MYCIN system for diagnosing infectious diseases. The calculation of uncertainty in MYCIN is based on the idea of Bayes' probability. At present, probability reasoning is being studied as a method for expert systems, but it goes without saying that the idea of fuzzy theory is that expert systems are essentially fuzzy rather than probabilistic.

With the coming of the 1980s, Dempster–Shafer theory was often used as a method for the processing of uncertainty in expert systems. In Dempster–Shafer theory, the value for the probability that expresses uncertainty is not a single numerical value; it is expressed as an interval. At this point it is not boiled down to probability rules, so the uncertainty of expert systems can be expressed more naturally than that of the MYCIN-type systems. However, even if we talk about uncertainty, in most expert systems it is little more than the application of outside evaluations of rules and facts in the end. In other words, it is the idea of matching logical truth values. For example, it is the external consideration of the degree of certainty or truth of the proposition "Towards morning a fine rain sprinkled down." But here there is no internal accounting of the problem of the ambiguity in meaning of "towards morning," "fine rain" and "sprinkled." The essential uncertainty of expert systems is actually the ambiguity of the concepts and words the experts work with. This kind of uncertainty is not something to be evaluated from outside the proposition; if it is not put into the proposition itself, it cannot be dealt with. The problem must be seen as what the meaning of "towards morning" is. Is it around 5 o'clock, around 8 o'clock or as late as 10 o'clock?

Above and beyond the fact that expert systems are systems that imitate the ways that the knowledge of such systems is expressed and used, methods for effectively constructing expert systems must accurately incorporate the ways in which experts perform. Generally speaking, the ways that experts do things involves the manner in which they bring together knowledge and their way of thinking, and this in its turn is nothing other than the way human beings think. The manner in which experts think can be said to be thinking that involves collection and management of knowledge for the solving of specific problems, but can be seen as a miniature of human thinking and humans in general. Human thinking processes involve characterization by means of all aspects of the obtaining of incomplete information and working with it, similarity inference, evaluation of complex objectives and making judgments and decisions in uncertain circumstances. If we ask how all of these activities are carried out in human thinking, the answer is through natural language. We can say that the everyday wisdom of human beings comes about because they can think in words. Given that the thinking of experts is much better than that of machines, the primary reason is that experts use language. Let us say the data for the relationship between x and y is incomplete. In this case human beings express and grasp incomplete information, as in, "When x is about 2, y is about 100; when x is about 7, y is about 120" or even more so in, "When x is small, y is large."

The assertion of fuzzy theory is that the essence of language is in its ambiguity—that is, fuzziness—rather than in logic. When we create a model of expert thinking, we cannot disregard the role and essence of language in that thinking. Fuzzy theory proposes a method that is close to the essence of expert systems in that it can handle the ambiguity in language and concepts. If there is something that can be called a logic in the human thinking processes based on ambiguous language, it must be an ambiguous logic. Fuzzy logic aims at descriptive language for the modeling of human thinking by means of a combination of fuzziness and logic. It is no mistake that the construction of fuzzy expert systems that make use of fuzzy theory makes for more effective, practical systems. There is actual proof in the field of control, in which there are already many concrete examples. Almost all of the expert systems for control are fuzzy expert systems for fuzzy control. Among these, there are not a few examples in which automation has been achieved for the first time using fuzzy control. There are few examples of fuzzy expert systems outside of control, but there are reports of fault analysis systems for locomotives and restaurant menu selection systems. Expert systems that do not make use of fuzzy theory are the main current, but of the almost 2,000 expert systems reported recently, only about 2% are actually being used. When viewed from the standpoint of practicality, the number of fuzzy expert systems has already surpassed the number of nonfuzzy expert systems.

Expert systems have three problems: (1) knowledge expression, (2) knowledge usage, and (3) knowledge acquisition. Fuzzy theory can contribute, a great deal to the solution of each of these three problems. As far as (1) is concerned, linguistic expression is the highest level, and real knowledge, rule knowledge, and judgmental knowledge can only be expressed in a form that includes ambiguity. And more than this, knowledge can be brought together in a manner that is easy to use, because human beings allow for ambiguity. The concept of fuzzy sets can be used for knowledge expressions, but fuzzy logic is naturally effective for (2). At the center is fuzzy inference, and fuzzy inference is inference by means of ambiguous rule knowledge using ambiguous information, which can be said to characterize the knowledge usage of experts. Knowledge is not limited to inference, and outside of inference are the large problems of evaluation and judgment. Fuzzy measures and fuzzy integrals are effective for evaluation and judgment models for expert systems based on experimental knowledge. For example, the Dempster–Shafer belief function, which is used for evaluation of rules and facts, is one form of fuzzy measure. The first problem with (3) is extracting the experiential knowledge of experts. Since at this time most knowledge is communicated in fragments in ambiguous language, fuzzy interpretations are necessary. Just taking rule knowledge alone, it is possible to acquire expert knowledge by means of a set-up for dividing up ambiguous cases for usage conditions. The second problem is methods for cases in which knowledge is not directly obtained from the mouths of experts, but their operations are indirectly observed, or when there are no experts and knowledge is obtained from experimentation and learning. Fuzzy theory can also be used for gathering data and information in the same ways that experts do in these cases. In other words, as was mentioned earlier, it is a qualitative gathering method as in "When x is small, y is large." The operations of experts can be accurately grasped using this kind of fuzzy expression. In addition, we can expect effective, high-level learning by means of the construction of learning models that are easy to understand on a linguistic level in cases where learning is undertaken.

Besides control, there are many fields in which fuzzy expert systems are effective, including fault analysis, medical diagnosis, scheduling, system planning, economic prediction, stock investment, sales predictions, management decision making, writing summary, and consultation areas such as legal consultation and travel consultation; among them are fields where nothing has been done, and others in which ideas with the versatility of fuzzy theory have been introduced in small, partial ways, but have had large effects. There are fields, such as management problems for small businesses, that are waiting for the appearance of fuzzy expert systems that can be used with laptop personal computers.

4.2 BUS SCHEDULING

4.2.1 The Bus Scheduling Problem

The planning for bus routes begins with an effective scheduling diagram that is matched to the demands of riders. The schedule diagram, the objectives of which are bus routes, is made up by operations specialists who give consideration to rider demands, traffic conditions, economics, and the working conditions of drivers. However, even with the free use of long experience and expert knowledge, the balancing and satisfying of the many reciprocal conditions requires a vast amount of time and effort, so it is a burden on scheduling specialists and one of the difficulties in the revision of schedules. To improve these conditions and in turn assist bus companies in the use of high-level information, an automatic bus scheduling system was developed.

This system solves a *combinatorial problem* based on the various restrictive conditions for bus scheduling, and by emulating the unconscious activities of specialists using heuristics, practical scheduling up to the standards of that done by hand was accomplished with a computer.

There were a number of difficulties in the automation of this office management process known as bus scheduling. Among these, we can note the fact that the conditions that form the basis of the system contain *ambiguity* (*fuzziness*). This problem is taken care of by means of *fuzzy sets*.

This system was a joint development by the Keihin Electric Railway Express Co., Toshiba Corp., and Toshiba Advanced System Corp.

(a) The Four Elements of Scheduling

Schedules vary from easy-to-follow ones to complicated ones, but it is easy to see that any schedule is made up of the following four essential elements: the distance and time necessary to get from the starting point, which is the depot, to the end point, the bus stop; a running interval; and first and last bus times that meet demand.

Distance and time required can be either directly or indirectly measured. Time required varies with traffic jams, during certain time periods within the day, and in some cases according to direction. In cases in which the required time changes drastically for weekdays, weekends and holidays or for different seasons, schedules are made for each. All in all, the different schedules are in most cases created in order to meet the fluctuations in rider demand.

As far as first and last bus schedules are concerned, first bus times for places that are more remote from the city center and are involved in the transportation of workers and students, for example, must be earlier in proportion to the distance, and last bus times must be proportionately later to meet demand. In these cases, the convenience for transferring to trains must be taken into consideration. The interval between buses can be computed from the anticipated number of riders for each time period and the capacity of the buses. For most lines the continuous length of the morning rush hour is shorter than that of the evening rush, and in addition, the hourly demand is greater; therefore, the interval between buses must be set shorter for the morning rush hour than for the evening rush hour. Generally, demand decreases during the daytime, but in shopping areas such as those around large department stores, the daytime running interval must take shoppers into consideration. For lines for which traffic jams are anticipated, allowances are made in required time, and furthermore, the time needed to turn the bus around again after the final stop (hereafter wait-to-turn) is lengthened, and an interval that can absorb delays, if they occur, is used. At suburban train stations where there are not so many buses, not only must the first bus timing be considered but also the transfer convenience of passengers during other time periods. The intervals must be set so that morning buses in particular arrive before trains leave and evening buses leave after trains arrive; in other words, the arrival and departure times are fixed. Aside from this, there are regions where connections with ferries are the main objective. Furthermore, there are cases in which connections with other buses must be considered. In areas with few buses, or in time periods of that sort, the greatest effort is made to keep the minutes of departure for each hour the same, so that riders can remember them easily.

In this way, scheduling specialists determine the four elements for the scheduling of departure and arrival times of buses at beginning and end-point depots.

Let us say that we have bus lines as shown in Fig. 4.1. Demand is large between station D and city E, and next comes that between city E and town F. The area in front of station D is small, and there is no space to stop buses to allow drivers rest periods. However, there is parking space

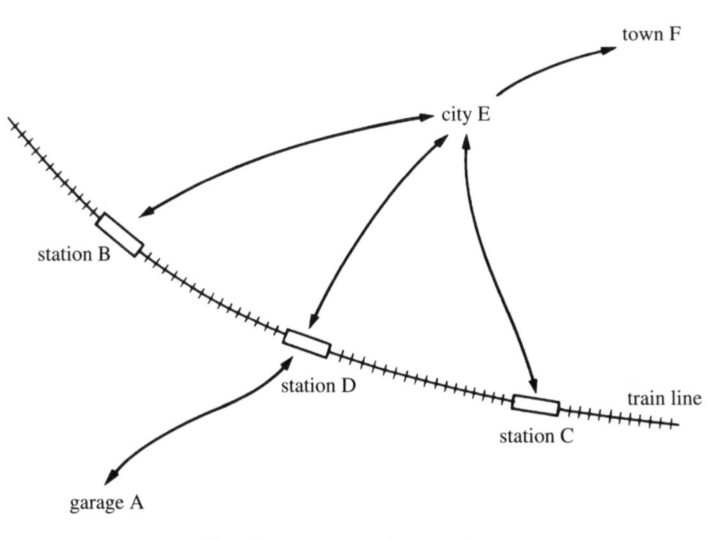

Fig. 4.1 Example of bus routes.

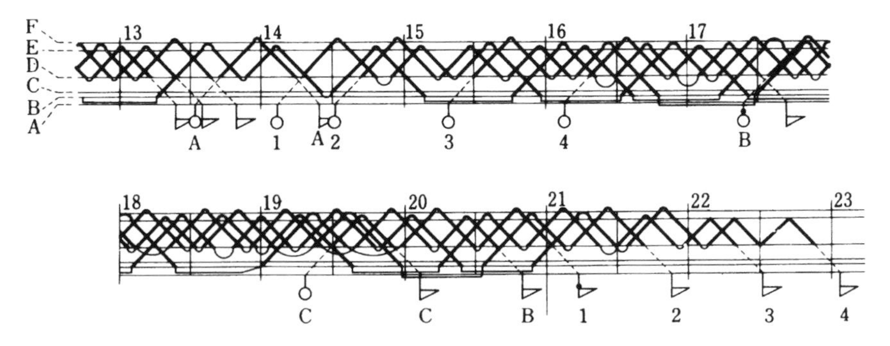

Fig. 4.2 Four elements of scheduling and schedule example 1.

near stations B and C, and they can be used as rest areas. Generally, if there is a depot convenient, buses are parked there and rest periods can be taken. However, since depot A is far from station D in this case, going back to the depot just for rest periods is avoided if at all possible, because of economic considerations.

First the four elements of the scheduling for these lines are determined. The four elements of the schedule are easily understood when expressed by the dark lines in Figs. 4.2 and 4.3. The four scheduling elements are the same in Figs. 4.2 and 4.3. These figures correspond to the afternoon of one day. The vertical axis is required time rather than distance. In this way, the tracks for vehicles traveling in the same direction are parallel line segments and are easy to see. The horizontal axis is time. A bus leaving depot A heads for station D. Since there is no foreseeable demand during this interval, it is just an out-of-service run, and it is denoted by a dotted line in the figure. From station D it heads for city E and provides service to passengers going to the city center and to town F, then heading to stations C and B. These intervals are operational runs carrying passengers and are

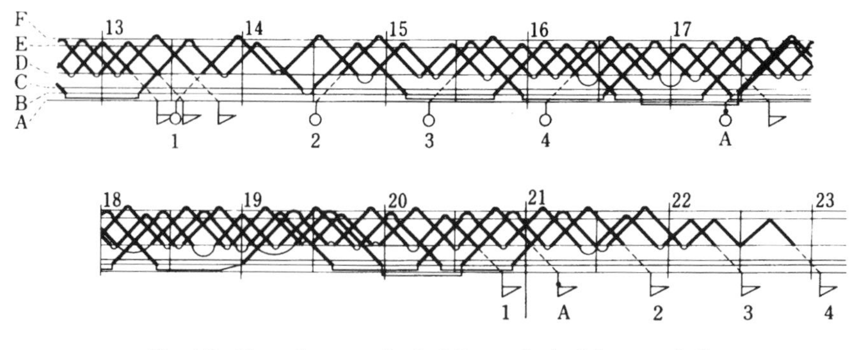

Fig. 4.3 Four elements of scheduling and schedule example 2.

denoted by solid lines in the figure. The path of the round trip is denoted by an inverted V shape. The vertex of the trip denotes the wait-to-turn situation. The four elements of the schedule are shown in this way by heavy solid lines in the figure, and they are expressed in the form of a large collection of trips. At this point the trips are not connected to one another. This situation gives the input conditions for scheduling.

(b) Scheduling Conditions

The trips that express the four elements of the schedule are matched and strung together, and they are summarized as man-days for each operator. In terms of Fig. 4.2, these are the schedule lineups numbered 1, 2, 3, and 4. This length must be kept within the fixed daily working time. If it goes over, there are cases in which overtime must be paid, which reduces proceeds, and over a certain amount it has an effect on safety. Conversely, if it is too short, the resources of operators and buses are not being used, and the economics are lost. There must be enough rest time in the workday. If the rest time is too short, working conditions deteriorate, and there is a loss of safety. If it is too long, it is uneconomical. With rest time, not just the length, but the way it is taken is also important. Rest time does not always have to be taken all at once; it can be divided into two segments, and it is good to allow for rest after some driving fatigue has accumulated. Even if it is divided up, there must always be at least one time that is long enough for a meal. In addition, the connections between trips and the wait-to-turn time for leaving the depot are added in above and beyond rest time. It is sufficient for the thinking behind the wait-to-turn time at the connections to be the same as that for the tops of the trips. In other words, appropriately short is ideal, but when traffic jams are anticipated, it is lengthened, so that fixed-time operation can be carried out. The rest is operation time. Naturally, the operating time must be long enough. The product of tying these trips to each other is called the work schedule.

The first element in determining the economics of scheduled operation is the necessary number of vehicles. The lower the necessary number of vehicles, the more reasonable it is. When a schedule is actually put into operation, each work schedule is apportioned an operator and vehicle. The method for this varies with the customs of each bus company, but the one day's bus and operator grouping is fixed—so, for example, a bus whose operator is on break cannot be used by another operator. However, when that operator's workday is over, that bus can be used by another operator. Thus, the number of work schedule lines at the time when that number is the greatest is the necessary number of vehicles for operation under that schedule. Finally, we can say that reasonable bus scheduling is one that makes possible a combination in which the working time and its elements of running time, rest time, and wait-to-turn time are appropriate and in which the work diagram that does this minimizes the maximum number of simultaneous lines.

It is sufficient to connect the trips together carefully following the preceding ideas, but it does not always go well doing this alone. Generally there is an extreme difference between the demand during the morning and evening rush hours and during the middle of the day. Above and beyond this, the peak demand is higher during the morning rush hour than in the evening. Therefore, in order to deal with the short time period of the morning rush, there is no way out of special scheduling that does not fulfill workday timing. In terms of Fig. 4.3, schedule A is such a schedule. These kinds of conditions cannot be avoided. Therefore, it is important that the necessary number of vehicles, inclusive of the number for this kind of special scheduling operation, should be kept as small as possible.

4.2.2 Automatic Bus Scheduling Functions

The trips that express the four elements of the schedule are arranged with a fixed goal, based on the various restrictive conditions; in other words, it is a construction that works from the parts to the whole. This means that the basic idea is similar to a CAD system for wiring diagrams.

There are many loose conditions for the combining of these trips. For example, since it is impossible to make a schedule in which the lengths of the work schedules, that is, the length of a day for one person, are all the same, there is in actuality no escape from having a concrete target value, and using values a little smaller and larger than that. Running time and rest time are the same. In other words, even if there are cases in which all the values miss the mark within a certain realistic permissible range, the main objective is the creation of a good schedule that is reasonable overall.

(a) Heuristics

Basically, scheduling is done by bringing together one schedule line, making the second line from the remaining trips, and repeating this process (Fig. 4.4). Therefore, this process constructs a tree, and the total number of groups becomes astronomically large. The calculation time for an exhaustive search is too long for this to be useful.

One method for raising the search efficiency and for avoiding the expansion of the number of groupings is called *heuristics*, which avoids searching all logically possible cases and is a method for searching intelligently for an answer within a realistic time range by making use of knowledge about the specific problem to be solved. Since the actual

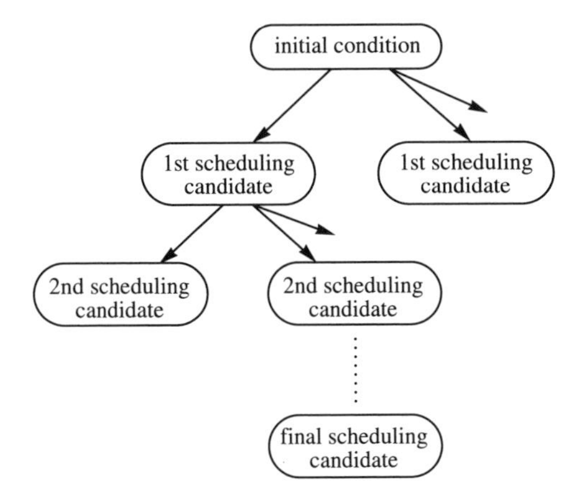

Fig. 4.4 Decision tree.

heuristic choices depend on the essential characteristics of the problem to be solved, there can be various types of heuristics even if we limit ourselves to bus scheduling, since labor agreements differ for different bus companies and we can assume that thinking about scheduling differs, too. However, it is generally important to make an effort to find a highly precise method, if at all possible.

First of all, the following three conditions are self-evident, but there are cases when a combination of these alone will reduce the numbers to a large extent.

(1) When a bus has returned to a depot, the trips should be connected so that it leaves from that depot on its next run.
(2) Combinations that make for taking rests at depots where there is no place to rest should be avoided.
(3) When it is rest time, a combination that uses the available trips to go to a depot that has a rest area should be given priority.

In addition, the following heuristics are worked in as a means for looking ahead:

(4) During rush hour, buses should be running, and during quiet times, combinations that allow for rests should be given priority.

If there are buses that are rested during the rush hour, other buses must run as substitutes. This invites an increase in the necessary number of vehicles required and is not economical. In addition, when there are many buses running leisurely, the other side of the coin is that rests are lengthened to that extent, and the results are unreasonable. When calculations are actually done, the number of vehicles necessary to get over the rush hour is a delicate determination, and there are cases in which condition (4) is the only weak point. Because of this we have the following condition:

(5) Combinations that make for rest periods during the rush hour should not be chosen. In other words, combinations that continuously connect the peak demand trips should be employed.

As can be seen from this, a good overall schedule cannot be made by accumulating schedules built up so that each one is optimized at its stage while ignoring the others. This is an essential complication of bus scheduling. The problem is just how far ahead we should look, and the following strategy is powerful under these conditions:

(6) When scheduling is done by combining "trips," choices are made so that the remaining trips can be combined easily and so that the number of overlapping trips is reduced.

Using the peculiarities of the problem in this way, the number of possible combinations is narrowed, but furthermore, it is necessary to employ some device for making the completed schedule appropriate in terms of working conditions. The reason for this is that even if, for example, we combine all of the trips in a way that reduces the required number of vehicles, it is not practical if driving is difficult because of unnatural combinations. Because of this the following condition is stipulated, and the quality of each schedule is improved:

(7) Priority is given to combinations of trips that are appropriate from the point of view of working conditions such as work time, driving time, and rest time.

The evaluation of the delicate points of whether or not working conditions are appropriate requires the judgment of experts, and at the stage in which each schedule is being put together, there is ambiguity that does not allow for unconditional determination. This point will be discussed in Section (b), which follows.

Various other methods are brought in. It is difficult to conceive intelligent heuristics and express them well in words, but inefficient trial and error is reduced by giving the utmost priority to combinations that are thought to have meaning, and an increase in the number of combinations can be avoided by early backtracking.

(b) Processing Ambiguity

From the standpoint of technologies used up to now, ambiguity was something that was not allowed. The basis was making handling easy through clarification with the utmost standardization. However, there is no escaping the fact that during the combining of schedules the problem of the *ambiguity* contained in the expressions of working conditions must be overcome when each schedule is judged to be desirable or not in the light of these conditions.

As in Table 4.1, for example, we can see even if the workday conditions call for 8 hours, examples like 8 hours 12 minutes or 8 hours 2 minutes are allowable, and in addition, this makes the overall form of the schedule fall into place. In other words, there is an ambiguity in the conditions for combining the trips that cannot be unconditionally determined beforehand, and the key is how this ambiguity is handled and used.

In this kind of case, it would be good to work with a tool that could handle the ambiguity itself. This is the concept of *fuzzy sets*. Speaking in terms of bus scheduling, what we have is the consideration of the fuzzy set for "the group of schedules with appropriate work times," expressing this well and using it. Let this be set A:

Set A: The set of schedules with appropriate work times.

We can think of operating time and rest time in the same way:

Set B: The set of schedules with appropriate operating times.

Set C: The set of schedules with appropriate rest times.

Table 4.1. Working Hours and Details

			Details			
	Schedule	Hours of Work	Driving Time	Rest Time	Wait-to-Turn	Preparation Time
Fig. 4.2	1	7:58	4:36	0:50	1:42	
	2	8:12	4:52	0:55	1:35	
	3	8:02	4:52	0:53	1:27	
	4	7:46	4:52	0:49	1:15	
	A	1:43	0:50	0:00	0:03	All
	B	4:16	2:36	0:00	0:50	0:50
	C	1:38	0:44	0:00	0:04	
Fig. 4.3	1	8:08	4:58	0:50	1:30	
	2	8:12	4:52	0:55	1:35	
	3	8:02	4:52	0:53	1:27	
	4	7:46	4:52	0:49	1:15	
	A	4:40	3:04	0:00	0:46	

Given this, if we consider the set of schedules with appropriate working conditions, Set H, we go with the idea of the intersection of these sets and get the expression:

$$\text{Set } H = A \cap B \cap C \cap \ldots: \quad \text{The set of schedules with appropriate working conditions.} \tag{4.1}$$

Set H is a fuzzy set. The degree to which one given schedule d satisfies the working conditions is the degree to which it belongs to set H. The function that expresses that degree is defined as $H(d)$, and fuzzy set H is expressed in terms of it. Function $H(d)$ is the *membership function* of fuzzy set H. There are various definitions for function $H(d)$, but if we follow a frequently used method, we can write it as

$$H(d) = A(d) \wedge B(d) \wedge C(d) \wedge \cdots. \tag{4.2}$$

given Eq. (4.1). In other words, this is the function that expresses the appropriateness of the work time, but we must consider the expression of the intuition and experience of scheduling specialists for dealing with the appropriateness of length of work time here. For example, we have the shape shown in Fig. 4.5. We get a hint for the concrete way this is obtained from the frequency distributions for work time in the many currently existing schedules created by hand. However, the soundness of this form of function should be judged good or bad in terms of the complete overall schedule. Functions $B(d)$ and $C(d)$, etc., can be defined in the same way for sets B and C, and they express the thinking about operating time and rest time, etc.

In the foregoing manner, the logical product on the right side of Eq. (4.2) can be calculated based on the membership functions for each. As far as heuristics are concerned, it is sufficient to determine the value for function $H(d)$ for schedule d that is made up of a certain combination of

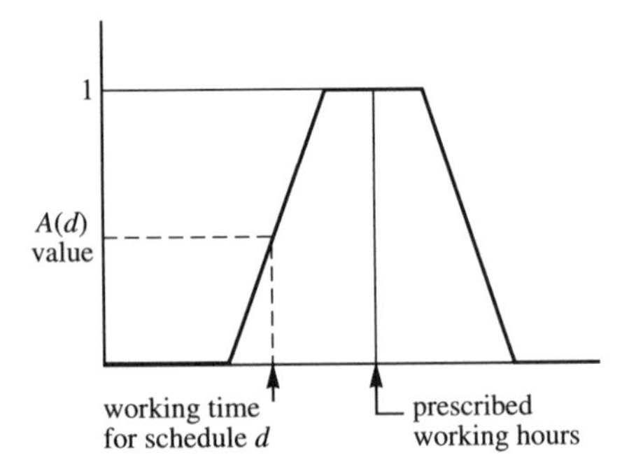

Fig. 4.5 Example of membership function.

trips and to give priority to combinations of trips for which this value is large.

4.2.3 Evaluation of Results

In cases where a number of scheduling specialists make schedules for the same line, the completed schedules are not all the same. Even if the necessary number of vehicles for these schedules are for the most part the same, the ways in which the trips are combined will differ a little among them. However, any of them will be fit for practical use. In other words, there is not a unique correct solution.

With this system, all of the possible combinations are not exhausted, and the solutions obtained are not guaranteed to always be the optimal ones. This is kept in mind, and from the possible combinations of trips obtained as schedules, a number of cases—ordered according to how low the necessary number of vehicles is—are output as solution candidates. In addition, if the computer results are reviewed by people who are scheduling specialists and fine-tuned by hand, there are many cases in which a possibility remains that they can be revised into more desirable schedules. In such cases, it is set up so that revisions can be made using tablet commands, while viewing the schedule on a CRT display.

(a) Actual Example

Figure 4.2 is one example of a solution that was among the schedule candidates found by the computer. This result received a good evaluation as a solution in itself, but it is also a good calculation result to use for this explanation, because it could be converted into the practical schedule in Fig. 4.3 through a simple manipulation when a specialist looked at it. In terms of the order of revision, schedule A in Fig. 4.2 was first connected to the beginning of schedule 1. Next schedules 1 and B in Fig. 4.2 were each divided in the neighborhood of schedule C. Finally, the order of the beginning of schedule B, schedule C, and the end of schedule 1 was rearranged, and on the other side, the beginning of schedule 1 was connected to the end of schedule B. The result is the form in Fig. 4.3. If Fig. 4.2 and Fig. 4.3 are compared, the necessary number of vehicles for each of them for this time period is six. However, Fig. 4.3 is more efficient than Fig. 4.2. The reason is the two back-and-forth transit trips between 14:00 and 14:30 and between 19:20 and just after 20:00 that are shown by dashed lines in Fig. 4.2 have been eliminated, and this makes for an operation configuration that is that much denser. Because of this there is an increase of one schedule for which the work time is a little over the regulation 8 hours, but it falls within the operating time regulation of 5 hours. The length of the special schedule for contending with the rush hour is also within the range that allows for expansion.

Schedules are for the most part created in this way. When viewed from the standpoint of ambiguity processing, the restrictive conditions, which are expressed ambiguously by the membership functions in Fig. 4.5, function well, and we can see that even if the regulations slip a little, the overall schedule is helpful in finding a candidate that is headed toward a solution that appeals to experts. Naturally it would be ideal if a solution like that in Fig. 4.3 would come out in the beginning, and if it were necessary, some kind of heuristics could be added, and the search precision could be raised by the addition of new evaluation items.

(b) Comparison of Time for Schedule Creation

In the case of doing all the work for scheduling by hand, one day's schedule is drawn out on a wide sheet of paper, the work operating and rest times totaled and investigated and the trips connected into a balanced whole through repeated trial and error.

The time this takes differs with the number of schedules, but if we consider a standard case in which there are 10 schedules satisfied by full workdays and three special schedules for peak times, the work done by hand by specialists normally includes about 5 hours for drawing in the "trips," and about 25 hours for connecting them and making it into a schedule. On the other hand, with this system, the inputting of the trips in dialogue form can be done in about 3 hours and the automatic production of the multiple schedule candidates completed in about 1 hour.

Actually the scheduling work is not over even after the trips have been tied together and the schedule completed. When the work is done by hand, it is first necessary to draw and produce a clean copy of the schedule diagram. In addition, the various numerical values necessary to actually operate the schedule have to be read off the chart and converted into signs and charts for use. The schedules posted at depots are one form of the registers made in this way. In addition, small running charts that tell the schedule and the route, which are placed beside the drivers' seats, are made by following each of the schedules made in this way from the departure from the depot to the return.

Besides this, the various kinds of registers that are necessary for the administration of bus operations are made one by one by totaling the changes repeatedly. In the present case, we have to look at about 70 hours to do this by hand. If it is done by computer it can be output on a printer or plotter in about 10 hours.

Looking at the totals, doing it by hand takes about 100 hours (14 days), and this system takes about 14 hours (2 days). The total amount of scheduling work for schedule revision is about 1/7 (Table 4.2).

Table 4.2. Comparison of This System and Manual System

This System		Manual System	
Inputting trips using comments viewed on CRT	3 hours	Drawing trips while considering the essentials of diagram construction	5 hours
Automatic schedule generation	1 hour	Connecting trips through trial and error	25 hours
Output of various lists, schedules and reports	10 hours	Creating various lists, schedules and reports	70 hours
Total	14 hours (2 days)	Total	100 hours (14 days)

The finished schedule governs daily operations in actual places from each branch. There the bus and operator groups are apportioned for each schedule following the working order, and the various registers are made for day-to-day additions and deletions of buses, delays, etc., but the automation of this field is already complete. Therefore, a great savings of effort has been achieved if we look at bus operation overall.

4.3 EVALUATION OF STRUCTURAL RELIABILITY

4.3.1 Introduction

From the time of their construction to now, existing structures have received some internal damage because of various external events such as earthquakes, typhoons and impact, and the result is that not a few have safety and serviceability problems. However, it would be impossible from an economic point of view to demolish and rebuild all of the damaged structures. Because of this, there is a requirement for accurately judging which structures should be repaired and which should be rebuilt. This judgment is easy with structures that are seen at a glance to be damaged, but it is very difficult in cases in which there is critical internal damage that is not readily apparent to the eye. In order to appropriately judge these kinds of complex damage conditions, it is necessary to evaluate the reliability of the structure from an overall point of view, mutually relating the various causes and other useful information (cause of the damage and state of progress, importance of the structure, various design and construction sources for the structure, local environment, etc.) with the damage visible on the surface.

However, the data that can be used for evaluation of structural reliability is insufficient in quality and quantity, and the progress and degree of damage must be inferred from investigations of various aspects of small amounts of ambiguous data, based on intuition and experience. Because of this, the evaluation of structural reliability has had to rely upon the intuition and engineering judgments of experienced specialists up to now.

Since there are not enough specialists to meet the demand in these days when the need for structural maintenance and repair has grown so much, it is impossible for all structures in which problems arise to be processed by experts. Therefore, a reliability evaluation system that would allow general structural engineers to achieve the same quality of evaluation results as experts would be desirable.

Given this background, a number of expert systems for structural damage assessment have been developed recently [1–8]. Special among these is the system called SPERIL, which, although it was developed about 10 years ago, already uses the idea of fuzzy sets in its inference process in order to treat the various kinds of ambiguity contained at various stages [9–11]. Other systems do not directly employ the concept of fuzzy sets, but most of them show the effects of, for example, operations with certainty factors. Besides this, a system for the evaluation of structural reliability that makes use of fuzzy logic has been developed [12]. In this section we will first introduce a pioneering structural reliability evaluation system and then give a detailed explanation of a fuzzy expert system developed by the authors for the evaluation of the durability of concrete slabs in bridge structures.

4.3.2 Structural Reliability Evaluation System Using FRIL

In the evaluation of the reliability of structures, how we quantitatively grasp the effects of uncertainty is important. However, the causes of uncertainty that we must consider, their origins, elements and other aspects are really diverse, and no matter what we do we must depend on human judgment for their comprehensive evaluation. In other words, we must make up for the lack of information and deficiencies of models by means of the general evaluation capacity of humans, who can grasp the essence of an object, even if it is unclear and vague.

In order to make use of this kind of human ability and to handle a wide range of structural reliability, an expert system that makes use of FRIL (fuzzy relational inference language) [13] has been proposed. With FRIL, the concept of fuzzy reasoning (also known as approximate reasoning) is used in order to avoid complicated calculations. Since this fuzzy reasoning is defined in a general form, it can also be applied in fields such as medicine, economics, and social problems.

This expert system is also basically constructed from a *rule base* and an *automatic fuzzy inference engine*, and it is set up so that the user inputs the necessary information each time it is used. The most characteristic feature of this system is the fact that the modus tollens is used in the inference system. Modus tollens means that for a proposition P (antecedent) $\rightarrow Q$ (consequent); if negation $Q*$ is given for Q, negation $P*$ is obtained. In this system P and Q are defined by fuzzy sets that have fuzzy truth values.

In order to explain this reasoning process using a simple example, let us consider this as a rule: If "S = the structure is built safely" then "NP = the structure has a small probability of failure for any limit state" is "τ_{abs} = absolutely true." In this instance, let us say that we obtain the information "the structure does not have a small probability of failure." We then derive the result that "this structure is not safe" is "τ = (true, for example)."

In the original system [14], a logical relations diagram was used in order to obtain a conclusion for S from the large number of rules, but in the new system, calculations have become more efficient using virtual rules. In addition, two different forms of inference—inference that only takes into consideration increases in belief according to success and inference that only handles decreases in belief due to failures—are used in order to increase the precision of inference.

4.3.3 Durability Evaluation System for Reinforced Concrete Slabs in Bridges

Among the bridges in existence, there are not a few that have undergone some damage and have safety and serviceability problems. In order to carry out maintenance on these bridge structures, it is necessary to evaluate their structural durability. However, not only is there quantitatively insufficient data to use for durability evaluations, but also a great deal of it contains some ambiguity, making it difficult to use. Thus a *production system* (fuzzy production system) in which a function that manipulates fuzzy sets has been added to the inference mechanism is used in this system in order to make for something that will hold up for use in actual maintenance facilities. However, this system uses something different from FRIL, modus ponens. The benefits generated by the integration of a manipulation system for fuzzy sets into the production system are, among others,

(1) Inference from ambiguous data and knowledge is possible.
(2) Natural language can be used in the rules.
(3) By means of (2), the knowledge acquired in the form of natural language through dialogues with experts can be input into computers as it is, so knowledge acquisition is easy.
(4) The number of pieces of knowledge can be kept to the minimum

necessary for evaluation because of the ability to include ambiguous knowledge.

In addition, this system was constructed on a 32-bit workstation, so it has an even better user environment. In the following we will give an outline of this fuzzy production system.

(a) Fuzzy Production System

Let us now consider the rule "If X is A, then Y is B." In production systems used up to now, for data "X is A," the rule could be used to draw the conclusion "Y is B," but for data in which X is A', which is slightly different from A ("X is A'"), a conclusion cannot be drawn from the rule. On the other hand, even with the data "X is A'," human beings can draw the conclusion "Y is B'." An inference mechanism similar to this kind of inference mechanism (known as approximate reasoning) that operates in the human brain can be realized using the concept of fuzzy inference. In other words "A'" is expressed by fuzzy set "@A'", the data rewritten as "X is @A'," "A" and "B" expressed by fuzzy sets "@A" and "@B," and inference carried out. The result is that the conclusion "Y is @B'" (@B' being a fuzzy set) can be obtained.

(b) Outline of the Durability Evaluation System

In this system, the following were carried out in order to precisely evaluate durability:

(1) Because experts in maintenance were met with many times and time was taken to conduct a dialogue with them, the knowledge peculiar to these experts was acquired. In particular, experts' experiential knowledge concerning the causes of damage to reinforced concrete slabs and degrees of progress of damage was acquired.
(2) It was possible to handle the ambiguity in the data obtained from inspections and the ambiguity in the knowledge obtained from experts through the introduction of fuzzy sets.
(3) In order to evaluate the durability of reinforced concrete slabs it can be considered necessary to infer the remaining life, and we attempted to estimate the remaining life of these slabs. The measures thought of for this were cause of damage, degree of damage, and progress of damage.

The durability evaluation is conducted in slab panel units. The evaluation progresses in the order shown in Fig. 4.6. First the type of damage in panel 1 is specified from the damage data obtained through inspection. In this system "cracks," "road surface damage," "reinforcing steel damage,"

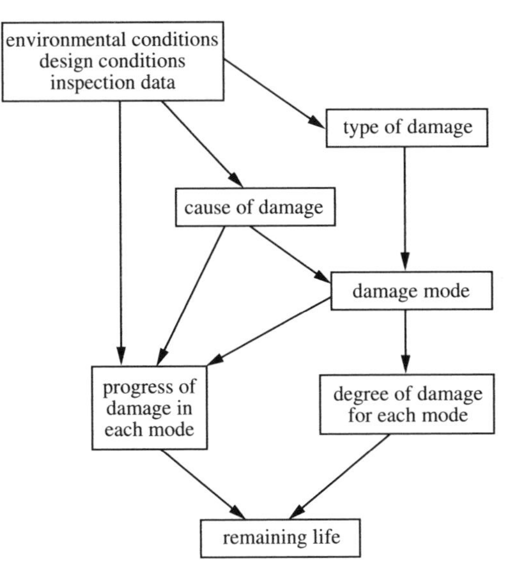

Fig. 4.6 Damage assessment process.

"concrete damage," and "structural damage" are set as the types of damage. Next the cause of the damage is inferred from various design and environmental conditions, etc. The considerations for causes of damage are given in Table 4.3. Generally multiple causes are inferred, but what is called the damage mode for each of these causes is considered. The damage mode is only made up of the types of damage that originate in each cause. After the damage modes have been identified from the causes of damage, the degree of damage for the damage mode is obtained. Furthermore, the progress of damage is found from the causes. The remaining life is then inferred for each damage mode using the degree of damage and progress of damage found in the foregoing manner for that mode and in addition the years of service of the slab.

(c) *Method for Estimating Remaining Life*

Next we will describe the method used in this system for inferring the remaining life. We assume that the slab has not undergone any harmful action that was not considered in its design, and the relation between years of service and degree of damage is stipulated by curve S-0 in Fig. 4.7. Furthermore, we assume that the straight lines S-1 through S-5 correspond to "large," "somewhat large," "medium," "somewhat small," and "small" degrees of damage. Here we will consider the inference of the remaining life of, for example, a reinforced concrete slab 20 years old with a

Table 4.3. Causes of Damage

Load	Extreme wheel load Impact Inappropriate positioning of support members
Design/structure	Insufficient rigidity due to insufficient slab thickness Insufficient main reinforcement Insufficient distributing reinforcement Defective load distribution cross members Additional bending moment due to uneven settling of main members
Construction	Poor concrete quality Insufficient compaction Poor curing process Insufficient processing of construction joints Insufficient covering
Other	Salt content Poor surface drainage Shifting of understructure

"somewhat small" degree of damage and a "somewhat small" progress of damage. The condition of this panel is given by point P in Fig. 4.8. As far as the "somewhat small" progress of damage is concerned, it can be assumed that the degree of progress of S-0 (similar for S-3) will come to dominate after 30 years, since the progress of damage for curve S-0 gets worse after 30 years. Therefore, as is shown in Fig. 4.8, the damage will progress along a straight line parallel to line S-4 from point P to R and from point R on along a straight line parallel to line S-4, and at point Q it reaches its upper limit. From the above we can find the remaining life by subtracting the horizontal coordinate of Q from that of P.

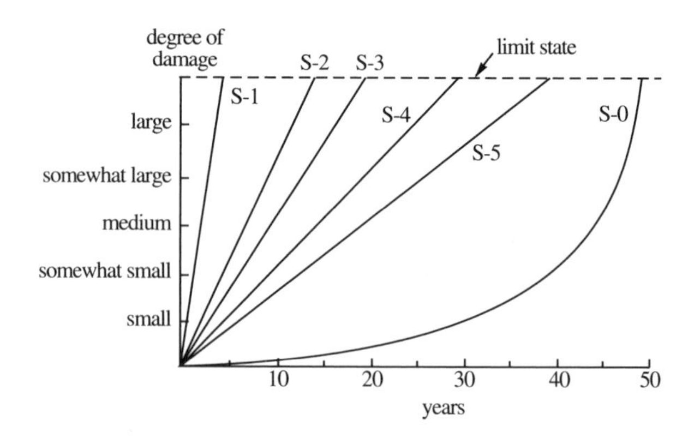

Fig. 4.7 Relation between damage and years elapsed.

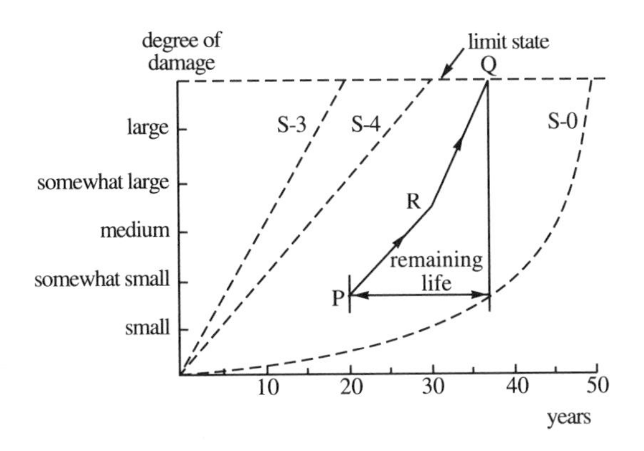

Fig. 4.8 Estimation of remaining life.

(d) Example Application

We can use a three-span Gerber girder bridge that was constructed in 1938 for an example application. Table 4.4 gives the various design and environmental conditions. Here we will give a simple explanation of process for carrying out the evaluation of P-1, one of the panels evaluated.

First, damage causes are inferred as in Table 4.5 from the design and environmental conditions, wheel loading position and inspection data for panel P-1 shown in Table 4.6. The rules carried out during the inference of cause (1) are shown in Fig. 4.9. The damage modes are identified from the damage that originates in this cause, and the degree of damage is found by means of evaluating the damage rate of the basic damage that makes up

Table 4.4. Bridge Design and Environmental Conditions

Type	Item	Data	Certainty Factor
Design conditions	Structural type	3-span Gerber plate girder (straight bridge)	1
	Design specification	Detailed recommendation for road construction (June 1926)	1
	Slab age	Old	Large
	Slab thickness	20 cm	1
	Length: 69.00 m		1
	Width: 12.95 m		1
	3 lanes		1
	1 sidewalk		1
Environmental conditions	Road type	Principal road	1
	Vehicle rate	Medium	Somewhat large
	Wheel location	Centered above support	Large

Table 4.5. Damage Cause Inference Results

	Damage Cause	Certainty Factor
Cause (1)	Extreme wheel load	Somewhat small
Cause (2)	Insufficient distributing reinforcement	Somewhat small
Cause (3)	Insufficient surface drainage	Somewhat small

the damage modes. Furthermore, the progression of damage is found from the causes and environmental conditions, etc. A table of these results would give us something like Table 4.7. If the remaining life of panel P-1 with the damage originating in each of these causes is inferred based on these results, the results shown in Table 4.8 are obtained. Since this bridge is situated in an area with an extremely large volume of traffic and the design specification applied is very old, 1926, we can easily understand the fact that "excessive wheel loads" is a cause. In addition, we can assume that the age of the application specification is a large factor in "insufficient steel reinforcement." Furthermore, there is a large possibility that drainage pipes are blocked from the inference of "insufficient surface drainage," and since the remaining life is as short as two years because of this, rapid repair is necessary.

4.3.4 Conclusion

In the foregoing we gave the outline of a practical fuzzy expert system for durability evaluations of reinforced concrete slabs for bridges, along with an example of application. Through the introduction of fuzzy theory, it is possible to use ambiguous natural language in rules and transplant the knowledge acquired from experts (expressed in natural language) as it is into the system. Furthermore, by means of using the concept of fuzzy inference in the inference mechanism, appropriate inference results can be obtained from ambiguous data and knowledge.

Table 4.6. Results of Panel P-1 Inspection

Damage Type	Damage Item	Condition	Certainty Factor
Cracks	Condition	Direction of width	Somewhat large
	Location	Centered under support	Somewhat large
		Under haunch	Medium
	Density	$1.72 \mathrm{~m/m^2}$	1
	Space	Large	Somewhat large
	Width	Medium	Somewhat large
Concrete	Free lime	Degree: medium	Somewhat large

```
(rule-1-1-2-2
If    (crack       configuration       direction of width)
      (crack       location       center of span       directly above)
      (wheel load       location       center of span       directly below)
      (design specification used       up to 1967)
then  (deposit (damage cause       use of extreme wheel load)
      (*cf time very-true = match)))


(rule-1-3-4-5
If    (crack       configuration       direction of width)
      (crack       location       haunch       directly under)
      (design specification used       up to 1967)
then  (deposit (damage cause       use of extreme wheel load)
      (*cf time fairly-true = match)))


(rule-1-5-3-6
If    (crack       configuration       direction of bridge axis)
      (crack       location       center of span       directly above)
      (design specification used       up to 1967)
      (wheel load       location       center of span       directly above)
then  (deposit (damage cause       use of extreme wheel load)
      (*cf time true = match)))
```

Fig. 4.9 Example of damage cause inference rules.

For this system, we were able to acquire expert knowledge valuable for evaluation through dialogue with experts in maintenance. Furthermore, by evaluating the durability of reinforced concrete slabs by estimating the remaining life from causes of damages, degrees of damage, and progress of damage, we were able to obtain evaluation results that are of practical use in various maintenance judgments.

Table 4.7. Results of Inference of Degree and Progress of Damage

Damage Cause	Damage Degree	Certainty Factor	Damage Progress	Certainty Factor
(1)	Medium	Small	Medium	Somewhat small
(2)	Medium	Small	Medium	Small
(3)	Large	Small	Somewhat large	Somewhat small

Table 4.8. Results of Inference of Remaining Life

Damage Cause	Remaining Life	Certainty Factor
(1)	5–10 years	Somewhat small
(2)	5–10 years	Small
(3)	2 years	Somewhat small

4.4 DECISION-MAKING SUPPORT

4.4.1 Introduction

H. A. Simon has said that human decision-making activities have the four aspects shown in Table 4.9 [16]. The various mathematical programming methods developed through the active research during the 1960s and 1970s, as well as the evaluation and utility theories, were ones in which evaluation functions based on strict formulations of problems were maximized (minimized), and we can probably say that they were related to support of the third and fourth aspects in Table 4.9, that is, [choice activities] and [reinvestigation activities]. World dynamics, ISM, DEMATEL, and cognitive maps, which attracted so much attention because of the need to grasp and solve wide-ranging business management, environmental, and international political problems, at times on a world scale, flushed out the important aspects of problems that cannot be expressed through strictly formulated models for numerical analysis through brainstorming, survey, and document analysis. Problem structures were expressed graphically to show the relations between these aspects [17]. The goal of these methods was to dig up competing proposals from the concerned parties participating in the problem—in other words, to bring about mutual formation in a shared arena, and we can say that this points toward the "establishment activities" for decision-making in the second aspect of Table 4.9. Strict formulation based on abstraction of the problem, creation of a good structure, and judgments of which items are important for the preceding aspects of decision-making first become possible with a grasping of the essence of the problem. This job corresponds to

Table 4.9. Four Aspects of Human Decision-Making (Simon)

1.	Discovery of decision opportunity	[information activity]
2.	Discovery of analysis of alternatives	[planning activity]
3.	Choice of alternative	[choice activity]
4.	Reinvestigation of choice	[reinvestigation activity]

the first item in Table 4.9, [information activities], and this for the most part originally depends on overall human judgment and heuristics. Recent technological innovations, especially the change to high-quality information that has accompanied the popularization of computers, has made dealing with excessive information that exceeds the limits that human beings can grasp unavoidable, with the result that the margin for reviewing the whole of a problem and thinking about it from a broad point of view has been lost. What is most necessary for decision-making problems today is a method that will aid judgments from a strategic human position by means of a summarized presentation of the context of the essence and changes in events related to the problem out of this convergence of data—that is, what can be thought of as the development of a support system for the first and second aspects of Table 4.9—[information activities] and [planning activities]—based on a grasping of the essence of the problem. In order to do this, the interaction between people and computers must come closer to the human side, and algorithms must be developed for information-processing functions based on human subjectivity, qualitativeness, and ambiguous circumstances. One of the things that gives us a foothold on this problem is Bartlett's model of human intellectual processes—the processes for creation of consistent circumstantial descriptions by means of "bottom-up activation of hypotheses using fragmentary information from the object" and "top-down verification of hypotheses for the object"—which has received so much attention since the beginning of the 1970s [18]. This model can provide good descriptions of the various characteristics of human knowledge and information processing such as "precise recognition and memorizing of event patterns," "rapid intuitive recognition searches," and "active information selection for decisions and judgments driven by expectation and anticipation."

Here we will introduce a strategic decision-making support system for political and social problems that is constructed by means of the introduction of primitive events for hypotheses in the preceding model, scripts that represent series of primitive events, and functions, which are based on fuzzy concepts and are similar to bottom-up activation, for the generalization of primitive events.

4.4.2 Strategic Decision-Making Support and Tiered Knowledge Expression Schema

We can say that through experience, human beings remember the essential meaning and characteristics of patterns that can be called prototypes and that come from groups of similar episodes; when they encounter an actual event they call up the appropriate pattern, recognize the meaning and importance of the event based on the circumstances, and determine the appropriate action. The expression of patterns of primitive events as a transition series of basic actions is called a script. This script idea was put forward by Schank and Abelson in 1977 [19], and it is based on the creation of prototypes of structures, actions, and phenomena from Schank's theory of conceptual dependency and the idea of expressing the relationships between them. In conceptual dependency theory, human intentions and the most basic activities are given as basic actions, but here the prototypical groups of primitive events shown in Table 4.10 are defined as the minimum syntax units necessary to describe theoretical series of political and social events. As is shown in Fig. 4.10, the system being introduced creates a script as a series of prototypical primitive events that can describe a similar series of actual primitive events, based on descrip-

Table 4.10. Primitive Events as Syntactical Units of Series of Social and Political Phenomenological Events

	(a) $A = B$	(b) $A \neq B$
$V(A) \overset{-S}{\leftarrow} P(X) \overset{C}{\rightarrow} D(B)$ (1)	[problem, [A], [X]] problem recognition	[inducement, [A, B], [X]] inducement of decision
$V(A) \overset{+S}{\leftarrow} P(X) \overset{C}{\rightarrow} D(B)$ (2)	[enablement, [A], [X]] enablement	[promotion, [A, B], [X]] promotion of decision
$D(A) \overset{C}{\rightarrow} P(X) \overset{+S}{\rightarrow} V(B)$ (3)	[success, [A], [X]] decision	[commitment, [A, B], [X]] contribution
$D(A) \overset{C}{\rightarrow} P(X) \overset{-S}{\rightarrow} V(B)$ (4)	[failure, [A], [X]] betrayal decision	[threat, [A, B], [X]] threat
$D(A) \overset{C}{\rightarrow} D(B)$ (5)	[motivation, [A], [ ]] decision recollection	[mental_trans, [A, B], [ ]] mental transfer
$D(A) \overset{T}{\rightarrow} D(B)$ (6)	[abandon, [A], [ ]] abandon decision	—
$V(A) \overset{a}{\leftarrow} P(X) \overset{b}{\rightarrow} P(Y) \overset{c}{\rightarrow} V(B)$ (7) $a \cdot b \cdot c = -1$	[trade_off, [A], [X, Y]] trade-off	[competition, [A, B], [X, Y]] conflicting advantages and disadvantages
$V(A) \overset{-S}{\leftarrow} P(X) \overset{b}{\rightarrow} P(Y) \overset{c}{\rightarrow} V(B)$ (8) $b \cdot c = -1$	[dilemma, [A], [X, Y]] complications of multiple objectives	[adversity, [A, B], [X, Y]] offsetting advantages and disadvantages
$V(A) \overset{+S}{\leftarrow} P(X) \overset{b}{\rightarrow} P(Y) \overset{c}{\rightarrow} V(B)$ (9) $b \cdot c \cdot = +1$	[concord, [A], [X, Y]] fulfillment of multiple objectives	[cooperation, [A, B], [X, Y]] agreement of advantages and disadvantages

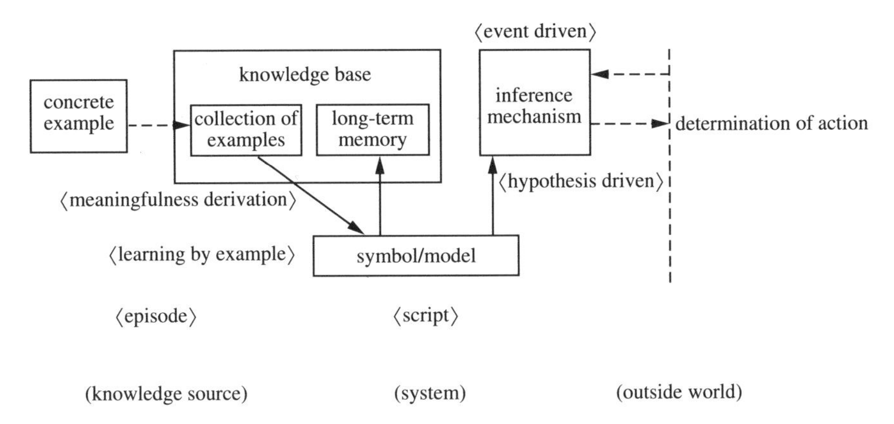

Fig. 4.10 Structure of strategic decision-making support system constructed.

tions of concrete events as episodes and the extraction of a series of concrete primitive events by means of matching them with prototypical primitive events. The script created corresponds to a hypothesis that expresses a decision-maker's prediction of circumstantial changes, and the system can provide support for a decision-maker's determination of actions to be taken by activating event-driven hypothetical scripts when a certain event occurs and giving predictions of circumstantial changes based on them.

Description of concrete events (episodes) concerning Japanese/American trade friction: "The inroads of the Japanese steel industry in the American market cause an increase of imports from Japan (P11 → P13), and American steel takes a blow (P13 → V12). The American steel industry brings a dumping suit to the Department of the Treasury, and because of this choice the Japanese situation becomes difficult (P13 → D11 → D12 → P14 → V13). The Treasury Department, which accepts the suit, establishes trigger prices, and Japanese industry moves toward having no alternative to self-regulation of exports (P14 → D13 → P15 → D14 → P16 → V17)." The cause-and-effect series for the events included in this description can be coded in the directed graph structure shown in Fig. 4.11a, using the three types of nodes (P: concrete event, D: actor decision, V: actor evaluation) and four types of links (C: occurrence, T: termination, S: positive or negative sentiment) [21, 22]. The structuring of these kinds of real events—that is, what can be called the surface knowledge structure—is summarized in the system of prototypical primitive events shown in Fig. 4.11b by means of the matching in Table 4.11. This summary structure is further summarized in the series of complex events shown in Fig. 4.11c and given a tiered structure by means of matching with the rules for creating

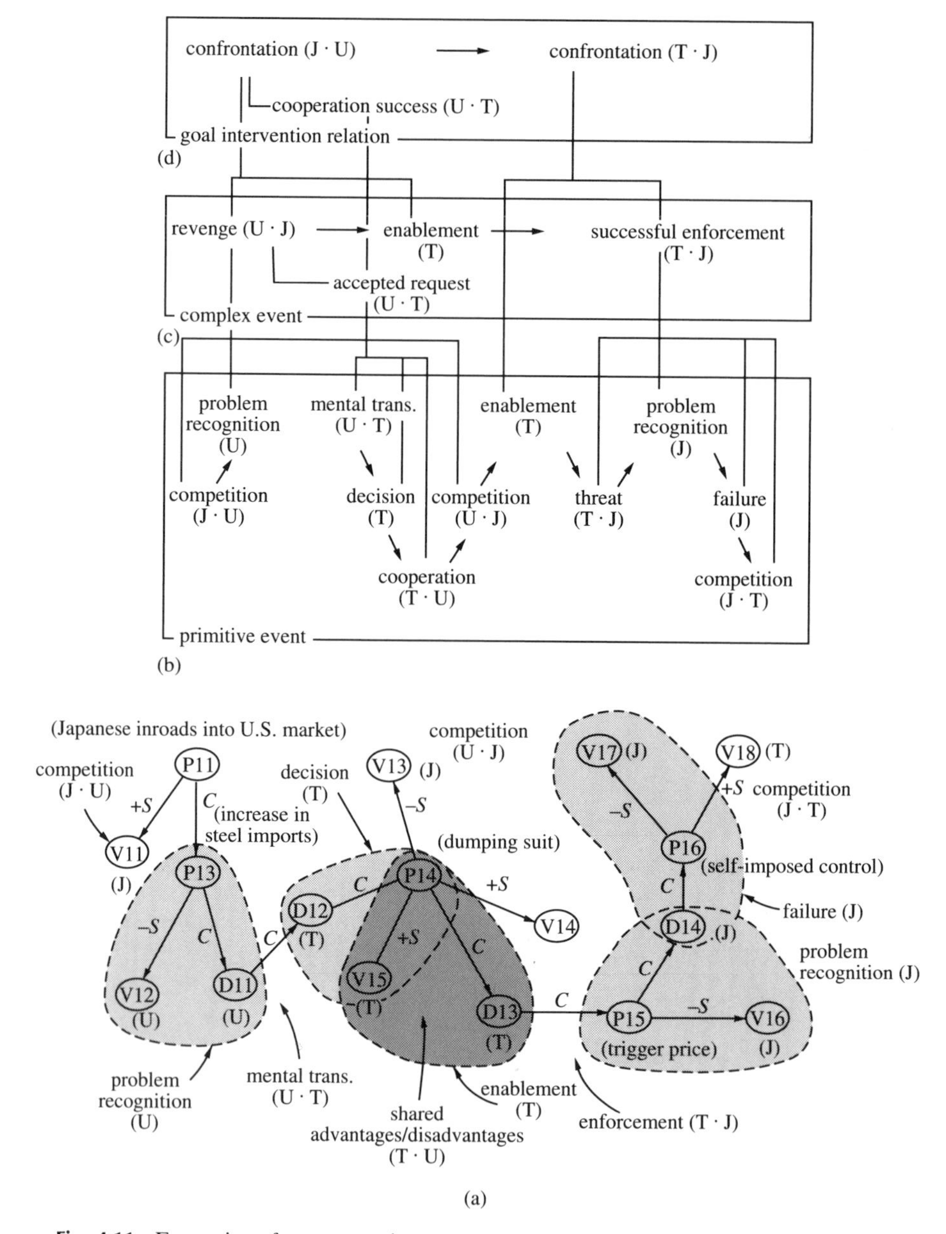

Fig. 4.11 Expression of structure using a cause-and-effect linking graph and schema hierarchical structural expression of concrete events (episodes). (a) Cause-and-effect network. J: Japanese steel industry; T: U.S. Treasury Department; U: U.S. steel industry. (b) Primitive event. (c) Complex event. (d) Goal intervention relation.

Table 4.11. Example of Rules for Creation of Complex Events

<table>
<tr><td>(a)</td><td colspan="2">intentional_resolution:
  problem → success* → concord
goal_pursuit:
  enablement → success* → concord
recovery:
  failure → problem → success*
....................</td></tr>
<tr><td></td><td>Consistency</td><td>compensation:
  competition → enablement → inducement → commitment
relief:
  inducement → success* → cooperation
accepted_request:
  mental_trans → success* → cooperation
....................</td></tr>
<tr><td>(b)</td><td>Contradiction</td><td>revenge:
  competition → problem → success* → competition
obstruction:
  promotion → success* → competition
rejected_request:
  mental_trans → success* → competition
successful_enforcement:
  threat → problem → failure* → competition
....................</td></tr>
<tr><td></td><td>Mutual contradiction</td><td>failed_request:
  mental_trans → failure* → adversity
....................</td></tr>
</table>

complex events in Table 4.11, which are recorded in the system as ordered sets of primitive events. Complex events, when they are participated in by two actors (column (b) in the table), are summarized in the highest-level structure shown in Fig. 4.11d—that is, a transition series of "goal intervention relations" as connotations of each of the relations, "cooperation (consistency)," "opposition (inconsistency)," "conflict (mutual contradiction)," "cooperation–success (consistency)," and "cooperation–failure (inconsistency, mutual contradiction)," between the actors, corresponding differently to content, main series and associated series complex events (for example, the accepted request in Fig. 4.11c) [23]. The high-level and low-level relational information for the primitive events, complex events and goal intervention relations are recorded in the system as slot valued schema p_ev( ), c_ev( ), and g_sq( ). Figure 4.12 shows the content of the three complex events in Fig. 4.11c, "revenge c_ev (12)," "enablement c_ev (14)," and "success–enforce c_ev (13)." The lower-level events of the has_part section—that is, the series of primitive events that form each

```
schema    c_ev(12)
   type      revenge
   actor     [US_Steel_Ind,Japan_Steel_Ind]
   content   [[Japanese,inroads,into,US,market],
              [dumping,suit]]
   pointer   a_part of([g_sq(1)])
             has_part ([p_ev(14),p_ev(15),c_ev(11),p_ev(16)])
             means_by(c_ev(11))

schema    c_ev(14)
   type      enablement
   actor     [Dept_of_Treasury]
   content   [dumping,suit]
   pointer   a_part_of([g_sq(1),g_sp(3)])

schema    c_ev(13)
   type      success_enforce
   actor     [Dept_of_Treasury,Japan_Steel_Ind]
   content   [[setting,Trigger,prices],
              [self-imposed,control,on,steel,exports]]
   pointer   a_part_of([g_sq(3)])
             has_part([p_ev(17),p_ev(18),p_ev(19) ,p_ev(20)])
```

Fig. 4.12 Complex event schema contents of "revenge," "enablement," and "success–enforce."

complex event—are recorded as values, and in addition, the a_part_of sections are recorded as goal intervention relations, which are a higher-level structure than these.

4.4.3 Process for Extracting Deep Structures of Problems through Schema System Interaction

Figure 4.13 is an example of multiwindow interaction. The system developed from the verb "revenges," which is included in question Q1 (window 1), recognizes that the point of scrutiny in the question is "revenge c_ev (12)." The primitive and complex events within this system are carried out with reference to their dictionary of event concepts (Table 4.12). Since the interrogative for Q1 in this system is "who," the content of the actor slot for c_ev (12) has the answer "US_Steel_Ind" (A1 of window (1), window (3) being the contents of the schema for the point of observation, and window (4) the tiered schema system inclusive of the point of observation). When "who" is replaced with "what" or "by whom," the system indicates the slot information within the schema for this point of observation as an

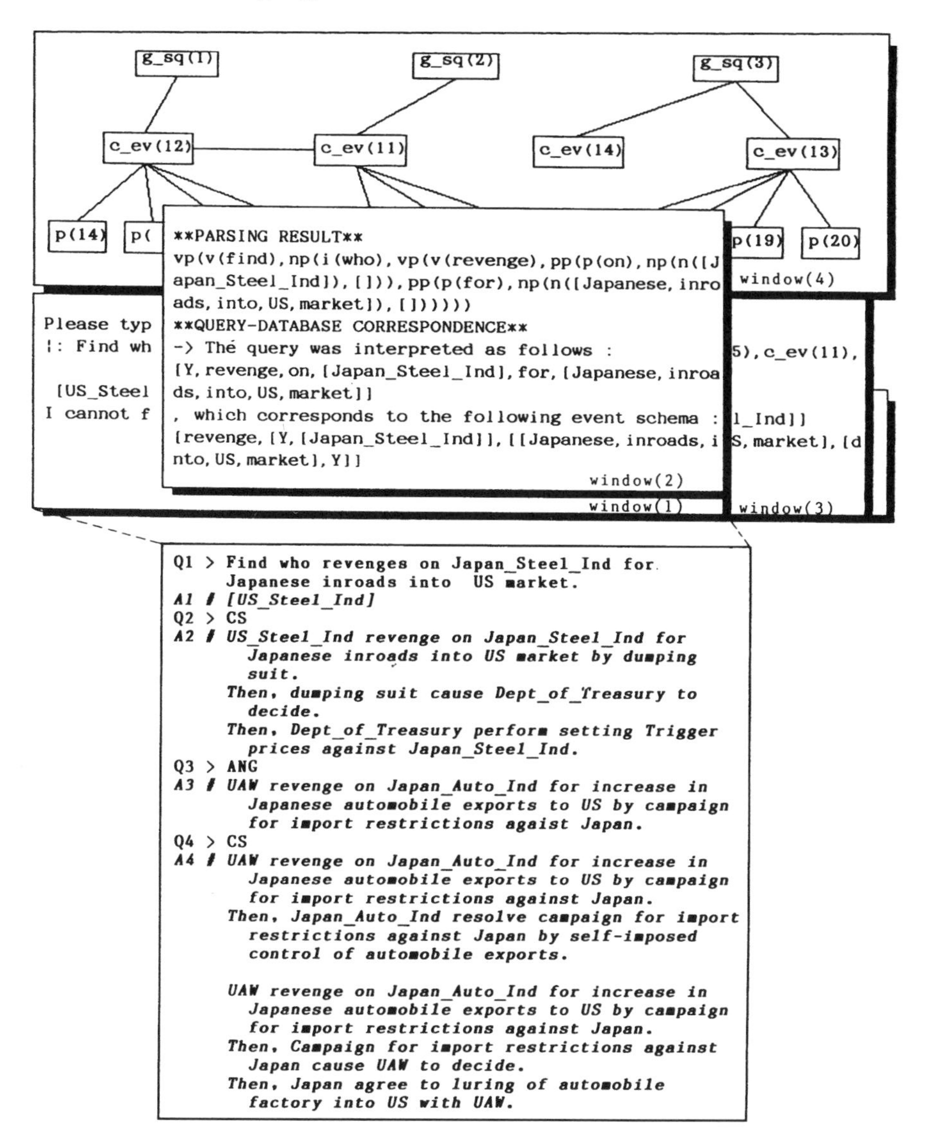

Fig. 4.13 Hierarchical structural schema knowledge and example of derivation of deep problem structure through interaction.

answer. In addition, for "why" and "how," the system activates the lower- or higher-level schema from the schema activated by the preceding dialogue, and the higher-level concept "why" and the lower-level concept "in what way" are answered. In addition to the natural-language interaction within this limited range, the system can display "transition summaries of preceding and following events" and extract "similar events," given ques-

Table 4.12. [Event Schema] ↔ [Natural Language Description] Conversion Table

```
type([A, succeed, in, B], [success, [A], [B]]).
type([A, contact, with, B], [mental_trans, [A, B], [ ]]).
type([A, agree, to, B, with, C], [accepted_req, [C, A], [B]]).
type([A, fail, in, B], [failure, [A], [B]]).
type([A, revenge, on, B, for, D], [revenge, [A, B], [D, C]]).
type([A, do, B, against, C], [revenge, [A, D], [B, C]]).
type([A, pursue, B, for, C], [goal_pursuit, [A], [B, C]]).
type([A, do, B, for, C], goal_pursuit, [A], [B, C]]).
type([A, suffer, from, B], problem, [A], [B]]).
type([A, spur, B], [enablement, [B], [A]]).
type([A, enable, B], [enablement, [B], [A]]).
type([A, motivate, B], [motivation, [B], [A]]).
type([A, cause, B], [enablement, [B], [A]]).
type([A, cause, B], [motivation, [B], [A]]).
type([A, compete, with, B, for, C], [competition, [A, B], [C, D]]).
type([A, recover, from, B, by, C], [suc_born_of_fail, [A], [B, C]]).
type([A, force, B, on, C], [successful_enforce, [A, C], [D, B]]).
type([A, perform, B, against, C], [successful_enforce, [A, C], [B, D]]).
type([A, cooperate, with, B, in, C], [cooperation, [A, B], [C, D]]).
```

tion command input AS (antecedent summary), CS (consequent summary), and ANG (analogize). The system started by question command CS in window (1) Q2 in Fig. 4.13 successively activates a series of complex event schema chains that started with c_ev (12)—which was activated by Q1, c_ev (12) (revenge), c_ev (14) (enablement), and c_ev (13) (success-enforce)—and it displays a summary of circumstantial transitions (window (1) A2) with this meaning by employing the relationship table, Table 4.12, in reverse order (event schema → natural-language expression). Furthermore, the system based on the ANG question command in Q3 finds and displays (two forms of circumstantial transition following from c_ev (25), which is summarized using the CS command in window (1) A3 and A4) events similar to the observed complex event c_ev (12) "revenge" that differ in actual content—in other words, schema c_ev (25) (trade friction surrounding automobile exports), for which the type slot value is "revenge." Along with recognizing circumstances similar to those of a deeper level (the revenge of the U.S. industrial sector on its Japanese counterpart surrounding "steel exports" and "automobile exports") from graphs of vast chains of causes and effects in surface information through this kind of dialogue, the decision-maker is given beneficial indications of the differences in the development of events in each case, for example "In contrast to 'steel exports' for which the hostility toward the Japanese steel industry progressed without change due to cooperation between the U.S. Treasury Department and U.S. industry, the Japanese auto industry solved the auto export problem with the U.S. auto industry peacefully through self-imposed restrictions," which should be investigated with caution when policy decisions are made.

4.4.4 Generalization of and Expectations for Schema Systems Based on Fuzzy Concepts and the Extraction of Predictive Information

Human decisions are to some degree based on "predictions and expectations" about the way situations develop, and this increases the effectiveness of human information processing in the real world with its vast amount of noisy information. As was discussed in Section 4.4.2, this function can be thought of as being carried out by the activation of prototypical patterns (scripts) of groups of similar concrete events (episodes) that have been accumulated through long-term experience. In our system, the formation of prototypes from actual primitive (complex) events and learning acquisition of scripts formed as systems of them is possible based on the concept of prototypicality, which can be thought of as one type of fuzzy concept [24].

The formation of a generalized event schema means coming up with general descriptions that are inclusive of groups of similar real events, from those groups. As is shown in Fig 4.12, the various attributes of primitive (complex) events are given in terms of slot values: event-type, actor, content, etc. As is shown in Fig. 4.14, in the actual world, a number of primitive events of the same type ("problem") for which other attributes (actor, contents) differ (the actor for e_2 being the U.S. Steel Co. and that for e_1 being the UAW) are extracted. As is shown in the figure, the slot values that are the same are transferred as is to the generalized event schema (E_1), and only for the ones that differ is the least general algorithm transferred, which replaces them with a higher-level concept (U.S. Steel Ind. for U.S. Steel Co. and UAW) by means of reference to the computer's internal high/low level relational knowledge map (Fig. 4.15). This is to avoid the fruitless creation of a generalized schema from concrete events with little similarity and the implications of the large number of unobserved events. When we let $RS(e_i)$ be the set of all attributes of primitive event e_i (for example, the actor attribute for e_i is UAW, if directly extracted, or the higher concepts for UAW as shown in Fig. 4.15, that is, the actors U.S. Auto Ind., U.S. Tech Ind. and U.S. Ind. are included), which is the same type as generalized event E_j, $CRS(E_j)$, the set of attributes with a majority of e_i $(i = 1, 2, \ldots)$ in common is an attribute of it. The E_1 in Fig. 4.14b is a generalized event for three primitive events, e_1, e_2, and e_3, of the same type. As with E_2 in the same

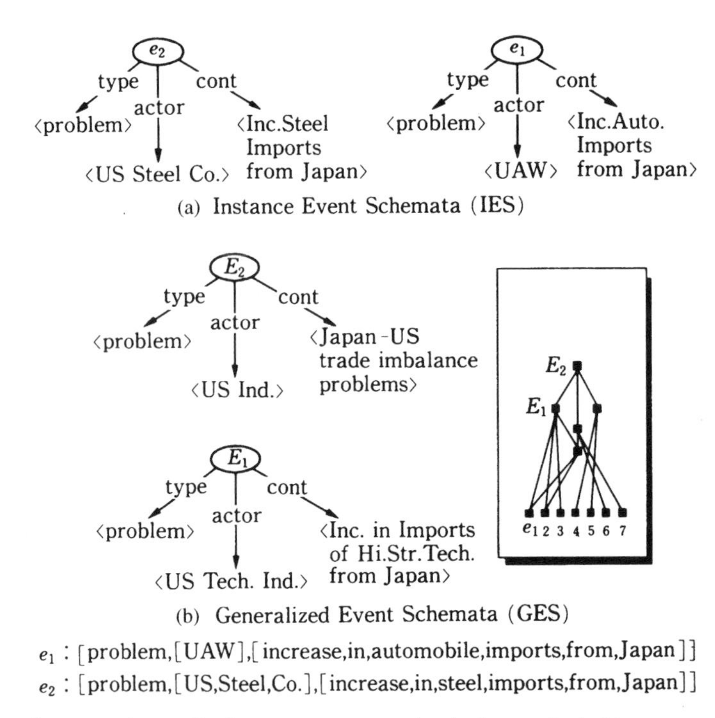

Fig. 4.14 Generalization process to upper-level schema of primitive events.

figure, the higher the level of generalization is, the more primitive events ($e_1, \ldots, e_7$) are included. The degree of attribution (prototypicality) of each event e_i in generalized event E_j is defined as

$$\mathrm{prot}\left(e_i | E_j\right) = \frac{\left(\mathrm{CRS}(E_j) \cap \mathrm{RS}(e_i)\right)^{\#}}{\left(\mathrm{CRS}(E_j) \cup \mathrm{RS}(e_i)\right)^{\#}}.$$

As is shown in Fig. 4.16, the higher the level of generalization (the larger the j in E_j) is, the smaller the prototypicality. Now, as is shown in Fig. 4.17, when events e'_1, e'_2, and e'_3 or e''_1, e''_2, and e''_3, which follow each of the primitive events e_1, e_2, and e_3 that are generalized in $E_1^{(1)}$, are generalized at the same level as $E_1^{(1)}$ in $E_1^{(2)}$, this generalized level transition pattern $S_1 : E_1^{(1)} \to E_1^{(2)}$ is created as a prototypical script that has a large possibility of being a typical pattern. The preceding scripts, which are based on generalized algorithms, are created on various generalized levels. When a concrete event e_t (either a primitive event or a complex event) occurs in the real world, as in Fig. 4.18, we let the system activate two levels of scripts, $S_1 : E_2^{(1)} \to E_2^{(2)} \to E_2^{(3)}$, $S_2 : E_1^{(1)} \to E_1^{(2)} \to E_1^{(3)}$

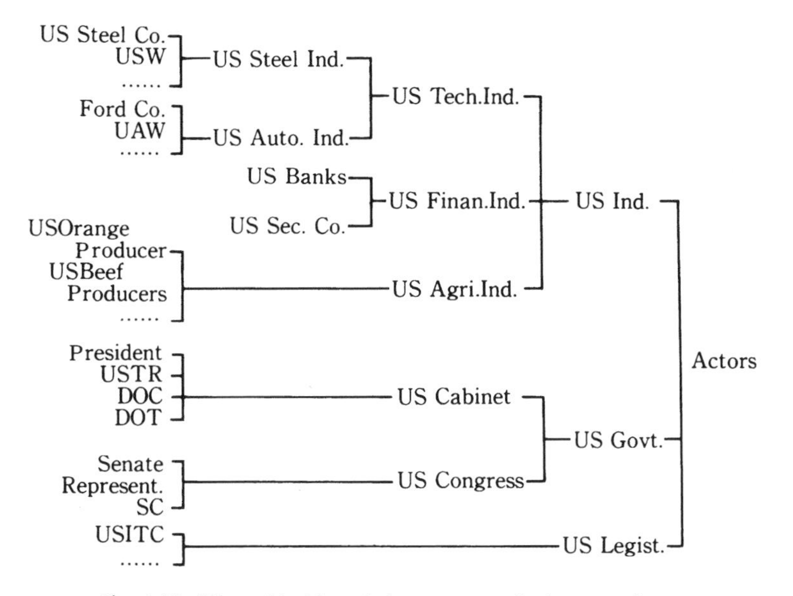

Fig. 4.15 Hierarchical knowledge structure of schema attributes.

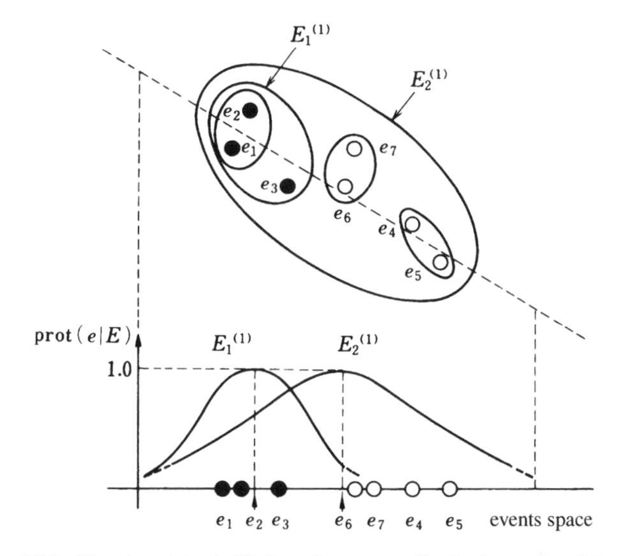

Fig. 4.16 Event prototypicality based on generalized schema for all levels.

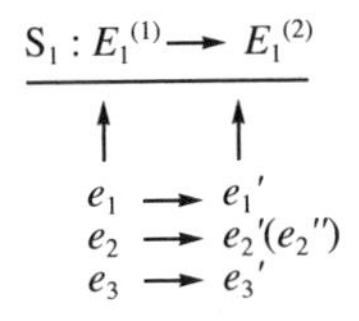

Fig. 4.17 Script as a generalized schema series.

and $S_3: E_1^{(1)} \to E_1^{(2)\prime}$. As can be seen from the figure, S_1 is a script for "protectionism" and S_2 is one for "free trade." When event e_t occurs, we say that the decision-maker assumes a transition scenario for what will follow, for example, $H_p: e_t \to e_{p1} \to e_{p2} \to \cdots \to e_{pk}$. The system calculates the prototypicality: prot$(e_t|E_j^{(1)})$ or max$_m$prot$(e_{pn}|E_j^m)$ for the trigger e_t and each of the assumed events e_{pn} $(n = 1, \ldots, k)$ that are input, based on $E_j^{(m)}$ (generalized event for constructing S_j), and using the average of these,

$$\text{prot}\big(H_p|S_j\big) = \frac{1}{k+1}\left[\text{prot}\big(e_t|E_j^{(1)}\big) + \sum_{n=1}^{k}\left\{\max_m \text{prot}\big(e_{pn}|E_j^{(m)}\big)\right\}\right],$$

considers the prototypicality of the decision-maker's scenario H_p based on each script S_j—in other words, the likeliness of the scenario (activation level of the script). Figure 4.19 shows the probability indication process

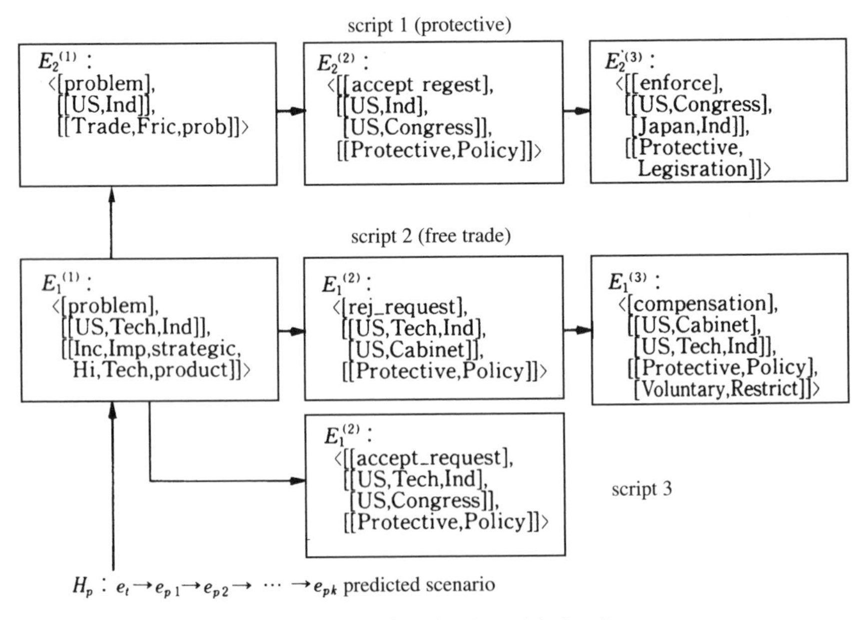

Fig. 4.18 Activation of scripts S_1, S_2 and S_3 by trigger event e_t.

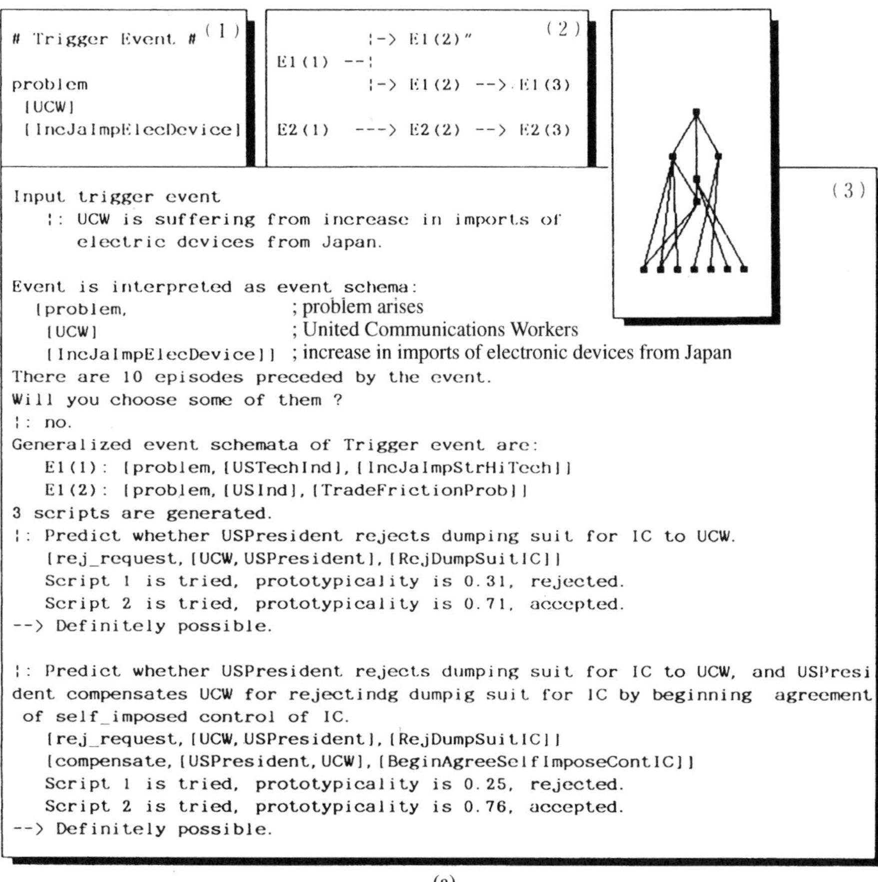

(a)

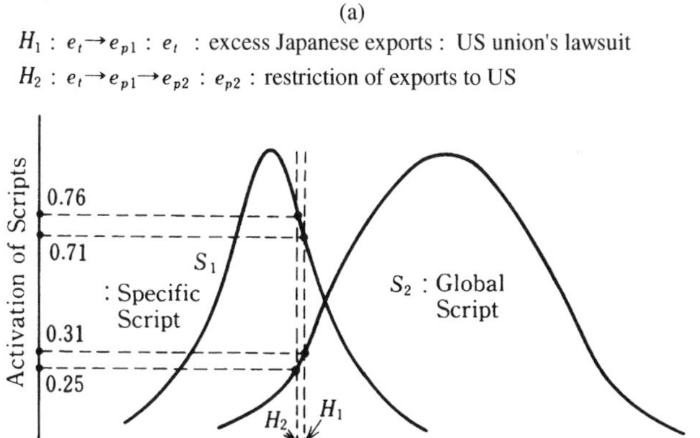

(b)

Fig. 4.19 System probability display process for scenarios H_1 and H_2, decision-maker's "expectations" or "fears and expectations."

based on the system's self-generated script for the decision-maker's anticipated or feared event transitions, $H_1: e_t \to e_{p1}$ (e_{p1} being "rejection of the UCW suit by the U.S. President": "rejected request") and $H_2: e_t \to e_{p1} \to e_{p2}$ (e_{p2} being "mandatory IC export limitations begun by President: compensation for U.S. unions": "compensation"), which follow the occurrence of trigger event e_t ("problem" surrounding "exports of electronic communications equipment by Japan"). As is shown in window (3) in the figure, the trigger event, which is input in natural language, is converted into the event schema in window (1) through a simple syntactical interpretation. The system displays the fact that it has a systemization of knowledge accumulated from 10 event series (scenarios) originating with events similar to e_t in tiered schema structures and poses the problem of whether all of them should be used in the generation of a script. In the figure, the decision-maker has answered that all of them should be used (window (3)). Window (2) is the resulting three scripts generated by the system, $S_1: E_2^{(1)} \to E_2^{(2)} \to E_2^{(3)}$, $S_2: E_1^{(1)} \to E_1^{(2)} \to E_1^{(3)}$, and $S_3: E_1^{(1)} \to E_1^{(2)\prime}$. The script associations are easier when they are more concrete—that is, when the generalization level is lower—and when they are constructed out of a large number of concrete examples. The system orders them $S_1 > S_2 > S_3$ based on the "ease of association" or "easiness to be reminded of." Figure 4.19b is a conceptual representation of the activation level of scripts S_j for the decision-maker's scenario H_p, which was found using prot$(H_p|S_j)$. The hypothetical scenario that shows the expectations of the decision-maker, $H_1: e_t \to e_{p1}$, is negated by prototypicality = 0.31, based on S_1 ("protectionist" script), but based on S_2 ("free trade" script), it is 0.71, and this shows a high enough activation level. Scenario $H_2: e_t \to e_{p1} \to e_{p2}$, which expresses the decision-maker's expectations and suspicions, has a lower activation level for S_1 and a higher one for S_2. Because the fact that there is a high enough activation level for S_2 is confirmed, the activation level of script S_3, which is next in the order, is not investigated. If a concrete primitive event that has already been extracted is input instead of hypothetical event series H_p, that event series will inevitably be displayed.

4.4.5 Conclusion

The way in which human beings bear the load of the huge amount of information from their environment within their limited information-processing capabilities and solve problems through judgments based on recognition driven by important hints and discretion that is not led astray by interfering information is very interesting. In this section we have discussed the role played by tiered schema structures for the equipping of computers with anticipatory knowledge, scripts, and fuzzy concepts in terms of creating an algorithm for the characteristics of humans' excellent information processing. The system introduced here gives contextual connections to fragmentary information from the outside and can provide assistance in choosing useful information that follows in a line of thought. The internal structured knowledge consists of prototypes, and the information from the real world that is supposed to be matched with these diverges from them to a certain extent. The degree of this divergence must be evaluated, and the ambiguity that depends on the line of thought must be quantified in order to confirm whether or not it is an instance of the prototype. The introduction of fuzzy concepts in this section made it possible to perform matching between the prototypical knowledge in the computer and reality.

4.5 MEDICAL DIAGNOSIS

4.5.1 Introduction

Unlike formulations, which could establish methods of solution using mathematical models and algorithms, handled by earlier programs, those for the objects of expert systems do not present easy problems; rather, it is more appropriate to think of them as being complex problems that can be solved using the knowledge that experts have obtained through long experience. For these kinds of problems, it is difficult to obtain enough of the knowledge necessary for solving them at the time the system is designed, and it is normal to go through the trial and error of adding, removing, and changing the knowledge to determine it. In this way, use in medical fields carries an extremely important meaning.

On the other hand, there have been a large number of reports of research on medical systems that make use of fuzzy reasoning. In other words, starting with the anticipatory research of Sanchez *et al.* [26], and coming up through that of Adlassing [25] and Tazaki *et al.* [27, 28], they have covered a variety of levels and fields of medicine.

In this section, the discussion will center on the results of research by Tazaki *et al.* [29–31], and we will explain a prediagnostic/inquiry system and a health checkup support system, both of which can be considered relatively close to being put into actual use.

4.5.2 Development of the Prediagnostic / Inquiry System [27, 31]

With the increasing demand for health maintenance in Japan and the changes in hospital structures, the early detection of adult diseases is carried out passively through extensive examinations. Because of this, the goal of this system is to stimulate the early detection and treatment of adult diseases by getting people to actively undergo examinations. The system that was developed is characterized by a lumping together of a fuzzy inference section that is written in a processing form of the C programming language, and knowledge matrices to carry out inference and make diagnoses.

(a) System Structure

(1) Hardware. This system is constructed using an engineering work station (EWS) that employs an American Motorola MC68000 as the main CPU, a color graphics terminal, and a console.

(2) Software. The software structure is shown in Fig. 4.20. The Japanese version of UNIX, which is a transplanted form of UNIX (System V), is used for the operating system, and a summary of the applications is given in Table 4.13. The following is a summary of the functions for each block:

(1) Screen Control Block: This is for making inputs in field units of the screen image possible, and it is made up of application interfaces, screen control processing and the routine groups for access. Because of this the user can easily perform the inquiry dialogue processing and knowledge inputting, without being aware of the UNIX system environment.
(2) Image Control Block: During the medical interview the patient visually confirms the diseased area, and this is what makes it possible to input that area using a mouse; it makes use of the graphics package by means of a standard GKS that takes independence from the EWS graphics section and the system's descriptive language into consideration.
(3) Database Management Block: This combines the knowledge base and the database. By making the knowledge within the knowledge base independent of the inference mechanism, the knowledge can be effectively used, and it is possible to use it as health maintenance data, independently of inquiry processing.
(4) Examination Interaction Block: The patient inputs symptoms and their degree in dialogue form, and this is software that makes it possible to obtain diagnostic results. This is a routine group that carries out examination inputs, display, inference, symptom checks and error processing.
(5) Knowledge Acquisition Support, Block: This makes batch processing of knowledge inputs by experts such as doctors possible, and it is made up of routine groups for messages and relations between symptom and diseases, recording of parameters, displays for each transaction, knowledge base checking, etc.

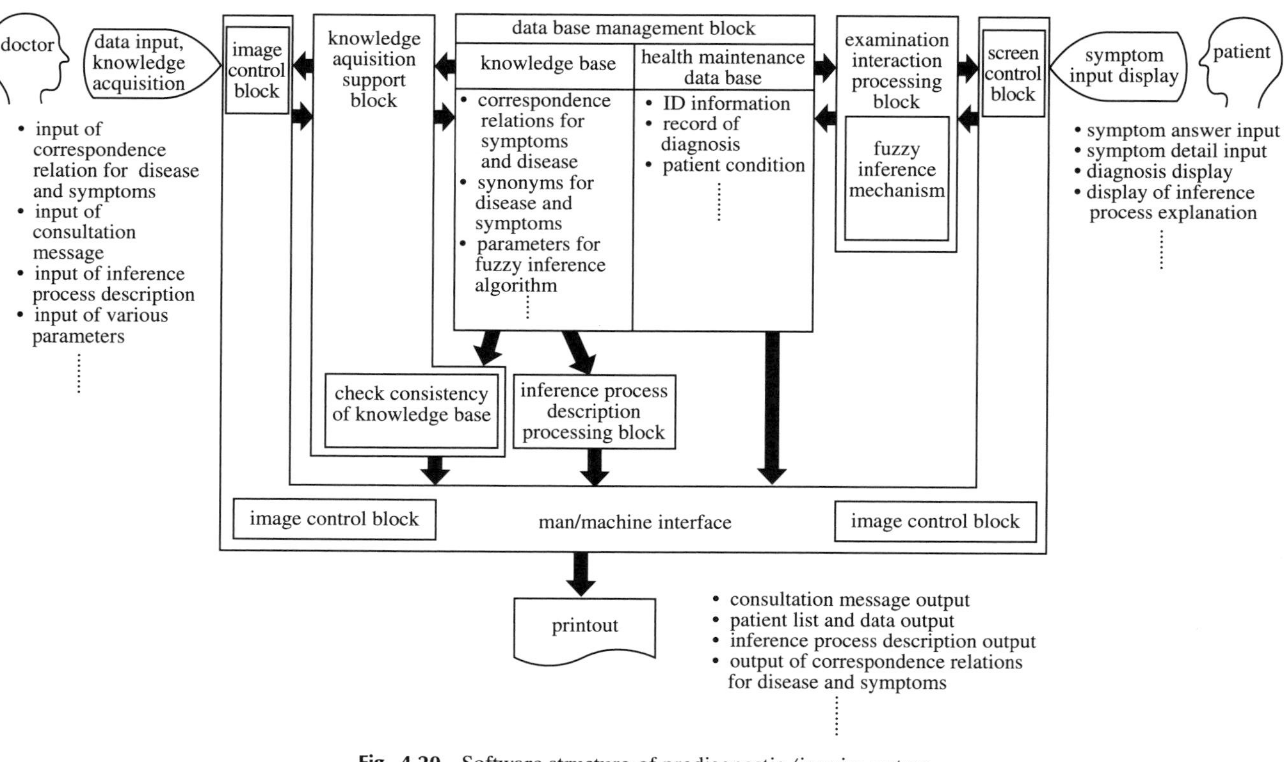

Fig. 4.20 Software structure of prediagnostic/inquiry system.

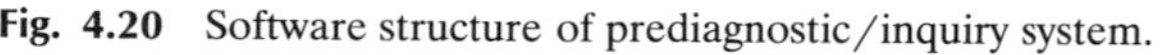

Table 4.13. System Specifications

<table>
<tr><td rowspan="2">Inference engine</td><td colspan="2">Theory</td><td>• Application of fuzzy theory</td></tr>
<tr><td colspan="2">Certainty</td><td>• Expression by seven-level word classification</td></tr>
<tr><td rowspan="4">Data base</td><td rowspan="3">Knowledge base</td><td>Number of disease systems</td><td>• 11 systems for adult heart, circulatory and respiratory systems</td></tr>
<tr><td>Number of diseases</td><td>• Approx. 300: esophagitis, diabetes, etc.</td></tr>
<tr><td>Number of symptoms</td><td>• Approx. 300: pain, nausea, etc.</td></tr>
<tr><td colspan="2">Health maintenance data base</td><td>• Accumulation of past records of diagnosis</td></tr>
<tr><td rowspan="4">Human interface</td><td colspan="2">Input system</td><td>• Chinese characters, Japanese syllabary, English alphabet, number input
• Function key selection
• Mouse positioning</td></tr>
<tr><td colspan="2">Screens for knowledge acquisition, etc.</td><td>• 22 basic record screens</td></tr>
<tr><td colspan="2">Screens for examinations</td><td>• 44 basic question screens</td></tr>
<tr><td colspan="2">Printer output</td><td>• 19 types of inference process explanations, etc.
• Color hard copies when required</td></tr>
</table>

(6) Block for Processing Inference Process Descriptions: This outputs lists obtained.

(b) *Establishing Knowledge Data for Inference*

When a doctor gives a medical interview, we can think of the process that is followed as being one in which he listens to the symptoms that the patient feels and considers the distance between them and the diseases that can be considered based on proof from the system of basic medical knowledge and on experiential knowledge. He then comes up with the result by asking questions about symptoms related to the relevant diseases. The inquiry procedure for the system that was developed based on this process is shown in Fig. 4.21. In this case, it is necessary to have general relations between symptoms for the knowledge data base, and in addition, based on these relations, the degrees of relations, viewing the disease from the symptom side and vice versa, are necessary.

If we consider the case in which the "disease" is a cold and the symptom is a cough as an example, the following would be an example of knowledge expressions:

(1) "cold" and "cough" are related (truth value = 0.9).
(2) "coughing indicates a cold" (truth value = 0.8).
(3) "colds are accompanied by coughs" (truth value = 0.6).

The truth value here is the certainty of the relation.

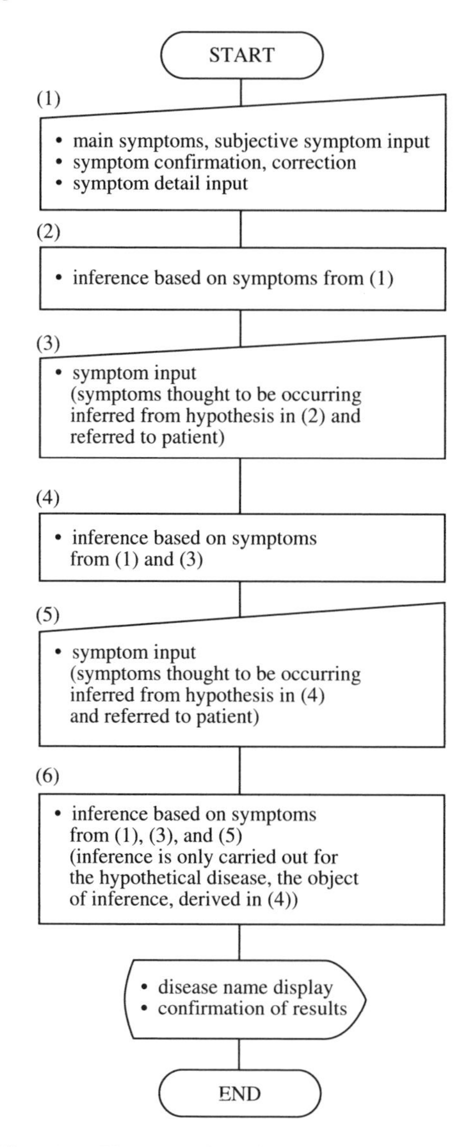

Fig. 4.21 Flowchart of prediagnostic/inquiry system.

By the way, the problem remains the degree to which the inference carried out by means of the fuzzy relational equations mentioned in the previous section is effective as a model, and it is related to the way in which the numerical values for fuzzy relation R are given. Actually, there are instances when they are too rough compared to the characteristics of the objects, with a construction using max–min operations, and because of this there is a loss in inference precision. Therefore, a more general inference replete with elasticity is made possible through a consideration of a model based on fuzzy correspondence with an extension of this. Based on the foregoing, propositions for the corresponding knowledge data are established within the fuzzy reasoning.

The sets for all of the diseases and related symptoms that are the objects of the system are as follows:

$$\text{Diseases: } X = \{X_i \ (i = 1, \ldots, m)\}.$$

$$\text{Symptoms: } Y = \{Y_i \ (i = 1, \ldots, n)\}.$$

Next, we let A_i, B_j, and R_{ij} be the following propositions:

A_i: disease i is present.
B_j: symptom j is observed.
R_{ij}: disease X_i is causally related to symptom Y_j.

Each of these contains uncertainty, and this means that they can be thought of as fuzzy sets. In addition, composite propositions P_j and P_{ij} are set up as follows:

$$P_j \triangleq \text{``}B_j \to \underset{i}{\mathrm{OR}}(R_{ij} \,\&\, A_i)\text{''}, \tag{4.3}$$

$$P_{ij} \triangleq \text{``}(R_{ij} \,\&\, A_i) \to B_j\text{''}. \tag{4.4}$$

Here the symbols are $\to$: implication, OR: disjunction, and &: conjunction. In other words, we let the certainty of "If symptom B_j is present, based on relation R_{ij} between symptoms and diseases, there is at least one disease A_j present" be P_j, and that of "If, based on relation R_{ij} between symptoms and diseases, symptom B_j will appear" be P_{ij}. According to this relation, propositions R_{ij}, P_j, and P_{ij} are established as knowledge for inference. At this point we can generally predict that the certainty of P_j will be high. However, it is not completely certain, because there is a possibility that the observer might have mistaken the symptom. In addition, that for P_{ij} is dependent on the depth of human caution and the observational conditions. Accordingly, we can generally predict that the certainty of P_j is greater than that of P_{ij}. A concrete example of R_{ij} is shown in Table 4.14. The certainties R_{ij}, P_j, and P_{ij} take the form of linguistic truth values (LTV), and these are given the seven levels in Table 4.15.

(c) Diagnostic Algorithm

The information obtained from patients and doctors is obtained in the form of LTVs that contain ambiguity, as was explained in the previous section, but it is necessary to convert them into numerical truth values

Table 4.14. Examples of Disease – Symptom Relations (R_{ij})

	Classification		Urinary Abnormality						
		Symptom	Increase in urine	Decrease in urine	Urine odor	Cloudy urine	Bloody urine	...	...
Classification	Disease	Number / Number	1	2	3	4	5	...	...
Kidney Diseases	Acute glomerulonephritis	1			RT		RT		
	Nephrotic syndrome	2		VT	VT				
	Chronic renal failure	3	PT		RT				
	Urethral calculus	4			PT		RT		
	Cystitis	5			PT	RT			
	Benign prostate hypertrophy	6					RF		
	⋮	⋮							
⋮	⋮	⋮							

Table 4.15. LTV Level

VT	Very true
RT	Rather true
PT	Possible true
PF	Possible false
RF	Rather false
VF	Very false
UN	Unknown

(NTV) in order to perform operations, and this quantification is carried out using membership functions. In this system there are seven levels of LTVs for expressions such as "a little" and "very much" for symptoms and "maybe" and "probably" for diseases and symptoms. At this time it is necessary to know what kind of membership value to take from each membership LTV and to set the level. These levels are called α-cut values, and at this time the value for A is expressed as A^α. In this instance A^α normally has one value, but in order to preserve the ambiguity of language, it is more natural to give them a range. For example, in a case in which an LTV is "UN" it will have a range of [0, 1]. In order to do this, all of the membership values are given as a range as

$$A^\alpha = \{x \mid \mu_A(x) > \alpha\}. \tag{4.5}$$

The relationships between LTVs, α-cut values, and membership values are shown in Fig. 4.22. However, each membership function and α-cut value is given in terms of coordinates or constants in the construction of the data base.

The inference algorithm is derived from Eqs. (4.3) and (4.4) from Section 4.5.2(b). Here we assume that P_j and P_{ij} are a fuzzy subset of LTV set V, which is probably true and convex. If we use the fuzzy modus ponens and fuzzy modus tollens in Eqs. (4.3) and (4.4), we obtain the following relations between symptoms and diseases:

For $\forall\alpha \in [0, 1]$,

$$\bigvee_i \left((R_{ij}^\alpha \wedge A_i^\alpha)\right)_l = (B_j^\alpha)_l - \left((\neg P_j)^\alpha\right)_u \bigvee 0, \qquad j = 1, \ldots, n, \tag{4.6}$$

$$(R_{ij}^\alpha \wedge A_i^\alpha)_u = (B_i^\alpha)_u + \left((\neg P_{ij})^\alpha\right)_u \wedge 1, \qquad i = 1, \ldots, m, j = 1, \ldots, n. \tag{4.7}$$

Here the symbol $\neg$ means the negation from fuzzy logic. In addition, 1 indicates lower bound (see the supplementary note on fuzzy logic operations). In Eqs. (4.6) and (4.7), the observed symptoms B_j and knowledge P_j, P_{ij}, and R_{ij} are given, and all of the diseases $\{A_i\}$ are found. A_i is

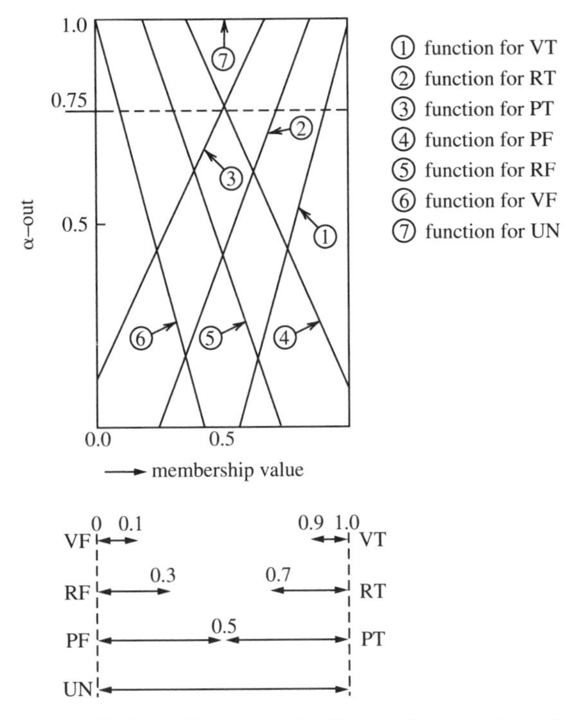

Fig. 4.22 Relationship between LTV, α, and membership value.

obtained by finding a common solution to Eqs. (4.6) and (4.7). Here the certainties of P_j^α, P_{ij}^α, R_{ij}^α, and B_j^α are defined by spatial values (lower bound and upper bound) as follows:

$$\begin{cases} P_j^\alpha = [p_j, 1], \quad P_{ij}^\alpha = [p_{ij}, 1], \\ R_{ij}^\alpha = \left[\gamma_{ij}(1), \gamma_{ij}(2)\right], \quad B_{ij}^\alpha = \left[b_j(1), b_j(2)\right] \end{cases}. \tag{4.8}$$

In addition, the distance between the symptoms and knowledge is given as follows:

$$\begin{aligned} Z_j^\alpha &\triangleq \left[\left(b_j(1) + p_j - 1\right) \vee 0, 1\right], \\ E_{ij}^\alpha &\triangleq \left[0, \left(b_j(2) + 1 - p_{ij}\right) \wedge 1\right]. \end{aligned} \tag{4.9}$$

Furthermore, we let the set of spatial values for knowledge and distance be as follows. In other words, for any i and j,

$$\overline{R} = \{R_{ij}^\alpha\}, \overline{Z} = \{Z_j^\alpha\}, E = \{E_{ij}^\alpha\}. \tag{4.10}$$

In addition, $R \in \overline{R}$, $Z \in \overline{Z}$, $E \in \overline{E}$ means that for each of any i and j,

$$\gamma_{ij} \in R_{ij}^{\alpha},\ z_j \in Z_j^{\alpha},\ e_{ij} \in E_{ij}^{\alpha}. \tag{4.11}$$

Therefore, the inverse problem for Eq. (4.6) is the problem of finding the following vector.

$$\overline{\alpha} = \left\{\alpha | {}^{\exists}R, {}^{\exists}Z, R \in \overline{R}, Z \in \overline{Z}, a \circ R = \overline{Z}\right\} \tag{4.12}$$

is a vector for the case in which these elements are spatial values. Using an algorithm for the inverse problem related to the fuzzy relational inequality, the solution is

$$\alpha_i^k = \left[\max_j \left(\inf w_{ij}^k\right),\ 1\right], \qquad i = 1, 2, \ldots, m. \tag{4.13}$$

Here $\overline{w}_{ij}^k$ is

$$\overline{w}_{ij}^k = \begin{cases} \overline{u}_{ij} & \text{for } {}^{\exists}i\colon \overline{u}_{ij} \neq 0 \\ \overline{v}_{ij} & \text{for all other } i\text{'s} \end{cases} \tag{4.14}$$

Here, $\overline{u}_{ij} = R_{ij}^{\alpha} \varepsilon Z_j^{\alpha}$, $\overline{v}_{ij} = R_{ij}^{\alpha} \tilde{\varepsilon} Z_j^{\alpha}$ (see supplementary note about ε and $\overline{\varepsilon}$, in Section (e) below).

On the other hand, the solution for Eq. (4.7) is given by finding the following vector:

$$\overline{\alpha}' = \left\{\alpha' | {}^{\exists}R, {}^{\exists}E, R \in \overline{R}, E \in \overline{E}, \alpha_i' \wedge \gamma_{ij} = e_{ij} \text{ for } \forall_i, \forall_j\right\}. \tag{4.15}$$

This solution is given by the following:

$$\alpha_i' = \left[0,\ \min_j \left(\sup(R_{ij}^{\alpha} \varepsilon E_{ij}^{\alpha})\right)\right], \qquad i = 1, \ldots, m. \tag{4.16}$$

Therefore, for any $k \in K$, the solution for Eqs. (4.12) and (4.16) is

$$\overline{\alpha}_i^k = \alpha_i^k \wedge \alpha_i', \qquad i = 1, \ldots, m. \tag{4.17}$$

Here, K is given by the following:

$$K = \left\{k | \forall_i : \bigcap_j \overline{w}_{ij}^k \neq \phi \text{ and } \alpha_i^k \wedge \alpha_i' \neq \phi\right\}. \tag{4.18}$$

If $K = \phi$, the solution does not exist. In this case the solution strategies that can be considered are

(1) Lower parameter α (α-cut value), which expresses the inference precision, and repeat inference closer to 0.
(2) Conduct a reexamination of observed symptom B_j, correct it with very precise information and perform inference again.

However, we can easily see that we will get a highly ambiguous result with (1), but it is not very effective. Therefore, we will investigate the implementation of (2).

(d) Evolutionary Diagnosis

In the previous section the case in which a solution for an inverse problem based on a given value for α existed. However, there are cases in which a solution cannot always be reached, as in when there is a solution for $\alpha = 0.6$, yet there is no answer for $\alpha \geqq 0.8$. In this type of case there is generally not enough information about symptoms, and there are cases in which it is better to make a diagnosis after more information has been obtained. Therefore, it is necessary to consider how this kind of symptom should be determined and by what method. This bears a relationship to being able to reach more accurate diagnoses in cases in which there has been an error in the initial data—that is, evolutionary diagnosis. Here this is carried out in the following manner.

First of all, we consider vector A, which gives the LTVs for the elements of the fuzzy set of diseases related to the symptom:

$$A \triangleq (A_1, \ldots, A_m). \tag{4.19}$$

We let the number i base vector for A be the following:

$$\underline{A}^i \triangleq (\text{c.f}, \ldots, \underline{A}_i, \ldots, \text{c.f.}), \qquad i = 1, \ldots, m. \tag{4.20}$$

Here, c.f. means "completely false."

This means that only the number i vector is A_i, and that the rest are all vectors that are c.f. In other words, we can say that there is only a possibility for the occurrence of disease number i, and that the rest are LTV vectors that are c.f. In addition, we assume here that the LTV items obtained for A^i are as follows:

$$L = \{\text{false}, \text{unknown}, \text{true}, \text{completely true}\}$$

If we use the fuzzy modus ponens and fuzzy modus tollens in Eqs. (4.3) and (4.4), we obtain the following for each for i and j:

$$\left(B_j^\alpha(\underline{A}^i)\right)_u = \left((R_{ij}^\alpha \wedge A_i^\alpha)_u + (1 - p_j)\right) \wedge 1, \tag{4.21}$$

$$B_j^\alpha(\underline{A}^i)_l = \left((R_{ij}^\alpha \wedge A_i^\alpha)_l - (1 - p_{ij})\right) \vee 0. \tag{4.22}$$

Here we let the elements for similarity (arithmetic mean expected value) $t(\underline{A}_i)$ for NTVs for predicted symptoms when $A = \underline{A}^i$ be the following vector:

$$t_j \triangleq \left(\left(B_j(\underline{A}^i)\right)_l + \left(B_j(\underline{A}^i)\right)_u\right)/2, \qquad j = 1, \ldots, n. \tag{4.23}$$

We define arithmetic mean b for the actual observed symptoms in the same way. Element b_j of this is given as follows:

$$b_j \triangleq \big(b_j(1) + b_j(2)\big)/2, \qquad j = 1, \ldots, n. \tag{4.24}$$

The direction of new observations is determined by considering the geometrical form of $t(\underline{A}^i)$ and b. In other words, the deviation between t_j and b_j is transferred to the determination of items for the symptoms that should be checked. The algorithm for this comes out as follows. First, the normal distance $D(\underline{A}^i)$ is considered:

$$D\big(\underline{A}^i\big) \triangleq \sum_{j=1}^{n} (t_j - b_j)^2. \tag{4.25}$$

$D(\underline{A}^i)$ is the expected value for the truth value when $A = \underline{A}_i$. However, this is a method for obtaining one distance, and we can think of various other meaningful distances. The minimum distance is obtained as follows:

$$d_i \triangleq \min_{A^i \in L} {}^i D\big(\underline{A}^i\big), \qquad i = 1, \ldots, m. \tag{4.26}$$

In other words,

$$d_i \leqq \min_{A^i \in L} \sum_{j=1}^{n} \Big(t_j\big((\text{c.f.}, \ldots, \underline{A}_i, \ldots, \text{c.f.}) - b_j\big)^2\Big). \tag{4.27}$$

Here we record $\underline{A}'_i$, which satisfies the following condition:

$$d_i = D(\underline{A}'(A'_i)), \qquad i = 1, \ldots, m. \tag{4.28}$$

d_i is the minimum value of $D(\underline{A}')$ in relation to $\underline{A}_i$ for disease i. At this time, we let the $\underline{A}_i$ that gives d_i be $\underline{A}'_i$. Next we calculate i^* such that

$$d_i^* = \min_i d_i. \tag{4.29}$$

d_i $(i = 1, \ldots, m)$ is determined for m diseases, and i^* is the number of the disease with the smallest d_i among them. From this, in relation to the truth value, base vector $\underline{A}^{1*}$ is determined for the disease for which the difference between the expected value and the observed value is the smallest:

$$\underline{A}^i \triangleq (\text{c.f.}, \ldots, \underline{A}_i^*, \ldots, \text{c.f.}). \tag{4.30}$$

This is for creating relative standards for expected values for truth values for diseases. In addition, we obtain $t(\underline{A}^{i*}(\underline{A}_i'^*))$, that is, t_j, which can be compared to b for each element.

From the preceding, we can indicate the items that correspond to the numbers for which the value of $t_j - b_j$ $(j = 1, \ldots, n)$ is large. Yet when two or more diseases are predicted as occurring simultaneously at this time, improvements are made by changing the groupings of $\underset{\sim}{A}^i$ base vectors.

(e) Supplementary Note

(1) Fuzzy Logic (Spatial Values). Let P and Q be fuzzy propositions, and let $\underset{\sim}{P}$ and $\underset{\sim}{Q}$ be LTV. When the α-level set satisfies the equation

$$\underset{\sim}{P}^{\alpha} = [p_1, p_2], \quad \underset{\sim}{Q}^{\alpha} = [q_1 \cdot q_2] \qquad \text{for } \forall\alpha \in [0, 1],$$

The combinations for fuzzy logic are given by the following relations:

$$\begin{array}{lll}
\text{Negation} & R_1 \triangleq \neg P, & \underset{\sim}{R}_1^{\alpha} = 1 - \underset{\sim}{P}^{\alpha}, \\
\text{Conjunction} & \underset{\sim}{R}_2 \triangleq P \,\&\, Q, & \underset{\sim}{R}_2^{\alpha} = \underset{\sim}{P}^{\alpha} \wedge \underset{\sim}{Q}^{\alpha}, \\
\text{Disjunction} & R_3 \triangleq P \text{ or } R, & \underset{\sim}{R}_3^{\alpha} = \underset{\sim}{P}^{\alpha} \vee \underset{\sim}{Q}^{\alpha}, \\
\text{Implication} & R_4 \triangleq P \to Q, & \underset{\sim}{R}_4^{\alpha} = \overline{(\neg \underset{\sim}{P})}^{\alpha} \oplus \underset{\sim}{Q}^{\alpha}.
\end{array}$$

Here we let

$$\begin{aligned}
1 - [p_1, p_2] &\triangleq [1 - p_2, 1 - p_1], \\
[p_1, p_2] \wedge [q_1, q_2] &\triangleq [p_1 \wedge q_1, p_2 \wedge q_2], \\
[p_1, p_2] \vee [q_1, q_2] &\triangleq [p_1 \vee q_1, p_2 \vee q_2], \\
[p_1, p_2] \oplus [q_1, q_2] &\triangleq [(p_1 + q_1) \wedge 1, (p_2 + q_2) \wedge 1].
\end{aligned}$$

(2) ε Composition and $\bar{\varepsilon}$ Composition. For two spatial values $[p, q]$ and $[r, s] < [0, 1]$, ε and $\bar{\varepsilon}$ are each defined as follows:

$$[p, q]\varepsilon[r, s] \triangleq \begin{cases} [r, 1] & \text{if } [p, q] \cap [r, s] \neq \phi \\ [r, s] & \text{if } p > s \\ \phi & \text{if } q < r \end{cases}$$

$$[p, q]\bar{\varepsilon}[r, s] \triangleq \begin{cases} [0, 1] & \text{if } p \leq s \\ [0, s] & \text{if } p > s \end{cases}$$

4.5.3 Development of a Health Checkup Support System [28 – 30]

As far as health maintenance on the job is concerned, the labor safety and hygiene laws require that periodic health checkups be carried out, and industry doctors devote much of their time to post-examination processing. With the recent variety and diversification of medical information, this has become even more complicated than in the past, and the workload of industry doctors is increasing.

In addition, even if the health maintenance systems used up to now have decision-making functions based on medical data bases, we cannot find one that gives support for the provision of information for health maintenance guidance. Because of this, a workplace health checkup support system was developed that has a function for automatically generating health maintenance comments in cases where lifestyle guidance is necessary even for healthy examinees, along with decision-making support for examination results.

The following are included in the inference function for the medical data base in this system:

(1) Judgments on primary and secondary examinations are made, and the results for patients that must be checked are set up so they can easily be corrected by a doctor.
(2) Test items for those who have to be reexamined after the primary examination are chosen.
(3) Health maintenance comments are generated from individual examination results.

By adding the preceding functions, a system in which the following are possible was created:

(a) The decision-making workload of industry doctors can be reduced, and more time can be spent on individual post-examination health maintenance guidance of the examinees.
(b) Even for large numbers of examinees, checkup decisions based on fixed standards and the generation of health guidance comments are possible.
(c) When decisions concerning health are made, they can be made with an accurate grasp of the examinee's professed symptoms, examination results, consultation findings, and, in addition, past medical history.

In addition, there are many problems with the ambiguity represented by the expressions of symptoms during the questioning of examinees within the data handled during inference. Furthermore, there are more examinees in health checkups than at hospitals, and even with clinical tests, there are many judgments on abnormal and borderline test results. In order to solve these problems, the knowledge for a two-valued information processing approach is massive and complex, and this creates a barrier to knowledge base revision.

Because of this, a system that can output judgmental results that correspond to degrees in examination results was created by employing fuzzy inference, which makes judgmental processing of ambiguous data and knowledge possible.

(a) System Structure

This system is constructed of a 32-bit engineering workstation (EWS), color bit-mapping display terminal, laser printer, optical character reader (OCR) and 8-inch floppy disk drive, and the EWS and all other terminals are connected by means of RS232C interfaces. In addition, the main memory of the EWS has its maximum of 20 Mb, and there is an external 172 Mb hard disk drive, a 50 Mb cassette tape streamer and a 5-inch (1.6 Mb) floppy disk drive for recording information.

The software configuration is shown in Fig. 4.23. The EWS is equipped with the UNIX system V operating system (OS) to which the 4.2BSD functions have been added, and it has multi-window graphics, Japanese language and communications functions. A summary of the operation methods for the system is shown in Table 4.16.

(b) System Functions

The system is constructed of the following function blocks.

(1) Examination Data Input / Output Processing. Individual data (ID) and examination data are input and checked from the optical character reader (OCR, floppy disk (FD), keyboard, and mouse), and along with this data are displayed and corrected.

(2) Knowledge Acquisition Support Processing. A maximum of 10 files and for each file unit, 500 rules, are possible. In addition, there is a maximum five-variable unit set-up possible for membership functions, and 25 functions can be set up for coordinated usage.

(3) Data Base Management. Unified management of ID inputs, examination data, and judgmental results is carried out by means of relational data base management functions. In addition, there is a file maintenance function that makes it possible to easily update items in examination data. Furthermore, examination data for a chosen year can be backed up on tape and past data can be referenced.

(4) Judgmental Support Processing for Checkup.

(1) Whether or not there is any question of illness is investigated based on primary judgmental rules from the primary examination data and past data for all examinees; if there is a question, the name of the illness and the certainty are inferred, and the steps to be taken are determined in correspondence with the degree. At this point,

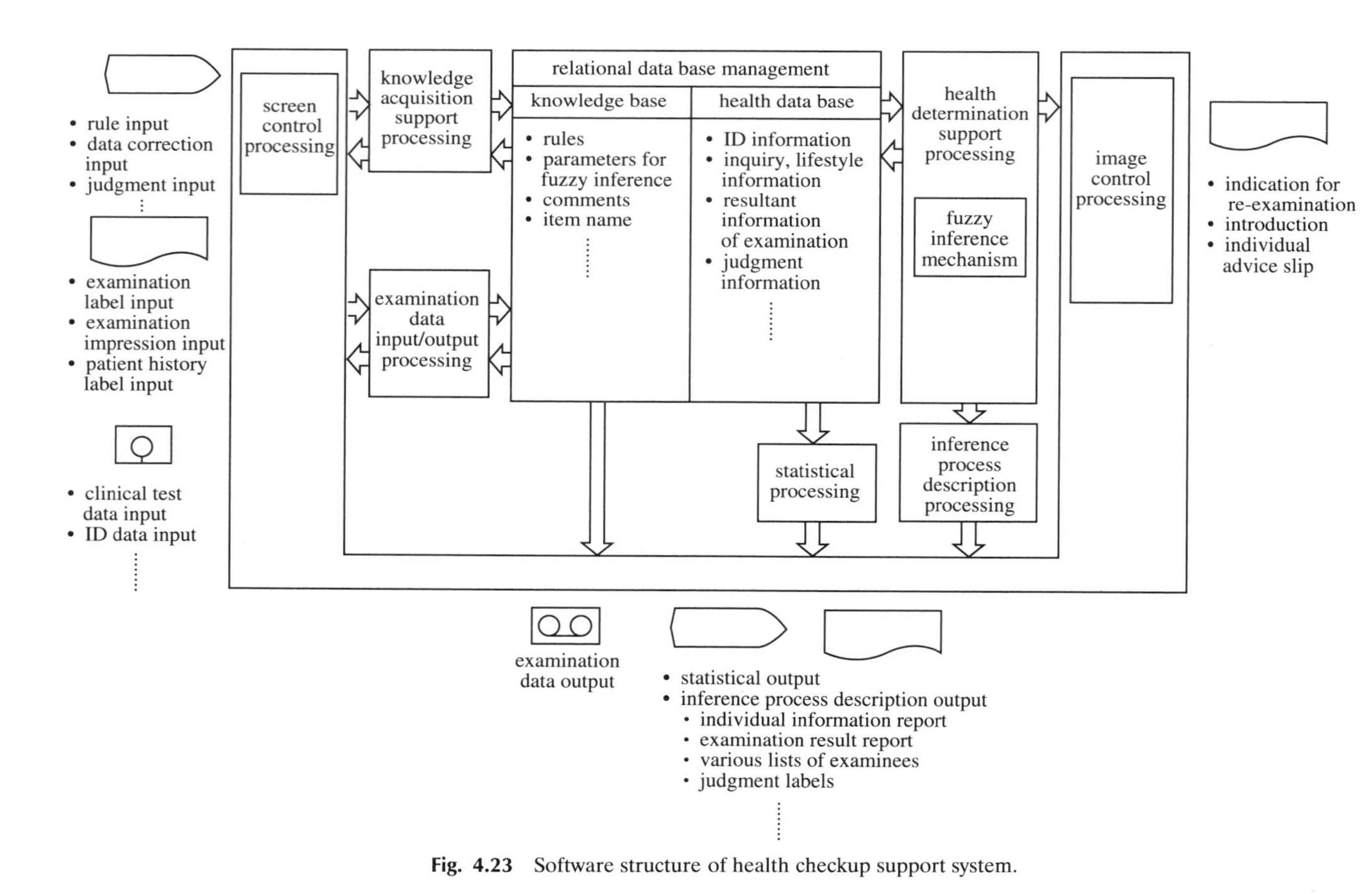

Fig. 4.23 Software structure of health checkup support system.

Table 4.16. System Specifications of Health Checkup Support System

System environment	OS	UNIX System V + 4.2 BSD
	Language	C programming language
Inference engine	Method	Forward inference with production rule
Knowledge base (main components)	Number of rules	500 rules: primary determination, secondary determination, health maintenance determination
	Range of determination	Obesity, anemia, blood sugar abnormality, white blood cell count abnormality, blood pressure, bloody urine, kidney function, liver function, hyperlipidemia, electrocardiogram, high blood uric acid level, stomach fluoroscopy, gall bladder, pancreatic disease, hemorrhage tendency, thyroid diseases, internal inflammation, chronic blood diseases, prostate diseases
	Output comment	Approximately 100 health management comments
	Membership function	Approx. 170
Data base	Method	Relational data base management
	Data	• Clinical test data (70 items for 16 areas such as urinalysis, blood count, etc.) • Impression data from image examinations (80 items for 4 test areas such as brain x-rays) • Observational impression data (20 items: heart noise, etc.) • Questioning data (90 items including self-asserted symptoms and patient history) • Examination data and determination data for past 3 years
Human interface	Input methods	• OCR (image examination impressions, observation impressions, question answers, etc.) • FD (ID, clinical test data) • Cassette tape (examination data from more than 3 years previous) • Mouse, keyboard (indicators, ID, knowledge input, etc.)
	Output methods	• Cassette tape (examination data storage) • CRT (various types of display using multiwindow and color graphics) • Laser printer (20 types of reports)

the confirmation and judgments of the doctor are obtained and final decisions are made concerning examinees who need reexamining. A reexamination notice that elucidates the items to be reexamined is printed for those who need reexamining, and individual advice slips on which health maintenance comments based on the steps to be taken are printed for those who do not require reexamination.

(2) Inference is carried out for those who are reexamined in the same way primary judgments were made, based on secondary judgmental rules, using secondary examination data, primary examination data, and past data. At this point the judgments and confirmation of a doctor are obtained, and final decisions about beginning treatment and close examinations are made. Introductions to medical agencies that contain clear explanations of checkup results and judgments are printed for those who require treatment or closer examination, and individual advice slips that contain health maintenance comments are printed for those who do not.

(3) Besides the advice slips, guidance notices and introductions output in (1) and (2), reports to the Labor Standards Agency, lists for the health office and company managers, and various lists for post-checkup processing by industry doctors are output.

(5) Statistical Processing, Processing of Explanations of Inference Process and Screen Control Processing.

(1) Statistical information, based on position, age, and sex, for checkup data is displayed.

(2) In order to make it easy to confirm the soundness of established knowledge, there is a detailed output of activated rules.

(3) Besides indicator input by means of the multiwindow function, system malfunction log screens, judgmental processing conditions screens, etc., can be referred to simultaneously. In addition, passwords are established in order to preserve data and privacy during operations.

(c) Inference and Knowledge Base

Production rules and fuzzy relations can be used as fuzzy inference methods for carrying out inference using ambiguous information, but since there is a tiered construction that establishes intermediate hypotheses for judgmental impressions and lifestyle guidance and makes final judgments, and since there is a necessity for it to be driven by data with a non-dialogue input, the system employs inference that uses forward chaining production rules. Since with chaining, ambiguity increases each time a rule is followed, this system is set up to produce results after two or three tiers.

In addition, in order for the input to be clinical tests and questioning information and the output to be degrees of illness, direct values and linguistic truth values are handled as intermediates. The inference mechanism stands as an independent subsystem as shown in Fig. 4.24; the fact base and knowledge base are inputs, and the certainty of the target and the inference process are its outputs.

The inference form is one that employs the fuzzy modus ponens, which is an extension of the modus ponens, one type of traditional deductive reasoning. This is expressed as follows:

$$\text{If } A_1 = a_1 \text{ and } (A = a \rightarrow B = d) \text{ then } B = b_1. \tag{4.31}$$

Here, we let a and b be fuzzy subsets in the universes U and V, and u and v be elements of these. In addition, the symbol $\rightarrow$ means implication.

If a_1—information about U for the fuzzy relation ($[A = a \rightarrow B = b] = R$) between u and v, which are objects that carry knowledge—is obtained, the proceeding equation gives as its result information concerning V from a_1 and R.

A fuzzy subset on the universe U is a set characterized by the membership function $\mu a: U \rightarrow [0, 1]$, and $\mu a(u)$ is the degree of membership of element u in a. Fuzzy set a is expressed in the following way:

$$a = \int U \mu a(u)/u. \tag{4.32}$$

Here, $\int U$ represents the combination $\mu a(u)/u$ over the total space of U. The / is a separator.

In fuzzy inference there must be a method for transforming fuzzy condition $A = a \rightarrow B = b$ into fuzzy relation R. In addition, conclusions

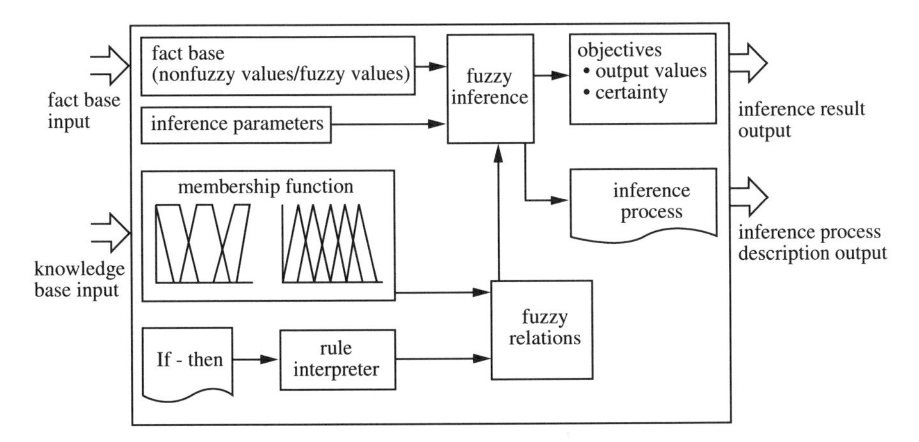

Fig. 4.24 Inference mechanism.

are obtained by carrying out compositions with real data and fuzzy relations. Several methods for transformation and composition have been generalized. In this system, the following method, for which divergence of solution will be small, is used for inference:

$$b_1 = a_1 \circ R = \int v \bigvee_u [\mu a_1 \wedge \mu R(u, v)]/(v). \tag{4.33}$$

Here b_1 is the conclusion and a_1 is real data.

In other words, operations are performed as follows in this inference section. Let us say that plural propositions have been established in the antecedents of the rules:

$$\text{If } A = a \,\&\, B = b \text{ then } C = c. \tag{4.34}$$

At this time, if $A = a_1$ and $B = b_1$ are given in the fact base, conclusion $C = c_1$ is given by the following equation:

$$c_1 = [a_1 \circ (a \rightarrow c)] \cap [b_1 \circ (b \rightarrow c)]. \tag{4.35}$$

Here $\circ$ is a max–min operation, and $x \rightarrow y$ is $\mu x(u) \cdot \mu y(v)$. The inference flow is shown in Fig. 4.25.

Next, the knowledge base that is used in inference is constructed of membership functions for the cases of making judgments on various inputs and outputs, rules, input and output ranges, and comments for final output. Membership functions are made up of function names (fuzzy labels) and membership values, and in this system the following fuzzy labels are used.

(1) These are used in antecedents with a maximum of five item-units for clinical test items other than qualitative data:

TRS (considerably small)	TPS (probably small)
MM (medium)	TBS (probably big)
TRB (considerably big)	

(2) These are used in antecedents for the qualitative data for clinical test items:

MM (−)	TPS (+ − less)	PB (+ −)
RB (+)	TPB (+bigger)	TPB (+ + bigger)
VB (+ + +)		

(3) This is used in antecedents for questions about symptoms.

LT1 (irrelevant, sometimes relevant, relevant)

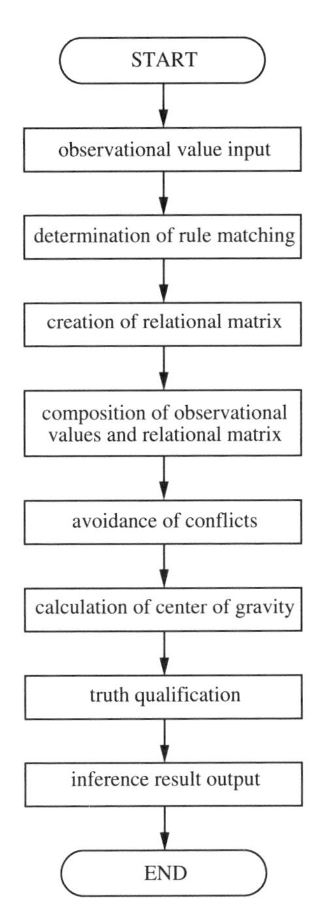

Fig. 4.25 Inference flow.

(4) These are used in conclusions for degree of illness and intermediate hypotheses:

CLA (classic) DEF (definite)
PRO (probable) POS (possible)
SUS (suspicious)

(5) These are used in antecedents and conclusions for image diagnostic impressions, observational impressions and intermediate hypotheses:

YES (present) NO (absent)

An example of a membership function for (1) is shown in Fig. 4.26 and an example of one for (4) in Fig. 4.27. The construction of rules is such

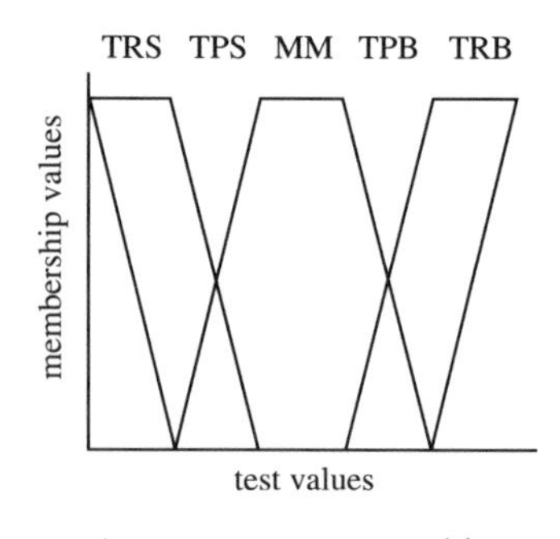

Fig. 4.26 Function for (1).

that there can be a maximum of 10 fuzzy propositions predicated in the antecedent and two in the conclusion in the following manner:

If input data item = fuzzy label,

then output item = fuzzy label.

In addition, it is possible for fuzzy labels, which represent membership functions, to establish negations and logical sums within the same proposition.

Rule files are constructed from three files, primary judgment, secondary judgment, and health maintenance guidance judgment; rule files are switched by designation.

An example of primary judgment rules in the area of liver function is shown in the following:

(1) If GOT = TPB then liver dysfunction = DEF
(2) If GPT = TPB then liver dysfunction = DEF
(3) If GGT = TPB then liver dysfunction = DEF
(4) If (GOT = < GPT) = YES then liver dysfunction = DEF

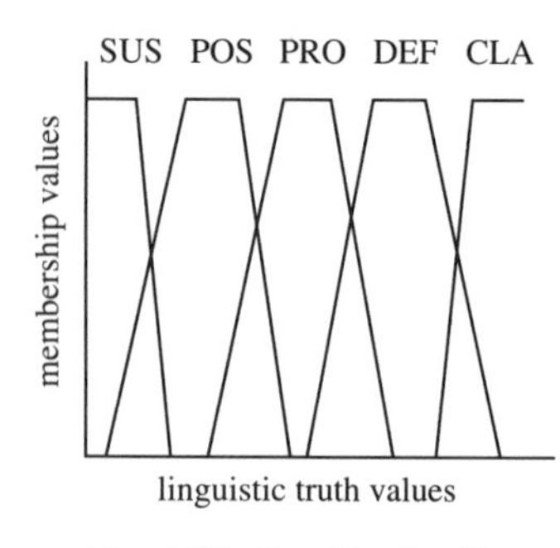

Fig. 4.27 Function for (4).

(5) If question (tire easily) = LT1 then liver dysfunction = PRO
(6) If GOT = MM & GPT = MM & (GOT = < GPT) = NO & previous GOT = TPB then liver dysfunction = PRO

(d) RDB (Relational Data Base), Etc.

In order to carry out effective unified maintenance of the variety of data for large numbers of people and a number of past years, this system uses a simple relational data base subsystem (RDB).

The RDB is one made with the objective of using it in an application program as a C programming language base, and it is constructed from a user interface, data base maintenance functions, and system utilities, as shown in Fig. 4.28.

The user interface function accesses the data base by organic combinations using application interface function groups, combination information, viewing information, various information tables for access conditions, conditional functions, and combined conditional functions.

The data base maintenance function carries out combinations of various relational file records by means of equations for designated combination conditions, and creates relational records according to viewing information.

System utilities include a data base for support and maintenance of the RDB, relational files, definition functions for records and the like, reconstruction functions for times when records and relational files are revised, backup functions that make use of cassette tapes, and restoration functions.

The concrete construction of the data base is as follows. Employee numbers and examination item numbers are accessed as indices (keys) (see Fig. 4.29).

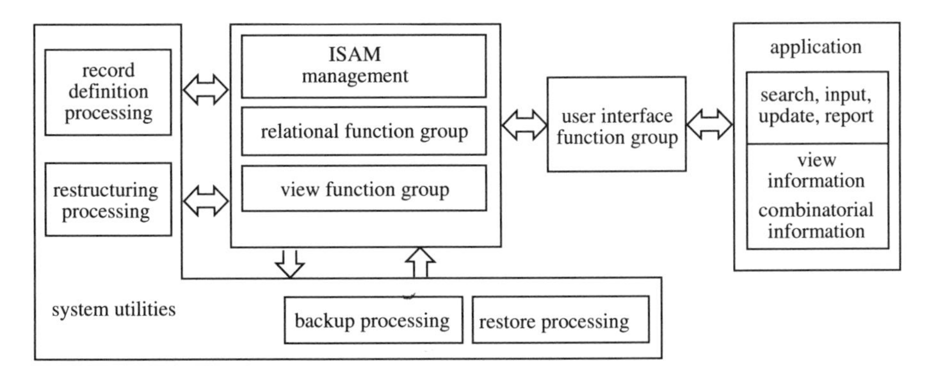

Fig. 4.28 Relational data base structure.

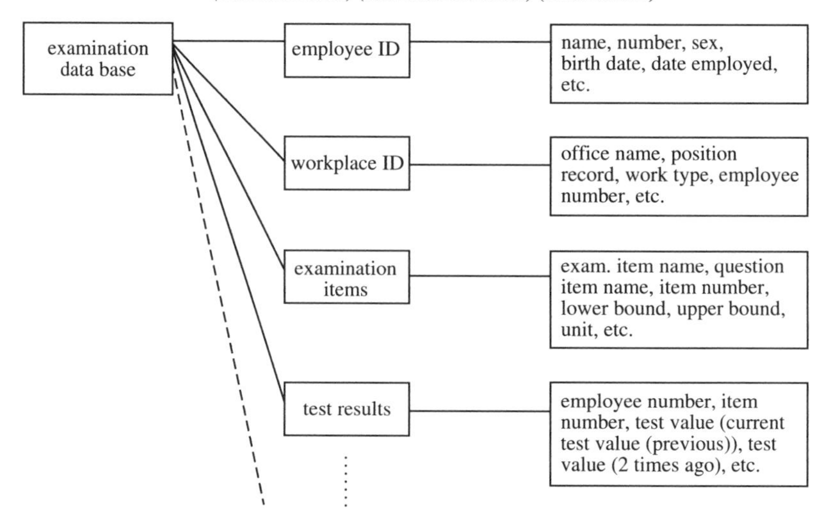

Fig. 4.29 Health checkup data base.

This system is set up in the company health management room, which holds the company doctor, health maintenance nurse, and responsible office workers, and the objective is support for regular health checkups. When the data base is constructed, the items for discriminating individuals are obtained on floppy disk (FD) from the host computer; examination impressions and question tables are on OCR forms, and clinical test results are obtained from medical instruments on FDs. Three past years' worth of these are converted to a data base, and data for the period that must be preserved according to law are stored on cassette tapes so that they can be retrieved at any time.

The set-up is on the scale of 5,000 examinees, but if the volume of the recording media is increased, this can be expanded. Furthermore, operations are carried out through menu selection with a mouse and operability is excellent, with such functions as automation of output size for reports.

4.5.4 Conclusion

The prediagnostic/inquiry diagnostic system that was developed was able to give accurate diagnostic results with a practical processing speed in tests in which symptoms appropriate to the data base were input. Hereafter, repeated tests with actual medical records are planned, and the results of these will be reflected in the system.

On the other hand, the data base system, with its inference function cannot only be considered as having great value for use in health maintenance, but also in other fields of medical information. Even though there are several issues remaining that should be investigated, such as membership functions and optimization processing for dropout, we can expect that the health checkup support system that was developed in this environment will be an excellent processing method, with ambiguity included in its functions.

REFERENCES

1. Ishizuka, M., Fu, K. S., and Yao, J. T. P., "Inference Procedure with Uncertainty for Problem Reduction Method," Technical Report, Purdue University, CE-STR-81-24 (1981).
2. Ishizuka, M., Fu, K. S., and Yao, J. T. P., "Rule-Based Damage Assessment System for Existing Structures," *Solid Mechanics Archives* **8**, 93–118 (1983).
3. Ogawa, H., Fu, K. S., and Yao, J. T. P., "Knowledge Presentation and Inference Control of SPERIL-II," *Proc. of Annual Conference of the Association for Computing Machinery*, San Francisco (1984).
4. Nakamura, H., Matsuura, S., Matsui, S. and Terano, T., "Life Term Prediction for Hydropower Steel Structures Based on Knowledge Engineering Techniques," *Proc. of Japan Society of Civil Engineers* **368 / I-5**, 301–310 (1986) (in Japanese).
5. Hayakawa, K. and Yokoi, N., "Expert System for Displaying Assessments of Concrete Cracks and Measures for Dealing with Them," *Nikkei Computer* 179–190 (1986) (in Japanese).
6. Mikami, I., Ezawa, Y., Tanaka, S., and Asakura, T., "Knowledge-Based Expert System for Damage Assessment and Repairing Method on Reinforced Concrete Slabs of Highway Bridges," *Journal of Structural Engineering* **33A**, 317–325 (1987) (in Japanese).
7. Furuta, H., Shiraishi, N., and Yao, J. T. P., "An Expert System for Evaluation of Structural Durability," *Proc. 5th OMAE Symposium*, Tokyo, pp. 11–15 (1986).
8. Shiraishi, N., Furuta, H., Umano, M., and Kawakami, K., "Structural Durability of RC Bridge Decks," *Proc. of Japan Society of Civil Engineers* **386 / I-8**, 285–291 (1987) (in Japanese).
9. Furuta, H., Fu, K. S., and Yao, J. T. P., "Applications of Knowledge Engineering—Expert Systems—in Structural Engineering," *Journal of the Japan Society of Civil Engineers* **70**, 28–33 (1985) (in Japanese).
10. Furuta, H., Fu, K. S., and Yao, J. T. P., "Structural Engineering Applications of Expert Systems," *Computer-Aided Design* **17**, 410–419 (1985).
11. Terano, T., Asai, K., and Sugeno, M., eds., *Fuzzy Systems Theory and Its Applications* (trans. from the Japanese), Academic Press, Boston (1987).
12. Blockley, D., Pilsworth, B., and Baldwin, J., "Structural Safety as Inferred from a Fuzzy Relational Knowledge Base," University of Bristol, Dept. of Civil Engineering Res. Rep. (1982).
13. Baldwin, J., and Zhou, S., "A Fuzzy Relational Inference Language," University of Bristol, Dept. of Eng. Math. Res. Rep. (1982).
14. Blockley, D., "Logical Analysis of Structural Failures," *J. Eng. Mech. Div., ASCE* **107**, 355–365 (1981).

15. Umano, M., "Fuzzy Set Manipulation System in Lisp," *Third Fuzzy Systems Symposium*, pp. 167–172 (1987) (in Japanese).
16. Simon, H. A., *The Science of Decision Making* (Inaba and Kurai, trans.), Sangyo Noritsu University Press (1979) (in Japanese).
17. Sawaragi, Y., *et al.*, *Participative System Approach*, Nikkei Kogyo Shinbunsha (1981) (in Japanese).
18. Bartlett, F. C., *Remembering: A Study in Experimental and Social Psychology*, Cambridge University Press, Cambridge (1932).
19. Schank, R. C. and Abelson, R. P., *Scripts, Goals, Plans and Understanding*, Lawrence (1977).
20. Schank, R. C., "Conceptual Dependency: A Theory of Natural Language Understanding," *Cognitive Psychology* **3**, 552–631 (1972).
21. Nakamura, K., Iwai, S., and Sawaragi, T., "Decision Support Using Causation Knowledge Base," *IEEE Trans.* **SMC-12**(6), 765–777 (1982).
22. Sawaragi, T., Iwai, S., and Katai, "Decision Making Support, Quantification of Extended Effects and Introduction of the Decision Maker's Subjective Judgmental Standards Based on Cause and Effect Relational Knowledge," *Studies from the Society for Measurement and Automatic Control* **22-6**, 629–636 (1986).
23. Sawaragi, T., Iwai, S., and Katai, "Creation of a Multilayered Structure of Cause and Effect Linked Knowledge for Grasping the Deep Structure of Societal Phenomena," *Studies from the Society for Measurement and Automatic Control* **23-9**, 977–984 (1987) (in Japanese).
24. Sawaragi, T. and Iwai, S., "Cognitive Stimulation for Intelligent Decision Support," in *Methodologies for Intelligent Systems* (W. R. Ras and Zemankova, eds.), pp. 362–369, Elsevier Science Publishing Company (1987).
25. Adlassing, K. P., "Fuzzy Set Theory in Medical Diagnosis," *IEEE Trans. S.M.C.* **SMC-16**(2), 260–265 (1986).
26. Sanchez, E., *et al.*, "Linguistic Approach in Fuzzy Logic of W. H. O. Classification of Dyslipoproteinemias," in *Fuzzy Set and Possibility Theory—Recent Developments* (Yager, ed.), pp. 582–588, Pergamon Press (1982).
27. Tazaki, E., *et al.*, "Development of Automated Health Testing and Services System via Fuzzy Reasoning," *Proc. IEEE Int. Conf. on SMC*, pp. 342–346 (1986).
28. Tazaki, E., *et al.*, "Medical Diagnosis Using Simplified Multidimensional Fuzzy Reasoning," *Proc. IEEE Int. Conf. on SMC* (1988).
29. Tazaki, E., *et al.*, "Medical Data Base System with Inference System (1)—Artificial Intelligence in Health Checkup Assessments," *Proceedings of the 7th Joint Conference on Medical Informatics*, pp. 377–380 (1988) (in Japanese).
30. Tazaki, E., *et al.*, "Medical Data Base System with Inference System (2)—Artificial Intelligence in Health Risk Assessments for Health Maintenance Information," *Proceedings of the 7th Joint Conference on Medical Informatics*, pp. 385–388 (1988) (in Japanese).
31. Tazaki, E., "Fuzzy Expert System," *Mathematical Science* **284**, 46–54 (1987) (in Japanese).

Chapter 5

Applications in Business

The management organizations of companies and government agencies conduct various kinds of business in order to carry out management and administration. When it comes to the management of the entire organization, managers and the staffs or workers that support them analyze both inside and outside information, refine programs that shed light upon management goals, and make the decisions of managers. After this, they continue supervision to see if management has been carried out according to plan. In addition, since organizations are run through hierarchical structures, general principles are based on these decisions, but division and department heads, plant managers and operations managers proceed with the more concrete detailed planning, decision-making, and supervision within the authority given to them. Here we will introduce the main aspects of fuzzy systems theory (fuzzy theory) that are used to make such planning, administration, and the business that goes along with them more efficient, and we will show the thinking behind applications. However, we will limit ourselves to a summary of the information included in the companion volume to this one, *Fuzzy Systems Theory and Its Applications* [1].

5.1 KEYS TO BUSINESS APPLICATIONS

5.1.1 The Present Situation in Business

In order to make accurate decisions and improve the efficiency of operations, top managers and supervisors make use of reports and information to conduct business and manage. The word "business" is used in Japan, but it has a broader meaning than business and its management; it includes activities such as operations, trade, planning, and meetings. Here we will use it in this broader sense, and in the following we will describe the current state of business.

The characteristics of business recently include moving toward larger-scale and more complex operations than in the past, with a great deal of change; it has become difficult to make predictions from the data accumulated in the past. Therefore, fuzzy theory has attracted a great deal of interest as a potential method for making business more accurate and effective. In other words, with fuzzy systems, generalized intelligent activities similar to those of humans, such as being able to carry out modeling of large-scale, complicated objects using insufficient data from those objects or creating rules from expert knowledge and experience received in natural language, can be imitated, and by means of these, flexible responses can be made to rapidly changing environmental conditions.

The changes in the state of economics and society in the world and Japan are common knowledge, and what has brought out these business characteristics is the change from rapid to slow growth in world economics since the oil shock. Despite the shrinking of markets, a large number of companies from the rapid-growth era have remained in the market and have come to the present through the development of heated sales competition. Because of this, choice on the user side is esteemed both in commodities and services, and there has been a shift to variety and flexibility in products along with smaller production quantities. Along with this, business has diversified and become flexible, but for competitiveness, there has been a move toward holistic approaches—vertical market, production, sales and distribution approaches and horizontal holistic approaches in which the parent company has a total business center that forms a nucleus. In other words, there has been a drive toward large-scale activities. Diversification, creation of flexibility, and holistic approaches make business even more complicated.

These circumstances extend to organizations other than business such as public offices and schools, and there is a demand for a powerful methodology or tool to cope with them. As a methodology that responds to this, fuzzy theory gives rise to data bases and various means for planning and management. Naturally it is possible to use older-style software and computers, but with high-speed fuzzy computers and efficient fuzzy software, the effects are dramatic.

5.1.2 Outline of Fuzzy Methods for Use in Business

If we take a large company as an example, business is composed of fields such as marketing, product development, personnel management, accounting, production, and distribution; the main work for these can be thought of as being the acquisition, planning (including evaluation and decision making), and management of information and data. Table 5.1 gives fuzzy methodologies for these kinds of fields. Besides the methodologies described in Sections 5.1 through 5.5 of the current chapter, ones in other parts of this book and in *Fuzzy Systems Theory and Its Applications* [1] (* in the table) are also chosen.

Next, we will give an outline of these methods following the order in Table 5.1.

(a) Fuzzy Data Base [1]

In many business situations, it is very convenient if there is a uniformly established and accumulated data base. In these instances, there is a great deal of ambiguous data in areas of human interface where people are directly involved, such as man/machine systems, natural language processing and decision-making; there is a need for fuzzy data bases, and some have been proposed, in order to actively use these in an effective way. As a means of doing this, the relational models of standard data bases are extended to fuzzy domains in most cases. Uses include decision-making systems for various levels of management and supervision, and expert systems are indispensable in every field.

(b) Fuzzy Structural Models [2]

The qualitative models that are used to graphically express the large-scale complicated systems of the various kinds of programming problems for the organizations and business of various levels of management are called *structural models*. These are used for the analysis, planning, and solution of management organization and programming problems, but since it is difficult to bring reality into two-valued, 0 or 1, expressions used for the correspondence relations between the sections of graphs, fuzzification is carried out by replacing these with membership functions. An example of

Table 5.1. Fuzzy Methods for Business

Goal \ Field		Marketing and Product Development	Personnel	Accounting	Production	Distribution	Other
Data accumulation		*Fuzzy data base					
Planning	Goal determination and alternatives	—					
Planning	Modeling	Fuzzy modeling for large-scale systems: (Fuzzy structural model; *Fuzzy regression model; Fuzzy GMDH)					
Planning	Analysis and evaluation	Fuzzy multivariate analysis (*quantification theories I–IV) *Fuzzy integral Fuzzy AHP					
Planning	Optimalization and decision-making	*Fuzzy mathematical programming Fuzzy multiobjective programming Fuzzy decision-making support system (DSS) Fuzzy multiattribute decision-making *Fuzzy statistical decision-making					
Control		—	*Use of fuzzy theory in behavioral science	Use of fuzzy theory for investment	Fuzzy production control *Fuzzy expert systems (processing, fault analysis) Fuzzy QC	—	

GMDH: group method of data handling; AHP: analytic hierarchy process; DSS: decision support system; QC: quality control.

this that has been proposed is FSM (fuzzy structural modeling) [3], and it is used in group decision-making.

(c) Fuzzy Regression Models [1]

Fuzzy regression models can be generated for programming problems for business predictions and the like, using data obtained from specialists. In these instances, fuzzy numbers are used for the regression model coefficients; they express the ambiguity of the data and can be thought of as information that shows possibility. This type of model is also discussed in Section 5.2.1.

(d) Fuzzy GMDH

GMDH is a method based on the principle of heuristic automatic systemization from input data for carrying out modeling of large-scale, complex, nonlinear systems with no prerequisite for knowledge concerning system structure for programming problems for business predictions and the like. Here, the model parameters are identified using fuzzy numbers, and fuzzy GMDH is a method for the modeling of ambiguous phenomena and systems. It is discussed in Section 5.2.2.

(e) Fuzzy Quantification Theory

Fuzzy quantification theory is an extension of quantification theory, which is based on survey data that involves human thinking and judgments and is used to statistically investigate the thinking and tendencies of groups or society, to take into account ambiguity. It is effective for market surveys made by companies and tendency surveys for product development. There are four types of fuzzy quantification theory, types I, II, III, and IV.

I is used for problems such as the investigation of the effect of the degree to which the surrounding area is a shopping center on sales of everyday products at a number of shops. In this case the degree of the area being a shopping center and the degree to which everyday products are handled are given by membership functions.

II is used for problems such as the investigation of what happens with the customer's willingness to purchase two different companies' equivalent products when viewed through evaluation items such as performance, price and style. In this case, the degree to which an evaluation item is considered and the degree of willingness to purchase each company's product are given by membership functions. Here real, naturally occurring conditions such as the differences in one customer's willingness to purchase two different companies' products are dealt with.

III is used for problems such as dividing the buying activities of young people into several patterns from survey data concerning places of purchase, product knowledge, and brands at the time of their purchases. In this instance, the degree of youth, as determined by age, etc., is given by a membership function, and they are divided into the following groups: those without product knowledge who concentrate on brand-name products and buy at department stores, those who usually buy all products at supermarkets, those who mainly buy brand-name products at specialty shops that have a great deal of product knowledge, and those who respect their own product knowledge and buy at specialty shops without being concerned with brand names.

IV stipulates the same type of pattern classification as III, but the way in which problems are handled is different. The relationships between several objects (products, works of art, characters) and the people (customers, enthusiasts, fans) who like them most are considered, and the strong similarity among people who like the same object and objects that are liked by the same people are thought about. In these cases, the groups that include the people that most like something and the objects of their admiration are considered; the degrees of membership are given by membership functions. The group activity patterns of these are then investigated.

(f) Fuzzy Integral

When new products are developed or the willingness to purchase products is investigated, it is necessary to know how customers evaluate products. Now, given a product price, a number of evaluation items (performance, human engineering characteristics, economy, etc.) are brought up, and the weight of these is measured using pair comparison or some other method. When a general evaluation value is determined for something, there is the problem of getting the sum of the different characteristics of each evaluation item when using the linear combinations used up to now. Fuzzy integrals are a way of trying to achieve overall evaluation operations that are close to the thinking and judgments of human beings. Here, additivity, which is used in the four-rule calculation method, is relaxed, and monotonic formulation is carried out; it is an attempt to get closer to the subjective characteristics of human beings. This method is not limited to things, and it is the choice for many types of evaluation objects such as that for characters and ways of thinking.

(g) Fuzzy AHP

Fuzzy AHP is the fuzzification of AHP (analytic hierarchy process), which was developed for use in the same types of evaluation as the fuzzy integral in Section (f). When there are evaluations of a number of objects, weights for each evaluation item are determined through pair comparison, and using this, the method for finding the overall evaluation value for each object is the same as for (f); however, weights are expressed in terms of possibility and inevitability measures, the condition of having one sum for each weight relaxed and eigenvectors used to find the weights. In addition, this method is not only for evaluation problems for management and business, but also for selecting optimal proposals from large numbers of alternatives in wide-ranging multiobjective programming problems and decision-making based on this. This method is described in Section 5.3.

(h) Fuzzy Mathematical Programming

Mathematical programming is often used for programming and decision-making at various levels of management and production, but in these cases clear objective functions and restrictions are used. In real problems, there is leeway in the functions that express profit and loss and the restrictions that express things such as possible investment moneys, and a better flowing consideration would often be possible if these were ambiguous. Various forms of fuzzy mathematical programming have been proposed to attend to this situation.

(i) Fuzzy Multiobjective Programming

In management and production programming and decision-making, there are many cases in which a solution that satisfies a large number of diametrically opposed objectives must be found. For example, the quality, price, and delivery of a product to be supplied to the market stand in opposition, and there is no way to satisfy them all at the same time, so there must be a compromise. Furthermore, when actual problems are dealt with, there is ambiguity in the objective functions and restrictions. The fuzzy AHP described in Section (g) and the fuzzy mathematical programming in (h) can be used for the preceding kinds of problems, but fuzzy multiobjective programming has been proposed as a method that gains a high level of similarity to real systems by identifying the membership functions that give the ambiguity of objective functions and restrictions through a trial-and-error process of continuing dialogue with the programmer or decision-maker. This method is described in Section 5.4.

(j) Fuzzy Multiattribute Decision-Making

As was mentioned in (f), (g), and (i)—fuzzy integrals, fuzzy AHP, and fuzzy multiobjective programming—when overall evaluation equations are linear combination equations in problems in which the most appropriate proposal is chosen from several alternatives that have a large number of items to be evaluated (attributes), multiattribute decision-making places importance on the way in which these are identified. Since in real problems the weight of each evaluation item and evaluation value is most often given in terms of a fuzzy number or linguistic value, fuzzy multiattribute decision-making was proposed as a method for these cases; it is described in Section 5.5.

(k) Fuzzy Statistical Decision-Making

Among the various kinds of decision-making there are for management, statistical decision-making is one in which, when the most appropriate action must be chosen from several possible actions, the expected effects of each are calculated, and the activity that gives the maximum value is chosen. In real problems actions take the form of ambiguities such as "If one must be chosen, I'd do this," and in addition, conditions are often given in ambiguities such as, "Things sell well." Fuzzy statistical decision-making extends statistical decision-making to such cases.

In the foregoing we have outlined the various methods that are used in business, placing importance on applications and characteristics, but there are other uses of basic fuzzy set theory and fuzzy reasoning in decision making and management, and these are shown in Table 5.1.

5.1.3 Characteristics of Business Applications

The points that differ from previous programming and management methods when fuzzy methods are used in business are the ability to bring in much of the judgment of the programmer or decision-maker and the ability to utilize ambiguous information such as language. For example, there is the use of membership functions, possibility, inevitability, and fuzzy numbers in things such as the determination of model coefficients and parameters, the weight of evaluation items and restrictions, and in correspondence to the ambiguity in information given, solutions can be given in ambiguous forms. Rather than forcing real systems and problems into similarities with simple models, modeling is carried out in a form that is true to nature with human participation, the answers obtained are given a range, and a form that people can easily understand is exhibited, a form that depends on their final judgments. In this way, fuzzy methods can be considered flexible methods that cooperate with people under the recent, ever-changing conditions of business. However, there are still many problems in the actual use of fuzzy methods that have yet to be solved, and there is a necessity for feedback from real experience for development.

5.2 MODELING OF LARGE-SCALE SYSTEMS

Economics problems, environmental problems, resource problems and the like are large-scale systems, and various variables are intertwined in a nonlinear form. GMDH [5] (group method of data handling), which was proposed by A. G. Ivankhnenko, is used for the modeling of this kind of nonlinear system. In this there is no prerequisite for knowledge about the structure of the system, and nonlinear system modeling can be carried out based on the principle of heuristic automatic systematization from input/output data. Here, for the most part, we will discuss fuzzy GMDH, in which the model parameters are identified using fuzzy numbers, and give some examples of applications. Prediction by means of fuzzy GMDH yields fuzzy numbers; therefore, it is suitable for modeling of ambiguous phenomena.

5.2.1 Interval Linear Regression Models

Fuzzy linear regression models and possibilistic linear regression models have already been formulated, but here we will explain *interval linear regression models*, which use interval concepts, as the simplest case of fuzzy numbers.

Interval linear systems are used as mathematical models for ambiguous phenomena. Here an interval linear system is a linear system in which the coefficients are intervals, so it can be expressed as follows:

$$Y = A_1 x_1 + \cdots + A_n x_n. \tag{5.1}$$

Here, x_i is an explanatory variable, and A_i is a interval. Using center α_i and width c_i, we can write

$$A_i = (\alpha_i, c_i). \tag{5.2}$$

A triangular fuzzy number A_i can be represented by center α_i and width c_i.

Interval output Y for Eq. (5.1) can be calculated as follows:

$$Y = (\Sigma \alpha_i x_i, \Sigma c_i |x_i|) = (\boldsymbol{\alpha}\mathbf{x}, \mathbf{c}|\mathbf{x}|). \tag{5.3}$$

Here, $\boldsymbol{\alpha} = (\alpha_1, \ldots, \alpha_n)$ and $\mathbf{c} = (c_1, \ldots, c_n)$ are row vectors, and $\mathbf{x} = (x_1, \ldots, x_n)$ is a column vector. To give a numerical example, we can calculate

$$(3,1) \times 2 + (4,2) \times (-1) = (2,4).$$

In other words, we get the interval with center 2 and width 4.

In addition, the inclusion relation between A_i and A_j is expressed by the following inequality:

$$\alpha_j - c_j \leqq \alpha_i - c_i,\ \alpha_j + c_j \geqq \alpha_i + c_i. \tag{5.4}$$

Let us discuss a method in which the modeling of an interval linear system yields an interval linear regression model. There are N samples of input/output data given, and these are expressed by (y_i, x_i), $i = 1, \dots, N$. Here y_i is output data or an observed variable, and $x_i = (x_{i1}, \dots, x_{in})$ is an input vector or explanatory variable. We let the estimated interval linear model be

$$Y_i^* = A_1^* x_n + \cdots + A_n^* x_{in}, \tag{5.5}$$

and we establish the following:

(1) Given observed value y_i is included in the estimated interval Y_i^*. In other words, for all i, $y_i \in Y_i^*$.

(2) The width of the estimated interval for the ith sample is $\Sigma_i c_i |x_{ij}| = \mathbf{c}|\mathbf{x}_i|$, and we wish to make this width small. Accordingly, we consider minimizing the total of the widths of the estimated intervals. In other words, this means finding the interval coefficient $A_i^* = (\alpha_i^*, c_i^*)$ that minimizes

$$S = \sum_{i=1}^{N} \mathbf{c}|\mathbf{x}_i| \tag{5.6}$$

Problems with requirements (1) and (2) can easily be transformed into the following linear programming problem:

$$\begin{cases} \min\limits_{A=(\alpha,c)} \sum\limits_{i=1}^{N} \mathbf{c}|\mathbf{x}_i|, \\ y_i \geqq \boldsymbol{\alpha}\mathbf{x}_i - \mathbf{c}|\mathbf{x}_i|, \\ y_i \leqq \boldsymbol{\alpha}\mathbf{x}_i + \mathbf{c}|\mathbf{x}_i|, \\ \mathbf{c} \geqq 0, \qquad i = 1, \dots, N. \end{cases} \tag{5.7}$$

To put it another way, by solving a linear programming problem in which (1) has been made into a restriction and (2) into an objective function, we can obtain the estimated interval $A_i^* = (\alpha_i^*, c_i^*)$.

Since we have returned the interval linear regression analysis to a linear programming problem, we can introduce expert knowledge concerning coefficients. For example, if there is vague knowledge about the number i coefficient and we let that be interval $B_i = (\beta_i, b_i)$, the estimated interval coefficient A_i should be included in B_i. In other words, we introduce the condition $A_i \subset B_i$. From Eq. (5.4) we can write this as

$$\beta_i - b_i \leqq \alpha_i - c_i, \qquad \beta_i + b_i \geqq \alpha_i + c_i. \tag{5.8}$$

If we solve the linear programming problem in Eq. (5.7) with the addition of this condition, we can find an estimated interval coefficient A_i^* that reflects expert knowledge.

From the preceding formulation, we can say the following:

(a) Since the given data is finite, a solution to the linear programming problem exists for any data given.

(b) As the number of pieces of data increases, the width of the estimated interval increases. When the number of pieces of data for analysis increases, it means that there is that much information, and this can be interpreted as meaning that the estimated possibilities are greater. This is antithetical to the characteristics of conventional linear regression analysis, in which the interval estimation becomes smaller as the number of pieces of data increases.

(c) When there is a dependent relationship between explanatory variables, the signs of the estimated coefficients are sometimes reversed in conventional regression analysis. If, in interval regression analysis, there is vague knowledge about a coefficient, that knowledge is introduced as a constraint condition in Eq. (5.8). Therefore, modeling that takes into consideration both expert knowledge and given data can be carried out.

(d) Our partial ignorance must be reflected in the modeling of ambiguous phenomena. Interval coefficients can be interpreted as expressions of partial ignorance.

(e) Since we can obtain an interval linear model that includes all given data, the data suggest possibility rather than the central tendency interpretations of conventional regression analysis.

5.2.2 Fuzzy GMDH

Next we will construct a fuzzy GMDH using an interval linear regression model. However, in order to make estimated interval Y a fuzzy number, we let it be a symmetrical triangular fuzzy number with center α and width c, and the membership function is expressed by $\mu_Y(y)$.

Now when observed data (y_i, x_i) are given, a number of partial linear models are constructed, and the method of obtaining an estimated equation for the model through a hierarchical combination of these is GMDH.

These partial linear models are called partial expressions, and here the following interval linear model is used as a partial expression:

$$Y = A_0 + A_1 x_p + A_2 x_q + A_3 x_p^2 + A_4 x_q^2 + A_5 x_p x_q. \tag{5.9}$$

Here, $A_0, \ldots, A_5$ are interval coefficients, Y the interval output $p, q = 1, \ldots, n$, $p \neq q$. When we let interval output Y be a fuzzy number, we let it be a triangular fuzzy number, and its membership function is expressed by $\mu_Y(y)$. Interval coefficients $A_0, \ldots, A_5$ are obtained by solving the linear programming problem in Eq. (5.7).

When a number of partial expressions have been obtained, it is necessary to establish an evaluation index for choosing the partial expressions for the best model. As an evaluation index for doing this, we define the following evaluation function:

$$J = \sum_{i=1}^{N} J_i^{(2)} \Big/ \sum_{i=1}^{N} J_i^{(1)}, \tag{5.10}$$

where we have

$$J_i^{(1)} = \mu_{Yi}(y_i), \qquad J_i^{(2)} = \mathbf{c}|\mathbf{x}_i|. \tag{5.11}$$

$J_i^{(1)}$ shows the degree of closeness of observed value y_i to the center of estimated interval Y_i. The higher the value for evaluation value $J_i^{(1)}$, the closer observed value y_i is to the center of estimated interval Y_i. $J_i^{(2)}$ shows the width of estimated interval Y_i. Therefore, the smaller the value of $J_i^{(2)}$, the smaller the width estimated, and this is considered good. In other words, we can say that a model with a smaller value of J bears a greater resemblance to given data (y_i, x_i).

Let us describe the fuzzy GMDH algorithm, as follows:

(1) *Step* 1: input variable x_{ij}, $i = 1, \ldots, N$, $j = 1, \ldots, m$, is determined for observed value y_i. If necessary, observed data (y_i, x_i) are normalized.

(2) *Step* 2: The correlation between observed value y and each input variable x_j is obtained, and only input variables x_j, $j = 1, \ldots, n$, $n \leqq m$, with large correlation coefficients are chosen.

(3) *Step* 3: Observed data (y_i, x_i) is divided into training data (afterwards TRD: N_t number of data) for obtaining an interval linear regression model and checking data (hereafter CHD: N_c number of data), which is used for choosing intermediate variables. The method of division is to make data for which the data number i is an integer multiple of 3 ($i = 3, 6, 9, \ldots$) and the last piece of data CHD; the rest are TRD.

(4) *Step* 4: Using TRD, partial expressions are found as follows for groups of two input variables x_p and x_q, and these are made up into interval linear systems:

$$Y_k = A_{0k} + A_{1k}x_p + A_{2k}x_q + A_{3k}x_p^2 + A_{4k}x_q^2 + A_{5k}x_px_q, \quad (5.12)$$

where Y_k, $k = 1, 2, \dots, n(n-1)/2$ expresses the kth partial expression, and interval coefficient $A = (\alpha, c)$ is obtained from the interval linear regression. Since estimated value Y_k is an interval, we let the value for the intermediate variable of the next level be

$$y_k = \alpha_{0k} + \alpha_{1k}x_p + \alpha_{2k}x_q + \alpha_{3k}x_p^2 + \alpha_{4k}x_q^2 + \alpha_{5k}x_px_q. \quad (5.13)$$

In other words, we only consider the center values of intervals.

(5) *Step* 5: CHD is transformed using Eq. (5.12) and interval coefficient A, which was obtained in step 4 using TRD. The similarity of partial expression Y_k to observed value y_i is evaluated using Eq. (5.10), and we let that value be J_k. r number of small evaluation values J_k are chosen as intermediate variables and are eliminated later. Using evaluation J_k, the threshold θ for that level is determined as follows:

$$\theta = \min_k J_k. \quad (5.14)$$

(6) *Step* 6: Using the intermediate variables $x_p = y_p$ and $x_q = y_q$ obtained in step 4, the partial expressions for the next level are constructed. Afterwards, steps 4–6 are repeated.

(7) *Step* 7: If, using evaluation values J_k from step 5, the threshold value θ_l for level l and the threshold value θ_{l+1} for level $l+1$ have the following relationship, the algorithm is terminated:

$$\theta_{l+1} = \min_s J_s \geqq \theta_l = \min J_k, \qquad l = 1, 2, \dots. \quad (5.15)$$

If the intermediate variables calculated for the previous layer are substituted in one after another, the estimated model is determined, and an interval linear model that is appropriate for the data is obtained.

5.2.3 Example of Fuzzy GMDH Applications

We will discuss the application [9] of fuzzy GMDH to dam-reservoir water-temperature prediction. Prediction of dam-reservoir water temperatures is very significant for the growth of wetland rice. Using time series data from July 1 to August 31, 1980 for temperature and water temperature for the Dai Ichi Yasaku Dam in Aichi Prefecture, an estimated model

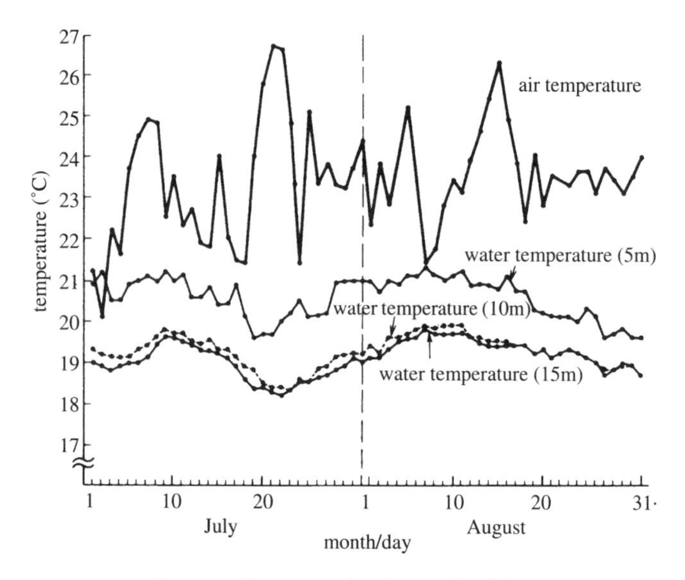

Fig. 5.1 Time series for observational data.

that expressed the structure of dam-reservoir water temperature was obtained from the July 1 to July 31 data. In addition, the estimated model obtained was used to predict water temperature for August. Figure 5.1 shows the data obtained, temperature, and water temperature (5 m, 10 m, 15 m). The structural equation for 10 m was

$$Y = A_0 + A_1 x_{21} x_{23} + A_2 x_{22} x_{25} + A_3 x_{22}^2 + A_4 x_{21} x_{23} x_{22}^2 + A_5 x_{21} x_{22} x_{23} x_{25} + A_6 x_{22}^2 x_{25}^2,$$

where $A_0 = (13.46, 0.3)$, $A_1 = (0.37 \times 10^{-2}, 0)$, $A_2 = (0.4 \times 10^{-2}, 0)$, $A_3 = (0.26 \times 10^{-2}, 0)$, $A_4 = (0.02 \times 10^{-4}, 0)$, $A_5 = (0.03 \times 10^{-4}, 0)$, and $A_6 = (0.77 \times 10^{-4}, 0)$, and in addition x_{2j} is observed values of water temperature (5 m) for x_{21}: previous day, x_{22}: 2 days prior, x_{23}: three days prior

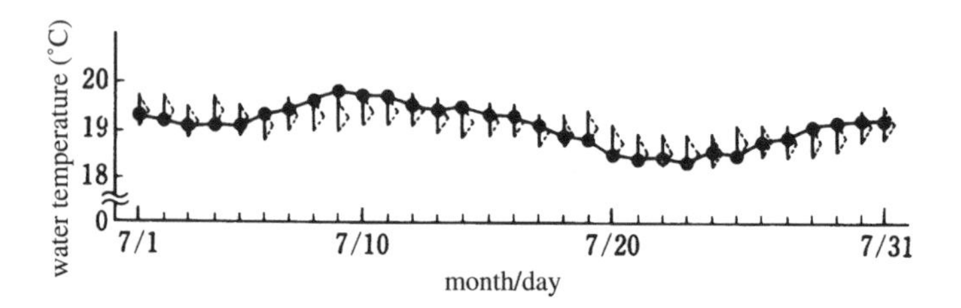

Fig. 5.2 Observational data and inferred fuzzy number (July, depth 10 m).

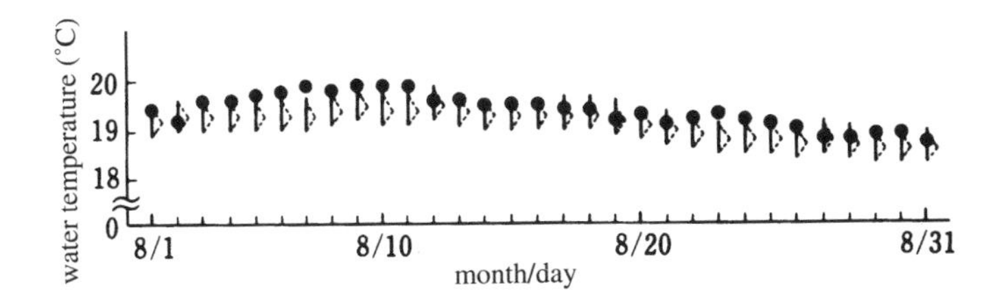

Fig. 5.3 Observational data and predictive fuzzy number (August, depth 10 m).

and x_{25}: 5 days prior. Figure 5.2 is a graph of the estimated values for water temperature at 10 m. In addition, Figure 5.3 shows the predicted values and observed values for water temperature at 10 m during August. The width of the predicted values for the period is small, 0.3°C, and of the 31 observed values 22 are included in the predicted values, so we can consider it to be a good model.

Finally, there have been experiments with using "if A then B" rules that are used in expert systems for the modeling of complex, large-scale systems [10], and this is a modeling method that makes use of the knowledge of specialists. Methods that use "if A then B" rules will probably be useful in the future for complex phenomena that cannot be interpreted by mathematical models alone.

5.3 MULTIOBJECTIVE EVALUATION

5.3.1 Fuzzy Evaluations Using AHP [11] (Analytic Hierarchy Process)

Evaluation is the clarification of the value of an object, based on a certain sense of value. In order to make this kind of evaluation, it is necessary to clarify the objectives and measurements of the attributes (function, performance, far-reaching effects, etc.) for the object. Since evaluation is the determining of the value of the object from the values for attributes, the weight of each attribute must be established. Evaluation objects are called alternatives.

Now we let there be n number of attributes, $x_1, x_2, \ldots, x_n$, and to these the decision-maker gives weights $\mathbf{W} = w_1, w_2, \ldots, w_n$. In this instance, the most preferred alternative y^* is expressed by

$$y^* = \left\{ y_j \middle| \min_j \sum_{i=1}^{n} w_i \cdot f_j(x_i) \right\}.$$

Here $f_j(x_i)$ is the value for the number i attribute for the number j alternative. In addition, the weight is standardized to $\Sigma_i w_i = 1$. Here, the problem becomes how to find weight $W = w_1, w_2, \ldots, w_n$. One method for this is the eigenvector method. Since this was proposed by Saaty in 1977, the problem-solving process, from creating hierarchies to final decisions, has been systematized as AHP (analytic hierarchy process).

Weight **W** in conventional AHP is standardized so that the sum is 1, but here we let **W** be a possibility measure on X. The conditions that **W** must satisfy are $0 \leqq w_j \leqq 1$, $i = 1, \ldots, n$, and that there must be at least one i such that $w_i = 1$. It is not necessary for $\Sigma_i w_i = 1$.

The decision-maker is asked, "What is the weight of attribute c as compared to d?" And a numerical value a_{cd}, like those shown in Table 5.2, that corresponds to the answer is obtained. We obtain Table 5.3. This shows an $n \times n$ matrix $[a_{cd}]$. In Table 5.3, it is assumed that $a_{cc} = 1$ and $a_{cd} = 1/a_{dc}$. If we let matrix $[a_{cd}]$ be **M**, the problem turns out to be the determination of eigenvector **W**, which corresponds to the maximum eigenvalue $\lambda_{\max}$ from the following equation such that $\max_i w_i = 1$:

$$(\mathbf{M} - \lambda \mathbf{I})\mathbf{W} = 0.$$

This can easily be found through a method that is used in the proof of the Peron–Frobenius theorem.

In AHP, weights are standardized so that the sum equals 1, and this means that it is an additive measure just like probability. In this system possibility and credibility measures, which are nonadditive measures from fuzzy theory, are used. In conventional AHP the substitutability and complementarity are not brought into consideration in evaluation from the additivity of the weight, but in this system, the decision-maker is shown two determinations that take into consideration both of these. *Substitutability* and *complementarity* have the same meaning as "substitutive" and "complementary" in the utility method. "Substitutive" means choosing alternatives that are especially excellent with regard to an attribute, and "complementary" means choosing ones that have no defects if at all

Table 5.2. a_{cd} Value: Correspondences to Linguistic Expressions

(compared to attribute d, attribute c is)	→	(a_{cd})
Equally important	→	1
Somewhat important	→	3
Quite important	→	5
Very important	→	7
Extremely important	→	9
Used as intermediate values		2, 4, 6, 8
$a_{cc} = 1$,	$a_{cd} = 1/a_{dc}$	

Table 5.3. a_{cd} Matrix Using Pair Comparison

	x_1	x_2	$\cdots$	x_n
x_1	a_{11}	a_{12}	$\cdots$	a_{1n}
x_2	a_{21}	a_{22}	$\cdots$	a_{2n}
$\vdots$	$\vdots$	$\vdots$		$\vdots$
x_n	a_{n1}	a_{n2}	$\cdots$	a_{nn}

possible. In their extremes these are max–max and max–min decisions, respectively. If we use nonadditive weight, we are relieved from the condition of independence among items that we have in the case of additive weight, and we can improve the rank reversal by adding a similar standard.

5.3.2 Evaluations Using Nonadditive Weights

One nonadditive measure is Dempster and Shafer's *Bel Measure* [12]. Bel measure expresses the degree of certainty of element $x \in X$ being a member of set A. This type of Bel measure is expressed by

$$\mathrm{Bel}(A) = \sum_{B \subset A} m(B), \qquad \forall A \subset X$$

using set function $m: 2^X \to [0, 1]$, which satisfies what is called the basic probability assignment

$$m(\phi) = 0, \qquad \sum_{A \subset X} m(A) = 1.$$

Set A such that $m(A) > 0$ is called the "focal element." Bel measure is defined using the basic probability, but it differs from probability in that it can express ignorance. Bel, for which $\mathrm{Bel}(A) = 0$ when $A \neq X$ and $\mathrm{Bel}(A) = 1$ when $A = X$, expresses complete ignorance.

In addition, *Pl measure* can be expressed by

$$\mathrm{Pl}(A) = \sum_{B \cap A \neq \phi} m(B), \qquad \forall A \subset X.$$

Bel is nonadditive, and so is Pl. The following relation arises between Bel and Pl:

$$\mathrm{Pl}(A) = 1 - \mathrm{Bel}(\overline{A}). \tag{5.16}$$

Another nonadditive measure is possibility measure Π, proposed by Zadeh. This is defined by

$$\Pi(X) = 1, \qquad \Pi(\phi) = 0$$

$$^{\forall}A, {}^{\forall}B, \quad \Pi(A \cup B) = \max(\Pi(A), \Pi(B)).$$

In addition, there is credibility measure N, which is a nonadditive measure and is the dual of Π. This is defined by

$$N(X) = 1, \qquad N(\phi) = 0$$

$$^{\forall}A, {}^{\forall}B, \quad N(A \cap B) = \min(N(A), N(B)).$$

N can be found from Π as

$$N(A) = 1 - \Pi(\bar{A}).$$

The relation corresponds to Eq. (5.16), a Bel and Pl equal to Π and N exist. Since focal element $A_1, A_2, \ldots, A_k$ are $A_1 A_2, \ldots, A_k$, we get

$$\text{Bel}(A \cap B) = \min(\text{Bel}(A), \text{Bel}(B)).$$

In this instance, since $\text{Pl}(A) = 1 - \text{Bel}(\bar{A})$, we get

$$\text{Pl}(A \cup B) = \max(\text{Pl}(A), \text{Pl}(B)).$$

In other words, Pl is the equivalent of a possibility measure. From the foregoing the basic probability can be found from possibility measure Π and then Bel and Pl can be found.

The expected value, using Bel and Pl, is given by the following Lebesque–Stieltjes integral. First, if we define upper bound distribution F^* and lower bound distribution F_* for real number function f on X as

$$F^*(v) = \text{Pl}(f \leqq v),$$

$$F_*(v) = \text{Bel}(f \leqq v),$$

and express the focal element by A, we find them with

$$E^*(f) = \int_{-\infty}^{\infty} v\, dF_*(v) = \sum_{A \subset X} m(A) \cdot \max_{x \in A} f(x), \tag{5.17}$$

$$E_*(f) = \int_{-\infty}^{\infty} v\, dF^*(v) = \sum_{A \subset X} m(A) \cdot \min_{x \in A} f(x). \tag{5.18}$$

Here E^* is the Bel expected value (upper bound) and E_* the Pl expected value (lower bound).

In the case of possibility measure Π and credibility measure N, an equivalent Pl and Bel also exist, so we can express the distribution functions of these measures as we did in Eqs. (5.17) and (5.18), if we employ the Lebesque–Stieltjes integral. In addition, a

$$\min f(x) \leqq E_*(f) \leqq E^*(f) \leqq \max f(x)$$

order arises. The equality holds on left- and rightmost sides when the focal element is only universal set X.

Let us consider w_i to be the degree of importance expressed by the possibility measure for the number i attribute. Now let us say that w_i has q number of values, $r_1 < r_2 < \cdots < r_q$; set A_l with q members is determined as

$$A_l = \{x_i | w_i \geqq r_l\}, \qquad l = 1, \ldots, q,$$

and we let basic probability m such that $r_0 = 0$ be

$$m(A_l) = r_l - r_{l-1}, \qquad l = 1, \ldots, q,$$

for focal element A.

Once the basic probability is determined, upper expected value E^* and lower expected value E_* are

$$E^*(f) = \int_0^1 v\, dF_*(v) = \sum_{l=1}^{q} (r_l - r_{l-1}) \cdot \max_{x \in A_l} f(x),$$

$$E_*(f) = \int_0^1 v\, dF^*(v) = \sum_{l=1}^{q} (r_l - r_{l-1}) \cdot \min_{x \in A_l} f(x).$$

The most preferred alternative, in the case of using Bel when we let f_j be the evaluation value for the number j alternative, is

$$y^* = \left\{ y_j | \max_j E^*(f_j) \right\},$$

and if in the Pl case we call it y_*, we get

$$y_* = \left\{ y_j | \max_j E_*(f_j) \right\}.$$

When $w_i = 1$ for all i, we get respective max–max and max–min decisions.

If the decision-maker prefers a substitutive determination, y^* is chosen, and for a complementary determination, y_* is chosen. A conventional AHP case will give an intermediate value.

5.3.3 Application: Used Automobile Selection

We will discuss a real example in which a pilot system was developed using a personal computer at a used car center in the Kansai Region of Japan. Since used cars, unlike new cars, add the usage conditions of the previous user, there are more attributes in the decision-making process for purchasing a used car than in that for a new one, and furthermore, the attributes are subjective things such as the dirtiness of the inside of the car. The customer looks around at as many cars as possible, and decides on the used car that is closest to what he wants. In reality, however, there is a limit to walking around and looking with customers scattered around the sales lot. As a policy for improving on past sales, the used car center put all of the cars on the lot into a data base, and introduced customers to cars that seemed to meet their desires through a computer by showing a photograph of the car on the monitor. There were three attributes for introducing the appropriate car in this system: car model, year, and price. Because of this, there was room for improvement as a system for supporting the original decision-making process, even though it was appropriate as a means for customer acquisition. Under actual conditions, there were many cases in which the salesperson introduced a car through conversations with the customer and his own experienced sense, as had been done all along.

At this point the authors developed a computer system using fuzzy theory, AHP, and a personal computer that could be used within the range of sales activities. The evaluation items that a customer uses when purchasing a used car are price, year, amount of exhaust, mileage, style, grade, interior damage, exterior damage, and equipment included. The content of these evaluation items is indicated in ways such as, for mileage, "between 10,000 and 30,000 km," or, for style, "sedan." Therefore, we let price, year, mileage, interior damage, and exterior damage (exterior appearance) be the former type of evaluation item and amount of exhaust, style, grade, and equipment be the latter; the car that the customer desires is first searched for using the latter items. Next, the authors determined the weight of each evaluation item using the eigenvector method, and overall evaluation was set up as the weighted mean of membership values and Bel and Pl expected values.

Now we will give one example. First, the weight of the evaluation items is calculated, and they are rearranged in order starting with the smallest; we get

$$W = (0.075, 0.215, 0.240, 0.706, 1.000).$$

In this case the order of evaluation items is year < exterior appearance < equipment < mileage < budget. Since there are five values for w_i, there are five A_1 sets, and the basic probabilities given to these are

$$A_1 = X = \{x_1, X_2, x_3, x_4, x_5\}, \qquad m(X) = 0.075,$$

$$A_2 = \{X_2, x_3, x_4, x_5\}, \qquad m(A_2) = 0.215 - 0.075 = 0.140,$$

$$A_3 = \{x_3, x_4, x_5\}, \qquad m(A_3) = 0.250 - 0.215 = 0.024,$$

$$A_4 = \{x_4, x_5\}, \qquad m(A_4) = 0.706 - 0.240 = 0.466,$$

$$A_5 = \{x_5\}, \qquad m(A_5) = 1.000 - 0.706 = 0.293.$$

Table 5.4. Points (Membership Values) for Each Attribute

Car Number/Make	Attribute: Year	Outside Appearance	Equipment	Mileage	Price
1701 Corolla	0.0	0.0	0.8	0.5	0.0
1718 Jimney	0.4	0.3	0.4	0.6	0.8
1719 Corona	0.0	0.7	0.8	0.6	0.3
1725 Bluebird	0.5	1.0	0.5	0.3	0.5
1738 Jimney	0.9	0.0	0.5	0.2	0.8
1749 Austere	0.0	0.0	0.5	0.7	0.0
1767 Bluebird	0.0	0.0	0.4	0.4	0.0
1780 Bluebird	0.0	0.7	0.9	0.9	0.0
1783 Corona	0.8	0.0	0.8	0.6	1.0
1785 Corolla	0.5	0.3	0.8	0.9	0.5

The membership values for each of the evaluation standards for the cars chosen and the overall evaluation points are shown in Table 5.4 and Table 5.5, respectively.

Car number 1793, which meets the requirements for the most important evaluation standard, "budget," has the first position under the additive and alternative evaluations. Even though its being a Corona is not such an important item, there is a fault in "outside appearance," so under complementary evaluation a Jimney, number 1718, takes the top position. In this way, using three different evaluation methods several alternatives are displayed, and it is easy to allow the customer to decide freely.

Table 5.5. Total Points and Ranking

Car	Additive Evaluation		Substitutive Evaluation		Complementary Evaluation	
Number/Make	Points	Ranking	Points	Ranking	Points	Ranking
1701 Corolla	0.244	10	0.425	10	0.0	7
1718 Jimney	0.632	3	0.800	3	0.589	1
1719 Corona	0.477	6	0.560	8	0.277	6
1725 Bluebird	0.485	5	0.608	6	0.359	7
1738 Jimney	0.505	4	0.808	2	0.333	5
1749 Austere	0.275	9	0.494	9	0.0	7
1767 Bluebird	0.328	8	0.565	7	0.0	7
1780 Bluebird	0.169	11	0.283	11	0.0	7
1783 Corona	0.749	1	1.0	1	0.588	2
1785 Corolla	0.639	2	0.783	4	0.457	3

5.4 MULTIOBJECTIVE PROGRAMMING

5.4.1 Multiobjective Programming with Ambiguous Information

Based on the diversification of societal demands in recent years, there are greater expectations for multiobjective programming, which takes into consideration a number of mutually conflicting objectives at the same time, than for single-objective mathematical programming, and there is a great deal of active research, not just on the theoretical side, but also on the practical side [13–16]. For multiobjective programming (MOP) problems that simultaneously optimize a number of mutually conflicting objective functions, there generally exists no (completely) optimal solution that optimizes a number of objective functions at the same time; therefore, in order to improve a certain objective function, a solution for which at least one other objective function must be sacrificed is introduced. This kind of solution is called a Pareto optimal or a noninferior solution, and it can be found by transforming the original problem by some means and solving the scalarizing problem obtained [13, 14, 16].

However, since there are generally an infinite number of Pareto optimal solutions, a decision-maker (DM), who is a human being in a real decision-making situation, must choose a reasonable final solution from the set of Pareto optimal solutions, based on his or her own preferences. However, preference functions that thoroughly reflect the preference structure of the DM are unknown at this point, and in addition, there are many cases in which it is difficult to directly identify one. Therefore the preference function for the DM is not identified in general; a preference solution for the DM is derived based on local preference information that can be obtained through interaction, and many of these so-called interactive methods have been proposed [13, 14, 16].

However, if we consider the ambiguity of human judgment, we can assume that the DM has a fuzzy goal such as "We want to basically keep each objective function under a certain value" for multiobjective programming problems, which are formulated as vector minimization problems. This kind of fuzzy mathematical programming was actually first proposed by Tanaka *et al.* [17] in 1974, but taking the multiobjective programming viewpoint, Zimmermann [18], in 1978, showed that we get a linear programming problem in cases of multiobjective linear programming in which the DM's fuzzy goals are expressed by linear membership functions, if we follow the maximizing decision [19] of Bellman and Zadeh.

However, in the past, fuzzy programming methods assumed that the DM would tacitly follow the maximizing decision, and interaction with the DM after the determination of membership functions for each objective function was not considered. In order to overcome this kind of problem, interactive fuzzy satisficing approaches—in which, after assessing the DM's fuzzy goals for each of the objective functions in the multiobjective programming problem, the reference membership values established by the DM are updated through interaction and a solution that satisfies the DM is derived—have been proposed in linear, linear fractional, and nonlinear cases, and examples of applications have also appeared [20–23].

However, in cases in which real decision-making situations with their complex interdependencies are formulated as multiobjective programming problems, we can generally, almost enough to say "always," predict that a large number of parameters will be included. Up to now, the experts who participated in the formulation of the problem would establish values for these kinds of parameters through some sort of subjective method or some method learned from experience, but this can be thought of as being insufficient for the modeling of real systems. In order to give experts' ideas about the parameters included in this kind of multiobjective programming problem more appropriate expression, rather than setting certain values by some sort of method as was done in the past, we can expect that it would be possible to have something more appropriately similar to a real system, if formulation as a problem in which the parameters included took the form of fuzzy sets were carried out. From this point of view, Tanaka *et al.* [24–26] have proposed a solution method that takes into consideration the ambiguity of coefficients based on linear programming methods, in cases in which the objective functions and constraints of multiobjective linear programming problems are expressed by trapezoidal fuzzy numbers. For multiobjective nonlinear programming problems that include fuzzy parameters, Orlovski [27] has shown that, based on the idea that these are problems for which a solution is found in which the degree of DM preference attains or is greater than a certain level, they can be transformed into ordinary multiobjective nonlinear programming problems, using the extension principle.

In these situations, there have been proposals not only for interactive decision-making methods [28–30] in which a solution takes into consideration the ambiguity of expert judgments concerning multiobjective programming problems that include fuzzy parameters, but also, for all cases—nonlinear, linear, and linear fractional—for interactive fuzzy satisficing methods for finding satisfactory solutions that take into consideration DM judgments in cases where the DM has ambiguous goals for each of the objective functions [31–33]. In addition, multiobjective programming problems that include fuzzy parameters have been examined in recent years by Luhandjula *et al.* [34] from the standpoint of problems that include possibility coefficients.

Since we cannot introduce all of this research in this section because of the limitations of getting it all on paper, we will only give a general outline of an interactive fuzzy satisficing method that simultaneously takes into consideration extension to cases that include fuzzy parameters in solution concepts for multiobjective programming problems, ambiguity in expert judgments concerning multiobjective programming problems that include fuzzy parameters, and DM fuzzy goals.

5.4.2 Pareto Optimality and α-Pareto Optimality

In general, multiobjective programming (MOP) problems are defined as problems in which a number of mutually conflicting objective functions are in some way optimized, under the given constraints, and formally they are defined as minimization problems for vectors such as the following:

$$\min f(x) = (f_1(x), f_2(x), \ldots, f_k(x))$$

$$\text{subject to } x \in X = \{x \in E^n | g_i(x) \leqq 0, j = 1, \ldots, m\}.$$

Here, x is an n-dimensional decision variable vector; $f_1(x), \ldots, f_k(x)$, the k number of mutually conflicting objective functions; $g_1(x), \ldots, g_m(x)$, the m number of constraint functions; and X the constraint set.

Since in MOP the objective functions are vectors, instead of an optimal solution for ordinal scalar objective functions, a solution in which the value for one objective function cannot help but deteriorate if the value for another is improved, in other words a Pareto optimality, is defined.

Definition 5.1 (Pareto Optimality). *When no $x \in X$ exists for $x^* \in X$, $f(x) \leqq f(x^*)$ and $f(x) \neq f(x^*)$, x^* is a Pareto optimal solution.*

However, from the point of view that, in contrast to the MOP used up to now, it is possible to have more appropriate formulation if each of the parameters for constraints and objective functions reflects the ambiguity of the human judgments of the experts that participate in formulation, the following kind of fuzzy multiobjective programming (FMOP), which includes fuzzy numbers, has come to be formulated:

$$\min f(x, \tilde{a}) = (f_1(x, \tilde{a}_1), f_2(x, \tilde{a}_2), \ldots, f_k(x, \tilde{a}_k)),$$

$$\text{subject to } x \in X(\tilde{b}) = \left\{x \in E^n | g_j\left(x, \tilde{b}_j\right) \leqq 0, j = 1, \ldots, m\right\}.$$

Here $\tilde{a}_i$ and $\tilde{b}_j$ are fuzzy vectors that include the number i objective function and the number j constraint, respectively, and for convenience they are written $\mu_{\tilde{a}i}(a_i)$ and $\mu_{\tilde{b}j}(b_j)$, respectively.

Since in FMOP the objective functions and constraints are characterized as fuzzy numbers, the Pareto optimalities that have been defined for MOP up to now cannot be used. Therefore, we introduce an α-level set such that the values for the membership functions for all of the fuzzy number vectors included are on α, and for convenience we write

$$L(\alpha) = \{(a,b) | \mu_{\tilde{a}i}(a_i) \geqq \alpha, \mu_{\tilde{b}j}(b_j) \geqq \alpha, i = 1,\dots,k, j = 1,\dots,m\}.$$

If we fix α at some permissible level value here, the parameters that include the objective functions and constraints have a certain width that depends on α; let us look at how this can be formulated as the following nonfuzzy α-multiobjective programming (α-MOP) problem:

$$\begin{aligned} &\min f(x,a) = (f_1(x,a_1), f_2(x,a_2), \dots, f_k(x,a_k)) \\ &\text{subject to } x \in X(b) = \{x \in E^n | g_j(x,b_j) \leqq 0, j = 1,\dots,m\}, \\ &\qquad (a,b) \in L(\alpha). \end{aligned}$$

In this instance, we can define an α-Pareto optimality that takes into consideration the ambiguity included for the α-MOP problem through a forthright extension of the ordinal Pareto optimality.

Definition 5.2 (α-Pareto Optimality). *When $x \in X(b), (a,b) \in L(\alpha)$ such that $f(x,a) \leqq f(x^*,a^*)$ and $f(x,a) \neq f(x^*,a^*)$ does not exist for an α-MOP, $x^* \in X(b^*)$ is called the α-Pareto optimality and (a^*,b^*) the α-level optimality coefficient.*

There have been proposals for interactive methods that bring out satisfactory solutions for the DM and take into account the ambiguity included in the problem, by upgrading α and reference (objective function) values from the set of this kind of α-Pareto optimalities through DM interaction [28–30].

However, if we consider the ambiguity of human judgment, it appears that the DM might have fuzzy goals for each of the α-MOP objective functions. In a minimalization problem, the DM might, for example, have a fuzzy goal (fuzzy min) such as, "I would like to keep objective function f_i below about A_i." This type of fuzzy goal is quantified by the DM's subjectively stipulating monotonically decreasing membership function $\mu_i(f_i(x,a_i))$, which concerns objective function f_i. For a maximization problem, DM's fuzzy goal (fuzzy max) of "I would like to keep objective function f_i above about B_i" is stipulated by means of monotonically increasing membership function $\mu_i(f_i(x,a_i))$, which concerns objective function f_i. In addition, as a more general case, cases in which the DM has a fuzzy goal (fuzzy equal) such as "I want objective function f_i to be about c_i" can be dealt with in the fuzzy approach. We let this kind of generalized α-MOP (G α-MOP) be expressed by the following [31]:

$$\begin{aligned} &\text{fuzzy min } f_i(x, a_i) && i \in I_1, \\ &\text{fuzzy max } f_i(x, a_i) && i \in I_2, \\ &\text{fuzzy equal } f_i(x, a_i) && i \in I_3, \\ &\text{subject to } x \in X(b), && (a, b) \in L(\alpha). \end{aligned}$$

Here $I_1 \cup I_2 \cup I_3 = (1, 2, \ldots, k)$.

For problems that have fuzzy equal objective functions, the following M-α-pareto optimality is defined, based on a magnitude relation for membership functions, rather than a magnitude relation for objective functions.

Definition 5.3 (M-α-Pareto Optimality). *When $x \in X(b)$, $(a, b) \in L(\alpha)$ such that $\mu(f(x, a)) \leqq \mu(f(x^*, a^*))$ and $\mu(f(x, a)) \neq \mu(f(x^*, a))$ does not exist for an $\alpha - MOP$, $x \in X(b^*)$ is called the α-level optimality and $(a^*, b^*) \in L(\alpha)$ is the α-level optimality coefficient. Here $\mu(f(x, a)) = (\mu_1(f_1(x, a_1)), \mu_2(f_2(x, a_2)), \ldots, \mu_k(f_k(x, a_k)))$.*

5.4.3 Interactive Fuzzy Satisficing Method

If the DM subjectively determines the membership functions for each of the objective functions, the decision-making problem for the fuzzy multiobjective programming problem can be defined as follows:

$$\begin{aligned} &\max \mu_D(\mu(f(x, a)), \alpha) \\ &\text{subject to } (x, a, b) \in P(\alpha), \qquad \alpha \in [0, 1]. \end{aligned}$$

Here $\mu_D(\cdot)$ is the conjunction function that expresses the DM's preference structure, and $P(\alpha)$ is the set of α-level optimality coefficients that correspond to the M-α-Pareto optimalities for Gα-MOP. If $\mu_D(\cdot)$ can be explicitly identified, this problem becomes an ordinary single-objective programming problem. However, since it is generally difficult to make a wide-ranging identification of $\mu_D(\cdot)$, it is necessary for the DM to find a solution that satisfies him through interaction.

Since, as is clear from the definition, the M-α-Pareto optimalities are normally a set of an infinite number of points, the DM must by some subjective judgment extract a satisfactory solution from the M-α-Pareto optimality set. Here the M-α-Pareto optimalities that are candidates for a satisfactory solution for the DM, which is close to being a min–max operation on the membership values $\overline{\mu}_i$ he has chosen, can be found by solving the following augmented min–max problem:

$$\min_{\substack{x \in X(b) \\ (a,b) \in L(a)}} \max_{1 \leqq i \leqq k} \left\{ \overline{\mu}_i - \mu_i(f_i(x, a_i)) + \rho \sum_{i=1}^{k} (\overline{\mu}_i - \mu_i(f_i(x, a_i))) \right\}.$$

Here ρ is a sufficiently small positive number.

When the M-α-Pareto optimal solutions obtained by solving the augmented min–max problem do not satisfy the DM, the DM must subjectively revise the values of reference membership value $\overline{\mu}_i$ and α, but the trade-off ratios among the membership functions in the current M-α-Pareto optimal solution and the trade-off ratio between α and membership functions can be thought of as useful information for doing this. Fortunately, this trade-off ratio information can easily be found by using Lagrange's multiplier for the constraint in the augmented min–max problem.

Based on the preceding discussion, an interactive algorithm for finding a solution satisfactory to the DM as guaranteed by M-α-Pareto optimality can be constructed as follows [31–33]. The steps marked by * indicate interaction with the DM.

(1) *Step* 1: For $\alpha = 0, 1$, calculate the individual maximum and minimum values for each objective function under the given constraints.
(2) *Step* 2*: Considering the individual maximum and minimum values for each objective function, the DM subjectively determines each membership function.
(3) *Step* 3*: The DM establishes a value for the initial α $(0 \leqq \alpha \leqq 1)$, and the initial reference membership value is set at 1.
(4) *Step* 4: An augmented min–max problem is solved for the α and membership values established, and the trade-off ratios between the corresponding M-α-Pareto optimality and each membership function and trade-off values between α and the membership values are found.
(5) *Step* 5: If the current M-α-Pareto optimal solution and α value are satisfactory, termination. If not, the reference membership value and α value are updated, considering the current membership function values and the trade-off information, and we return to Step 4.

5.4.4 Conclusion

In this section we have given an outline of an interactive fuzzy satisficing method in which solutions satisfactory to the DM are found within a set of M-α-Pareto optimalities that take into consideration the DM's fuzzy goals after having transformed a fuzzy multiobjective programming problem that was formulated in consideration of the ambiguity of expert judgment into a nonfuzzy α-multiobjective programming problem, in order to bring the ambiguity of human judgment into multiobjective programming. For numerical examples, application examples, and other methods that we did not have space for here, please see the corresponding references. In the future, we hope for improvements in this method through being able to formulate fuzzy multiobjective programming problems that reflect actual situations with the cooperation of experts in various fields and through an accumulation of actual applications of the method described in this section.

5.5 MULTIATTRIBUTE DECISION-MAKING

5.5.1 Introduction

In this section we will describe the method for using a fuzzy model to identify a multiattribute decision-making process. Decision-making activities are the actions taken in order to choose an appropriate alternative from those for realizing a certain goal. Activities such as (1) determining the set of alternatives, (2) evaluating alternatives, and (3) comparing alternatives are discussed in decision-making theory. When there are several objectives to be realized in making a decision, the decision-making is known as multiobjective decision-making.

The more difficult the decision-making, the more necessary it becomes to have the knowledge of experts. The problem then becomes just how to make a model of the decision-making processes of experts. The determination of an evaluation structure for selecting alternatives and the establishment of selection standards are especially important. In general, the evaluation of alternatives is carried out based on several attributes of the object, and this kind of decision-making is called multiattribute decision-making.

The given attribute information can be divided into numerical data, qualitative data, and linguistic data, and methods of decision-making for each of these have been proposed; however, in cases in which people follow their intuition in making decisions, linguistic expressions come closer to the human sensibility than numerical expressions. Here we will concentrate our discussion on decision-making methods that use linguistic expressions.

5.5.2 Structure of Multiattribute Decision-Making

(a) Statistical Approach

Methods for identifying evaluation structures for multiattribute decision-making based on numerical information can be divided into those that make use of probability and those that make use of possibility. In other words, in cases of multiattribute measurement and evaluation of alternatives that have the addition of probabilistic error (noise), probabilistic statistical methods are employed to identify the structures. In particular, regression analysis [37, 48] is often used for identification of linear structures. On the other hand, discriminant analysis is used in cases for which the evaluations of alternatives for decision-making are given in classes or groups [36].

(b) Possibility Approach

However, if we think of the data given as containing no noise or error, we must take into consideration all of the possibilities that are indicated by the data. In other words, we must understand the values for the evaluations obtained for each of the alternatives as possibilities. Possibilistic regression analysis [39, 40], into which the concept of possibility measure has been introduced, is used in the modeling for this.

Let us consider a simple example in order to gain an understanding of the concept of possibility. For example, if we take the example of the number of times a person bathes each day, he or she probably bathes every day. However, the probability of bathing 10 or more times a day is generally close to 0. This is the concept of statistical probability. However, when we consider the bathing possibilities for this person, it is possible that he or she could bathe 10 or more times. This figure expresses the possibility.

In order to express this kind of possibilistic concept, Zadeh defined possibility measure using and interpreting fuzzy sets [45].

(c) Multiattribute Evaluation Structure

The identification problem for the multiattribute decision-making structure that we are dealing with here means determining an evaluation structure for the multiattribute decision-making from multiattribute data X_i $(i = 1, 2, \ldots, k)$ given in Table 5.6 and alternative evaluations Y. In

Table 5.6. Multiattribute Evaluation Data

Alternative Number	Alternative Evaluation	Evaluation of Alternative Attributes				
j	Y	X_1	$\cdots$	X_i	$\cdots$	X_k
1	y_1	x_{11}	$\cdots$	x_{i1}	$\cdots$	x_{k1}
2	y_2	x_{12}	$\cdots$	x_{i2}	$\cdots$	x_{k2}
$\vdots$	$\vdots$	$\vdots$	$\cdots$	$\vdots$	$\cdots$	$\vdots$
j	y_j	x_{1j}	$\cdots$	x_{ij}	$\cdots$	x_{kj}
$\vdots$	$\vdots$	$\vdots$	$\cdots$	$\vdots$	$\cdots$	$\vdots$
n	y_n	x_{1n}	$\cdots$	x_{in}	$\cdots$	x_{kn}

other words, we assume that multiattribute evaluation is carried out by the linear equation

$$Y = A_i \cdot X_1 + A_2 \cdot X_2 + \cdots + A_k \cdot X_k, \tag{5.19}$$

and this is the determination of the weight of each attribute.

In Table 5.6, X_{ij} is the value for attribute i for alternative j. Y_j indicates the evaluation of alternative j. Let us say that there are values for j from 1 to n and for i from 1 to k. Each value is obtained as a numerical value or a linguistic expression based on their respective problems.

Here, we determine the coefficient A_i for linear multiattribute evaluation equation (5.19) that best estimates the evaluation of the alternative for the given object. In order to make writing the problem simple, we introduce the vector expression

$$Y' = [y_1, y_2, \ldots, y_n],$$

$$\mathbf{X} = \begin{bmatrix} x_{11} & \cdots & x_{i1} & \cdots & x_{k1} \\ \vdots & & \vdots & & \vdots \\ x_{1n} & \cdots & x_{in} & \cdots & x_{kn} \end{bmatrix},$$

$$A' = [a_1, a_2, \ldots, a_k].$$

In order to explain possibilistic regression analysis, we will use triangular fuzzy numbers. Triangular fuzzy number A is

$$\mu_A(x) \begin{cases} 1 - \dfrac{|a - x|}{c}; & a - c \leqq x \leqq a + c \\ 0; & \text{others} \end{cases}.$$

A is a fuzzy number with center a and width c. Here we will write $A = (a, c)$.

The equation for possibilistic linear multiattribute evaluation can be expressed by

$$Y = A_1 \cdot X_1 + A_2 \cdot X_2 + \cdots + A_n \cdot X_n,$$

and using the extension principal, its membership function can be calculated by

$$\mu_Y(y) = \begin{cases} 1 - \dfrac{|y - \mathbf{x}^t\mathbf{a}|}{\mathbf{c}^t|\mathbf{x}|}; & \mathbf{x} \neq 0, \\ 1; & \mathbf{x} = 0,\ y = 0, \\ 0; & \mathbf{x} = 0,\ y \neq 0. \end{cases}$$

Here, $x = (x_1, x_2, \ldots, x_n)$, $a = (a_1, a_2, \ldots, a_n)$, and $c = (c_1, c_2, \ldots, c_n)$; x^t indicates a transposition of vector x. However, y and A_i are fuzzy numbers. In addition, for y such that $c^t|x| < |y - x^t a|$, $\mu(y) = 0$.

In order to determine this kind of possibilistic evaluation function, a measure for minimizing the width of the possibility,

$$S = c_0 + \cdots + c_n,$$

is used. The possibilistic evaluation equation for this problem is determined by solving the following linear programming problem:

$$\min_{a,c} S = \min_{a,c}(c_0 + \cdots + c_n).$$

Here,

$$(1 - h)\Sigma c_j \cdot |x_{ij}| + \Sigma a_j \cdot x_{ij} \geqq y_i,$$

$$(1 - h)\Sigma c_j \cdot |x_{ij}| - \Sigma a_j \cdot x_{ij} \geqq -y_i, \qquad i = 1, \ldots, n.$$

Here h is a value on [0, 1] and indicates the congruence of the possibilistic regression model.

5.5.3 Decision-Making Models that Use Linguistic Expressions

In the following we will deal with the evaluation of attributes and alternatives that use linguistic expressions. First we will explain the handling of linguistic expressions that are called linguistic variables and make use of fuzzy sets.

(a) Linguistic Variables and Dictionary

Let us define the descriptive adjectives that express linguistic evaluations of attributes and alternatives and the linguistic variables that make correspondences with alternatives and attributes. Let Cartesian product $S \times A$ be the range of definition of set S of alternatives and set A of attributes for alternatives, and we will call the variables that make up the range of values for set D of descriptive adjectives linguistic variables $L(\alpha, i)$. The following are examples of linguistic variables:

$$L\ (\text{product A, popularity}) = \text{“good”}$$

$$L\ (\text{product B, design}) = \text{“bad”}$$

$$L\ (\text{product C, performance}) = \text{“extremely good”}$$

Here, linguistic variable $L(\alpha, i)$ shows the evaluation of alternative α concerning attribute i. These linguistic variables give evaluations of attributes and alternatives.

We can define the meanings of linguistic expressions made from descriptive adjectives like "good" and "bad" using fuzzy sets on interval [0, 1] [42]. These fuzzy numbers are written U "good" and U "bad." If we use possibility concept II, the possibility of popularity for product A can be expressed by fuzzy number U "good" for the value "good" for L (product A, popularity). We get

$$\begin{aligned} \text{II(product A, popularity)} &= \text{II(degree of popularity of product A)} \\ &= U\,\text{“good”}. \end{aligned}$$

Dictionary D for descriptive adjectives like "good" and "bad" is created by drawing correspondences between descriptive adjective L_i and fuzzy number U_i. In this instance, the correspondence between L_i and U_i is formed based on the concurrence of experts. In other words, if we obtain linguistic evaluation L_i ($i = 1, 2, \ldots, k$), its fuzzy evaluation is found as U_i, using dictionary D, which has been constructed by experts.

(b) Structure of Linguistic Evaluation

In multiattribute decision-making, evaluation of the attributes and characteristics of objects concerned with alternatives is carried out first, and next, the alternatives are evaluated based on these evaluations of attributes. Here the problem is constructing a method for identifying the process by which experts carry out evaluations of alternatives, based on the characteristics and attributes of evaluation objects.

Rephrasing this mathematically, we have to discuss the identification of linguistic evaluation process F;

$$Z = F(L_1, L_2, \ldots, K_k) \tag{5.20}$$

for the determination of linguistic expression Z for the evaluation of the alternative through linguistic variable $L_1, L_2, \ldots, L_k$ for the attribute.

We construct a model of linguistic evaluation process F from the following four parts (Fig. 5.4):

(1) The dictionary for drawing correspondences between descriptive adjectives and fuzzy numbers on interval [0, 1].

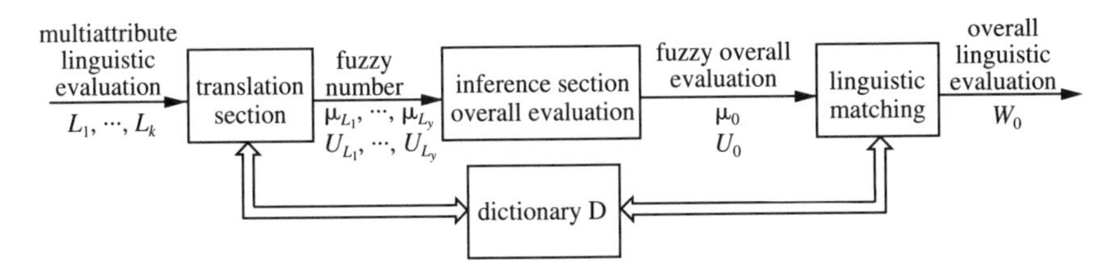

Fig. 5.4 Evaluation structure using linguistic information.

(2) Linguistic translation for translating linguistic variable value L_i for attribute i to fuzzy number U_i, based on dictionary D.

(3) Inference of overall evaluation from the fuzzy evaluation equation for determining fuzzy number U_0 for overall evaluation of the object, based on the fuzzy number for the attribute evaluations.

(4) Linguistic matching for matching dictionary D and the fuzzy overall evaluation and determining the appropriate linguistic expression. In this part, linguistic expression Z for the overall evaluation of the object is determined.

(c) Linguistic Translation and Linguistic Matching

Here we define the fuzzy number on interval [0, 1] that corresponds to the descriptive adjective with a triangular fuzzy number (a, c_1, c_r). a is the center, and c_1 and c_r are the width of ambiguity. At this time, the dictionary that gives correspondences between descriptive adjectives and fuzzy numbers is as follows (Table 5.7). When the attribute evaluation for a given object is L_0, (a, c_1, c_r) is retrieved for the fuzzy number from dictionary D, during linguistic translation. Fuzzy overall evaluation V is obtained from these fuzzy numbers $U_1, \ldots, U_k$ by means of fuzzy evaluation function f. Here, a descriptive adjective that corresponds to fuzzy overall evaluation U_0 must be searched for, in order to obtain the linguistic expression. In other words, by bringing fuzzy overall evaluation U_0 and the data base of dictionary D together, an appropriate word is determined.

Table 5.7. Dictionary

Descriptive Adjective	Fuzzy Number
Very high	(1.00, 0.20, 0.00)
High	(0.75, 0.15, 0.15)
Average	(0.50, 0.15, 0.15)
Low	(0.25, 0.15, 0.15)
Very low	(0.00, 0.00, 0.20)

This linguistic matching operation is defined by min–max operation

$$Z_0 = W_0 \in D \| [\max\{ \mu w_0(T) \wedge \mu_v(t)\} \\ \geqq \max\max\{ \mu_w(t) \wedge \mu_v(t)\}] \qquad (5.21)$$

Here $W_0 | P(W_i)$ means W_0 such that it satisfies condition P. μ_{U0} is the membership function of U_0, and in addition, μ_{Wi} is the membership function for word W, which is included in dictionary D.

Example 1. Let us consider the example of decision-making for the marketing for the selling of paintings. In order to evaluate target markets, two attributes, the living standard of people living in various cities and their cultural concern, are considered. Table 5.7 shows the dictionary for the words that are used.

Let us process this for fuzzy evaluation function f given by

$$V = f(U_{\text{Standard of Living}}, U_{\text{Cultural Concern}}) \\ = 0.3\ \mu_{\text{Standard of Living}} + 0.7\ \mu_{\text{Cultural Concern}} \qquad (5.22)$$

Here 0.3 and 0.7 are fuzzy numbers (0.3, 0.1, 0.1) and (0.7, 0.1, 0.1), respectively. Now let us say that a certain city's population is evaluated as a living standard and cultural concern of "very high" and "low." We have to evaluate this city as a market for the selling of paintings. The fuzzy numbers that correspond to "very high" and "low" in the dictionary in Table 5.7 are

$$U(\text{"very high"}) = (1.00, 0.20, 0.00), \\ U(\text{"low"}) = (0.25, 0.15, 0.15),$$

respectively. We can obtain the matching overall evaluation, fuzzy number (0.475, 0.255, 0.245), by substituting these into Eq. (5.22). By carrying out linguistic matching with this fuzzy number and the dictionary, we obtain the expression "normal," which corresponds to this fuzzy number (Fig. 5.5). The result is that the evaluation of this city as a market is "normal." Among these, the policies with the highest evaluations are employed.

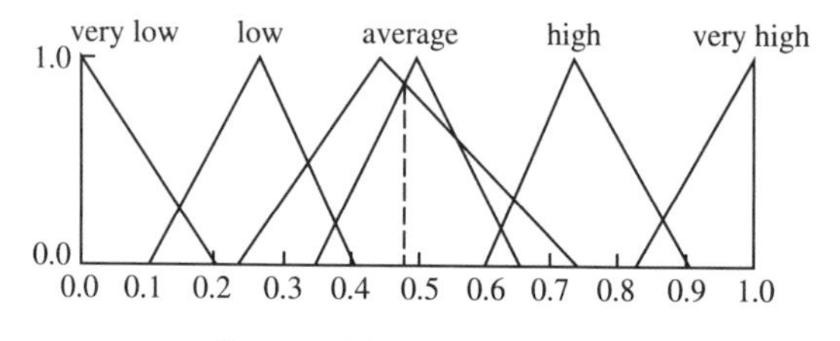

Fig. 5.5 Linguistic matching.

(d) Determination of the Fuzzy Multiattribute Evaluation Function

Above, we constructed the process for linguistic evaluation by experts. Here, we will discuss the method for determining fuzzy evaluation equation f for the linguistic evaluation model. Fuzzy evaluation function f is a function for converting the k number of fuzzy numbers that express attribute evaluations for the object into a fuzzy number for the overall evaluation of the object. Here we must determine this fuzzy evaluation function f in such a way that it imitates the evaluation process of experts, based on training data ω ($\omega = 1, 2, \ldots, n$) obtained from the expert evaluation process. In cases for which the variables are also fuzzy numbers, we cannot just solve a simple linear programming problem. A heuristic method is shown in Fig. 5.6.

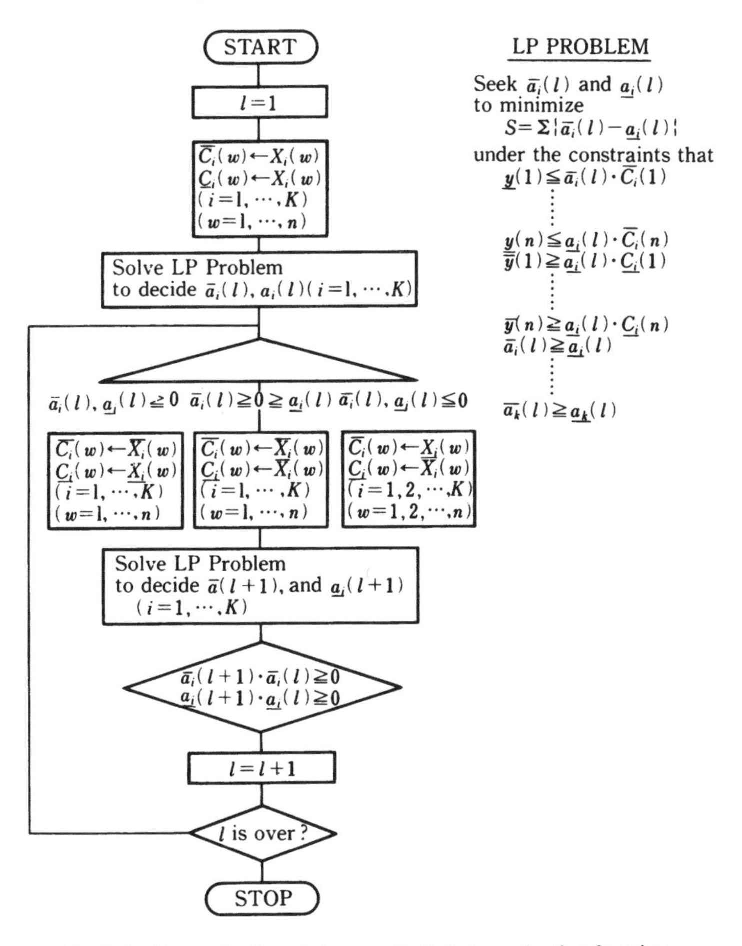

Fig. 5.6 Determination of fuzzy multiattribute evaluation function.

5.5.4 Conclusion

Here we have described identification methods for multiattribute evaluation structures for alternatives in the decision-making process. As far as linear multiattribute evaluation is concerned, possibilistic regression analysis is effective for cases in which attribute information is given in numerical values, and in addition, linguistic multivariate analysis, in which linguistic matching is introduced into multivariate analysis, is effective in cases in which they are given as linguistic expressions. In terms of applications of multiattribute decision-making using possibilistic regression analysis, the authors have used it in the identification of numbers of employees [43]; in addition, multiattribute decision-making that makes use of linguistic regression analysis has found applications concerning structural safety [1, 35, 41]. Chikio Hayashi's quantification theory and fuzzy quantification theory [1] are effective for evaluation structures for qualitative information.

REFERENCES

1. Terano, T., Asai, K., and Sugeno, M., eds. *Fuzzy Systems Theory and Its Applications* (trans. from the Japanese), Academic Press, Boston (1987).
2. Terano, T., *Introduction to Systems Engineering*, Kyoritsu Publishing (1985) (in Japanese).
3. Amagasa, M., *Theory of System Structure: Fuzzy Theory as a Foundation*, Moriyama Books (1987) (in Japanese).
4. Hayashi, C., *Quantification Methods*, Toyo Keizai Shinbun Publishing (1972) (in Japanese).
5. Tamura and Kondo, "Recent GMDH Methodology and Applications," *Operations Research*, 104–111 (February 1987) (in Japanese).
6. Tanaka, H., Hayashi, I., and Asai, K., "Formulation of Fuzzy GMDH," *Systems and Control* **30**(6), 581–587 (1986) (in Japanese).
7. Tanaka, H., Watada, J., and Hayashi, I., "Three Formulations for Fuzzy Linear Regression Analysis," *Transactions of the Society of Instrument and Control Engineers* **22**(10), 1051–1057 (1986) (in Japanese).
8. Tanaka, H., "Fuzzy Data Analysis by Possibilistic Linear Models," *International Journal of Fuzzy Sets and Systems* **24**, 363–375 (1987).
9. Hayashi, I., and Tanaka, H., "The Fuzzy GMDH Algorithm by Possibility Models and Its Application," *Fuzzy Sets and Systems* **36**, 245–258 (1990).
10. Takagi, T., and Sugeno, M., "Fuzzy Identification of Systems and Its Application to Modeling and Control," *IEEE Trans. SMC* **15**(1), 116–132 (1985).
11. Saaty, T. L., *The Analytic Hierarchy Process*, McGraw-Hill (1980).
12. Shafer, G., *A Mathematical Theory of Evidence*, Princeton University Press, Princeton, New Jersey (1976).
13. Chankong, V. and Haimes, Y. Y., *Multiobjective Decision Making: Theory and Methodology*, North-Holland (1983).
14. Seo, F. and Sakawa, M., *Multiple Criteria Decision Analysis in Regional Planning: Concepts, Methods and Applications*, D. Reidel Publishing Company (1988).
15. Sakawa, M., *Optimization of Linear Systems: From Single Objective to Multiobjective*, Morikita Publishing (1984) (in Japanese).

16. Sakawa, M., *Optimization of Linear Systems: From Single Objective to Multiobjective*, Morikita Publishing (1986) (in Japanese).
17. Tanaka, H., Okuda, T., and Asai, K., "On Fuzzy Mathematical Programming," *J. Cybernetics* **3**, 37–46 (1974).
18. Zimmermann, H. J., "Fuzzy Linear Programming and Linear Programming with Several Objective Functions," *Fuzzy Sets and Systems* **1**, 45–55 (1978).
19. Bellmann, R. E. and Zadeh, L. A., "Decision Making in a Fuzzy Environment," *Management Science* **17**, 141–164 (1970).
20. Sakawa, M., Yano, H., and Yamine, T., "An Interactive Fuzzy Satisficing Method for Multiobjective Linear Programming Problems and Its Applications," *IEEE Trans. on Systems, Man and Cybernetics* **SMC-17**, 645–661 (1987).
21. Sakawa, M., Yano, H., and Yamine, T., "Interactive Fuzzy Satisficing Method for Multiobjective Fractional Programming Problems," *Trans. Inst. of Electronics and Communication Engineers* **J69-A**, 32–41 (1986) (in Japanese).
22. Sakawa, M. and Yano, H., "An Interactive Fuzzy Satisficing Method Using Augmented Minimax Problems and Its Application to Environmental Systems," *IEEE Trans. on Systems, Man and Cybernetics* **SMC-15**, 720–729 (1985).
23. Sakawa, M., Narazaki, H., Konishi, M., Nose, K., and Morita, T., "A Fuzzy Satisficing Approach to Multiobjective Pass Scheduling for Hot Tandem Mills," in *Toward Interactive and Intelligent Decision Support Systems*, Vol. 1 (Y. Sawaragi *et al.*, eds.), pp. 363–373, Springer-Verlag (1987).
24. Tanaka, H. and Asai, K., "Formulation of Linear Programming Problems Using Fuzzy Functions," *Systems and Control* **24**, 351–357 (1981) (in Japanese).
25. Tanaka, H. and Asai, K., "Fuzzy Solution in Fuzzy Linear Programming Problems," *IEEE Trans. on Systems, Man and Cybernetics* **SMC-14**, 325–328 (1984).
26. Tanaka, H. and Asai, K., "Fuzzy Linear Programming Problems with Fuzzy Numbers," *Fuzzy Sets and Systems* **13**, 1–10 (1984).
27. Orlovski, S. A., "Multiobjective Programming Problems with Fuzzy Parameters," *Control and Cybernetics* **13**, 175–183 (1984).
28. Sakawa, M. and Yano, H., "Interactive Decision Making for Multiobjective Linear Programming Problems with Fuzzy Parameters," *Transactions of the Society of Instrument and Control Engineers* **22**, 162–167 (1986) (in Japanese).
29. Sakawa, M. and Yano, H., "Interactive Decision Making for Multiobjective Linear Programming Problems with Fuzzy Parameters," in *Large Scale Modeling and Interactive Decision Analysis*, Proceedings, Eisenach, GDR (G. Fandel *et al.*, eds.), pp. 88–96, Springer-Verlag (1986).
30. Sakawa, M. and Yano, H., "Interactive Decision Making for Multiobjective Linear Fractional Programming Problems with Fuzzy Parameters," *Cybernetics and Systems: An International Journal* **16**, 377–394 (1985).
31. Sakawa, M. and Yano, H., "Interactive Fuzzy Satisficing Method for Multiobjective Nonlinear Programming Problems with Fuzzy Parameters," *Trans. Inst. of Electronics and Communication Engineers* **J68-A**, 1038–1046 (1983) (in Japanese).
32. Sakawa, M. and Yano, H., "An Interactive Fuzzy Satisficing Method for Multiobjective Linear Programming Problems with Fuzzy Parameters," *Proceedings of the IFAC/IFORS Symposium, Large Scale Systems: Theory and Applications*, Zurich, Switzerland, pp. 497–502 (1986).
33. Sakawa, M. and Yano, H., "An Interactive Fuzzy Satisficing Method for Multiobjective Linear Fractional Programming Problems with Fuzzy Parameters," in *Toward Interactive and Intelligent Decision Support Systems*, Vol. 2 (Y. Sawaragi *et al.*, eds.), pp. 338–347, Springer-Verlag (1987).

34. Luhandjula, M. K., "Multiple Objective Programming Problems with Possibilistic Coefficients," *Fuzzy Sets and Systems* **21**, 135–145 (1987).
35. Hinkle, A., Watada, J., and Yao, J. T. P., "Linguistic Assessment of Fatigue in Welded Structures," Conference of NAFIPS, Louisiana, July 2–4 (1986).
36. Matsubara, *Fundamentals of Decision Making*, Asakura Books (1977) (in Japanese).
37. Takemura, ed., *Handbook of System Techniques*, Nihon Riko Publishing Group (1981) (in Japanese).
38. Takeuchi and Yanagii, *Fundamentals of Multivariate Analysis*, Toyo Keizai Shinjoho-sha (1972) (in Japanese).
39. Tanaka, H., Uejima, and Asai, K., "Linear Regression Model Using Fuzzy Functions," *J. Operations Res. Soc. of Japan* **25**(2) 162–174 (1982) (in Japanese).
40. Tanaka, H., Watada, J., and Hayashi, I., "Three Formulations of Fuzzy Linear Regression Analysis," *Studies from the Society for Measurement and Automatic Control* **22**(10), 1051–1057 (1986) (in Japanese).
41. Watada, J., "Linguistic Evaluating Methods," Lecture Materials for the 1984 Ambiguity Science Symposium, Society for the Study of Ambiguity Science, 9/24, Osaka (1984) (in Japanese).
42. Watada, J., Fu, K. S., and Yao, J. T. P., "Linguistic Assessment of Structural Damage," Technical Report, Purdue University, CE-STR-84-30 (1984).
43. Watada, J., Tanaka, H., and Shimomura, T., "Identification of Learning Curve Based on Possibilistic Concepts," in *Applications of Fuzzy Theory in Human Factors* (W. Karwowski and A. Mital, eds.), Elsevier, Amsterdam (1986).
44. Watada, J., "Identification of Multiattribute Evaluation Structures," *Ryugaya University Social Science Research Report*, No. 17 (1987) (in Japanese).
45. Zadeh, L. A., "The Concept of a Linguistic Variable and Its Application to Approximate Reasoning, Parts 1 & 2," *Information Sciences* **8**, 199–249; **8**, 301–351 (1975).
46. Zadeh, L. A., "Fuzzy Sets as a Basis for a Theory of Possibility," *Fuzzy Sets and Systems* **1**, 3–28 (1978).

Chapter 6

Fuzzy Computers and Software

When fuzzy theory is applied to actual problems, computers and their software are necessary as effective tools. The digital computers that have been used up to now can be used for this goal, but operations are slow, and they are especially inappropriate for control. Because of this, dedicated computers for fuzzy operations are being developed, and they have become partially operational. In addition, software for fuzzy operations has been created through extensions of Prolog and Lisp. Here we will center on introducing the present state of fuzzy computers and this software.

6.1 FUZZY COMPUTERS

6.1.1 Introduction

Human beings have made highly useful digital computers through the union of Boolean algebra (two-valued logic) with truth values of only 0 and 1 and switching devices (vacuum tubes and transistors) in an ingenious method known as the *stored program system*. These digital computers, through their robustness toward inherent noise and the expandability of systems, have built up the huge information-processing system society of today. During this time, the internal signal processing systems of computers have not emerged from the stored program system, and there have been no changes in their basic operations from their original development to today. The condition of being able to rapidly and deterministically process large amounts of information without mistakes is more than enough to show humans' weak points and incompleteness.

The greater the importance given to this tremendous capacity, the more occasions arise in which we find something lacking. This is especially obvious in the field of artificial intelligence. In this field, importance is placed more on development of logic than on numerical calculations, and the *if/then form of inference* is made much of. The inference currently carried out in artificial intelligence is based on a method of data matching called *unification*. Therefore, matching among pieces of deterministic data (data with clear outlines) is possible, and dedicated program languages and hardware have been developed for this. However, the information handled by human experts does not always have clear outlines. Rather, there are many cases in which most of the information does not have clear outlines. For example, there are cases such as, "*If a high fever persists for several days or more*, there is a *possibility* of pneumonia," and "*If the decrease in blood sugar is slow* after meals, insulin secretions are *low*." It is impossible to carry out unification between inputs and knowledge bases made up of this kind of *ambiguous linguistic information*. There is a requirement for methods that carry out "*soft matching*" between ambiguous information and ambiguous information and produce very appropriate conclusions, even if they are ambiguous. *Fuzzy reasoning* is certainly something that makes this possible, and it will probably take over from the methods employed in artificial intelligence up to now.

Digital computers are all-purpose machines, and if programs are devised, they can carry out fuzzy reasoning. The possibility of something and its being optimal are completely separate problems. And because of this, there is a desire for the appearance of hardware dedicated to fuzzy reasoning—in other words, a *fuzzy computer*. This is certainly a way to dissolve the dissatisfaction that people have with digital computers in one stroke, and since the internal structure of data, hardware architecture, and inference algorithms are totally different from those of inference machines up to now, these can be called *sixth-generation computers*. In the following we will discuss the important parts of the construction of fuzzy computers, their actual architecture, and the outlook for the future, etc.

6.1.2 Fuzzy Inference (Similarity Inference) and Fuzzy Computers

Fuzzy inference is the derivation of a new conclusion from inference rules stored in a knowledge base (just let these be called knowledge) and given facts, but it differs from normal inference in that all of the variables in the propositions are fuzzy variables; that is, they are constructed from ambiguous information. This can be written as follows:

$$\begin{array}{ll} \text{(knowledge)} & \text{If } x \text{ is } \mathbf{A}\text{, then } y \text{ is } \mathbf{B} \\ \underline{\text{(fact)}} & \underline{x \text{ is } \mathbf{A}'} \\ \text{(conclusion)} & y \text{ is } \mathbf{B}' \text{ [forward fuzzy inference].} \end{array} \tag{6.1}$$

In addition, we can think of the following kind of reasoning:

$$\begin{array}{ll} \text{(knowledge)} & \text{If } x \text{ is } \mathbf{A}\text{, then } y \text{ is } B \\ \underline{\text{(fact)}} & \underline{y \text{ is } \mathbf{B}'} \\ \text{(conclusion)} & x \text{ is } \mathbf{A}' \text{ [backward fuzzy inference]} \end{array} \tag{6.2}$$

Here **A**, **A′**, **B**, and **B′** are pieces of linguistic information with ambiguous outlines (*fuzzy sets*). Carrying out the inference in (6.1) and (6.2) would be extremely difficult using traditional (Boolean) logic.

Ambiguous linguistic information **A** is expressed by a characteristic function such as the one in Fig. 6.1a, which is known as a *membership function*. For convenience this is sampled as shown in Fig. 6.1b, and it is expressed in terms of the following set of ordered pairs:

$$\mathbf{A} = \{(x_i, a_i)\}, \qquad x_j \in X. \tag{6.3}$$

Here X is the total space and a_i is the degree to which x_i is a member of set **A** ("tall" in this case). For simplicity, a_i is a value from the space [0, 1]. Therefore, fuzzy set **A** can be seen as a vector and can be denoted as follows:

$$\mathbf{A} = (a_1, a_2, a_3, \ldots, a_i, \ldots, a_m), \qquad 0 \leqq a_i \leqq 1. \tag{6.4}$$

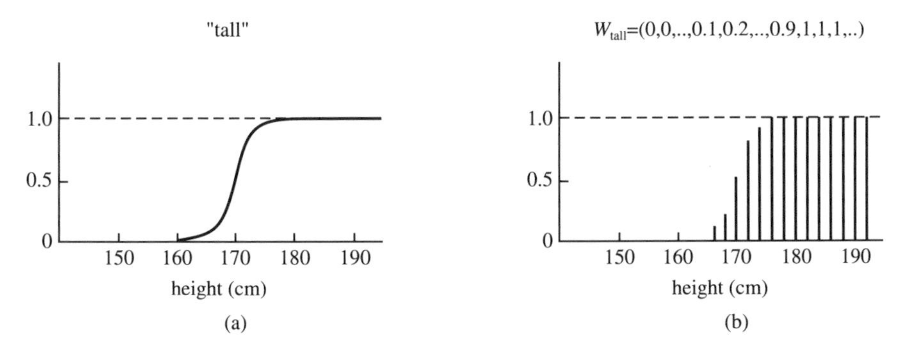

Fig. 6.1 (a) Membership function for the word "tall" and (b) distribution expression from sampling.

Here, m is the number of elements, and it is finite. In the same way, the other piece of linguistic information **B** can be expressed as follows:

$$\mathbf{B} = (b_1, b_2, b_3, \ldots, b_j, \ldots, b_n), \qquad 0 \leqq b_j \leqq 1. \tag{6.5}$$

Let **A** be the cause and **B** the effect, and we can then define a matrix that expresses the cause-and-effect relationship between **A** and **B**; this is called the *fuzzy relation* from **A** to **B**. In other words:

$$\mathbf{R} = \begin{bmatrix} r_{11} & \cdots & r_{in} \\ \vdots & r_{ij} & \vdots \\ r_{m1} & \cdots & r_{mn} \end{bmatrix} \tag{6.6}$$

L. A. Zadeh found the inference conclusion **B′** in (6.1) using the following equation [1]:

$$\begin{aligned} \mathbf{B}' &= (b'_1, b'_2, b'_3, \ldots, b'_j, \ldots, b'_n) \\ &= (a'_1, a'_2, a'_3, \ldots, a'_j, \ldots, a'_m) \circ \begin{bmatrix} r_{11} & \cdots & r_{in} \\ \vdots & r_{ij} & \vdots \\ r_{m1} & \cdots & r_{mn} \end{bmatrix} \\ &= \mathbf{A}' \circ \mathbf{B}. \end{aligned} \tag{6.7}$$

Alternatively,

$$b'_j = \bigvee_i a'_j * r_{ij}. \tag{6.8}$$

Here $*$ is the MIN or algebraic product; L. A. Zadeh explicated the MIN, and its appropriateness is shown by the fact that it is still widely used today.

Many operations have been proposed for fuzzy relations. L. A. Zadeh proposed the following types of fuzzy relations [2, 3]:

$$r_{ij} = (a_i \wedge b_i) \vee (1 - a_i), \tag{6.9}$$

$$r_{ij} = 1 \wedge (1 - a_i + a_j). \tag{6.10}$$

E. H. Mamdani proposed a different type of fuzzy relation as shown in the following, and its appropriateness is shown by practical applications [4]:

$$r_{ij} = a_i \wedge b_j. \tag{6.11}$$

Mizumoto *et al.* have proposed many types of fuzzy relations and evaluated them by certain standards, obtaining more appropriate fuzzy relations for our subjective reasoning [5, 6].

The authors have defined a non-Neumann computer (non-stored program system) that performs the forward fuzzy inference in (6.1) and the backward fuzzy inference in (6.2) in parallel using a dedicated "fuzzy" circuit, rather than a digital circuit, as a *fuzzy computer*. In other words, since ambiguous linguistic information **A**, **A′**, **B**, **B′**, etc., used in (6.1) and (6.2) can be expressed in terms of membership functions like those in Fig. 6.1, we can sample this as in Fig. 6.1b, and let it be a vector formed from m elements (each of a value 0–1), as in

$$\mathbf{W}_{\text{tall}} = (0, 0, \ldots, 0.1, 0.2, \ldots, 0.9, 1, 1, 1, \ldots). \tag{6.12}$$

Instead of the normal binary word, this is called a *fuzzy word*. Each truth value is given a correspondence to, for example, 0 V to 5 V or 0 μA to 100 μA, and 10 fuzzy words are handled by m signal lines (data bus). A basic fuzzy computer is made up of a "*fuzzy memory*," which stores ambiguous linguistic information, and a "*fuzzy inference engine*," which uses that information to carry out fuzzy inference, and according to need, a "*defuzzifier*" is added. Since each of these three blocks must handle continuous quantities of electric signals, a dedicated fuzzy circuit must be designed.

6.1.3 Fuzzy Building Blocks [10]

(a) MAX and MIN Circuits

Fuzzy logic functions are different from two-valued logic and many-valued logic in that, essentially, an infinite number can be defined, but only between 10 and 20 types are widely known [11]. Since of these the ones frequently used in fuzzy inference are the *fuzzy logical sum* (MAX) and *fuzzy logical product* (MIN), we will discuss the circuits for carrying them out here.

Electronic circuits can be divided into current-mode circuits and voltage-mode circuits, depending on the signal mode that is handled in them. In current mode, the information signal is expressed in terms of the direction of current and its size, and I^2L (integrated injection logic) is a representative example. On the other hand, in voltage-mode circuits, the information signal is expressed in terms of the polarity of the voltage and its size, so normal digital circuits and analog circuits take this form. In current-mode circuits, addition and subtraction can be set up with simple lead connections, and there is the excellent characteristic that no extra transistors or other circuit components are necessary. However, this has the defect of one circuit only being able to drive one load (a fanout of 1). On the other hand, voltage-mode circuits give a wealth of fanout, but they are not appropriate for addition and subtraction. In addition, current-mode circuits resist the effects of power-source voltage changes.

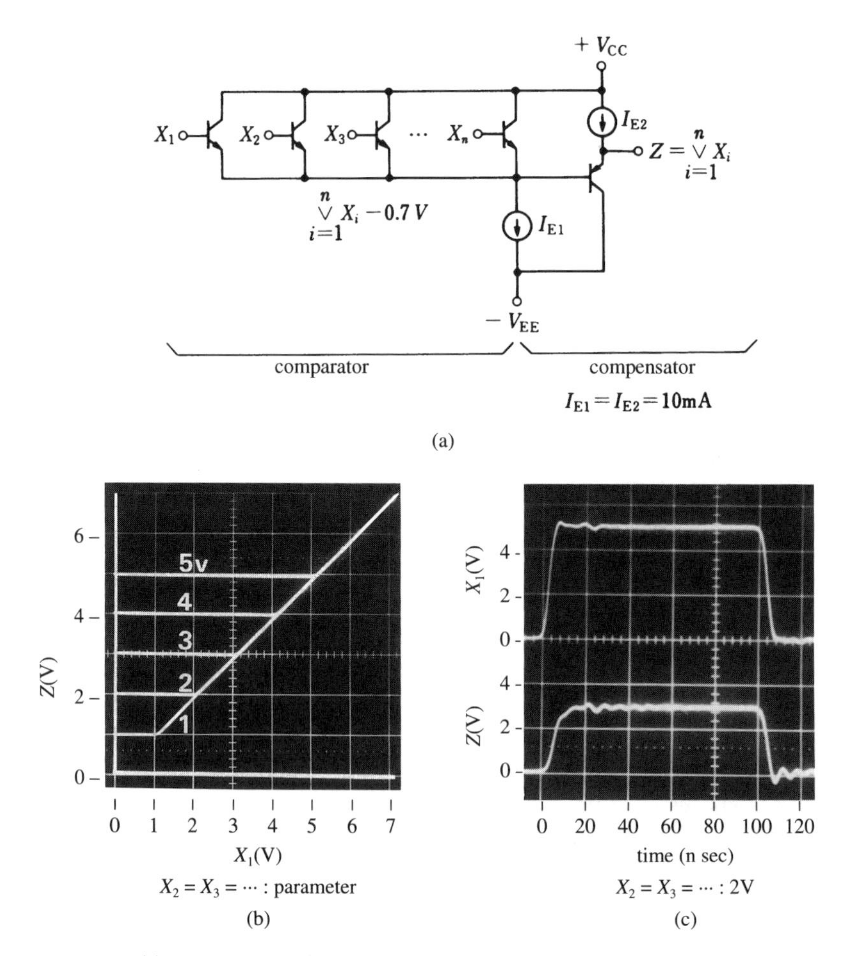

Fig. 6.2 (a) MAX circuit, (b) input/output characteristics, and (c) transition response.

Here we employ a voltage-mode circuit. Figures 6.2a and 6.3a show a MAX circuit and a MIN circuit, respectively [8, 14]. In this voltage-mode circuit, the truth values of 0–1 correspond to 0 V to +5 V. The MAX and MIN circuits are made up of comparators and compensators. The n input signal voltages that correspond to truth values (grades) are compared, and (maximum value −0.7 V) or (maximum value +0.7 V) appears at the emitter of an emitter-coupled NPN transistor array or an emitter-coupled PNP transistor array, respectively. Since the transistors of the comparator are all coupled at the emitter, Fig. 6.2a and Fig. 6.3a are called *emitter coupled fuzzy inference logic gates* (ECFL gates). The emitter voltage of 0.7 V at the comparator emitter junction is corrected at the next stage, the

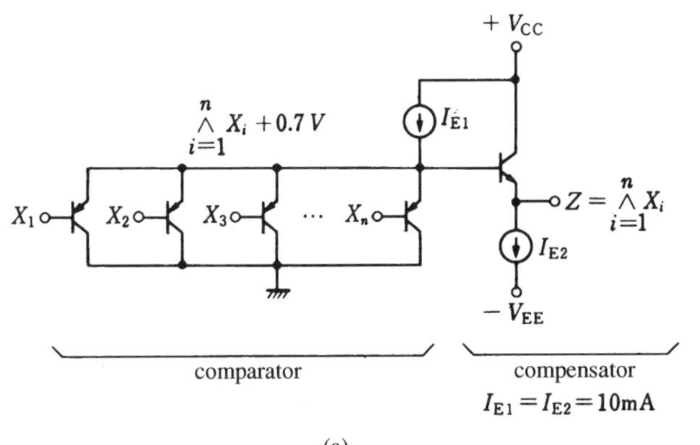

(a)

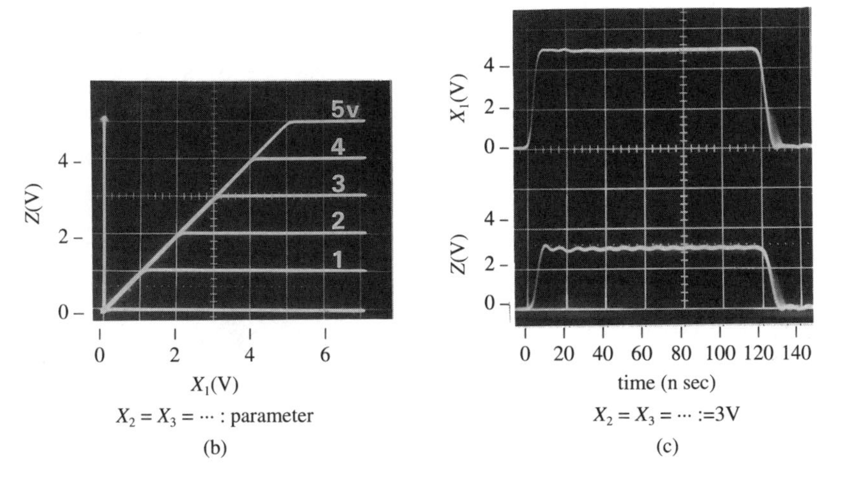

Fig. 6.3 (a) MIN circuit, (b) input/output characteristics, and (c) transition response.

compensator (an emitter follower driven by I_{E2}). By setting it up this way, the maximum and minimum input values appear at output Z in Fig. 6.2a and Fig. 6.3a, respectively. These are shown in Fig. 6.2b and Fig. 6.3b. The compensator not only corrects the voltage shift from the comparator, but also simultaneously compensates for the thermal drift of the emitter junction voltage. With a surrounding temperature varied over a range of $-55°C$ to $125°C$, voltage variations are less than $\pm 0.6\%$ F.S. In practice this is absolutely no problem. The transient responses for the ECFL gates are shown in Fig. 6.2c and Fig. 6.3c. From the figures, we can see that the response speed is 20 ns or less. In order to eliminate resonance and ringing caused by loads in real circuits, a resistor of about 100 Ω is placed in series with each base, but this is no problem because a construction in which this amount of resistance takes no space can be made using the base area pattern during integration.

The major characteristics of ECFL gates are large fanouts and resistance to changes in source voltage. The ECFL will operate in a completely normal fashion, with variations of $+V_{CC}$ in a range of +6 V to +51 V and $-V_{EE}$ in a range of −1 V to −46 V, and this circuit has a robustness to which binary digital circuits cannot compare.

(b) Fuzzy Memory Devices [7]

Fuzzy memories save membership functions like that shown in Fig. 6.1a or sampled membership functions like that in Fig. 6.1b. In other words, they save and retrieve information in single units of fuzzy words like that in Eq. (6.12). Figure 6.4 shows the structure of a *fuzzy memory device*. Here the ambiguous information handled by the memory contains the following seven items:

"negative large (NL)"
"negative medium (NM)"
"negative small (NS)"
"approximately zero (ZR)"
"positive small (PS)"
"positive medium (PM)"
"positive large (PL)"

NL, NM, NS, ZR, PS, PM, and PL are "labels" attached to linguistic information. With the fuzzy memory discussed here, the user inputs the membership function to the PROM section; the 000 to 111 labels (PL: 111, PM: 110, PS: 101, ZR: 100, NS: 011, NM: 010, NL: 001, NOT SPECIFIED: 000), which have been saved in external two-valued memory, are transformed into seven-bit words using a 1 of 8 decoder, and pass transistor array-controlled analog voltages (0 V–5 V correspondences to truth values 0–1) output to 25 output leads (output words). A photograph of an actual prototype chip is shown in Fig. 6.5. The production process was an Al gate pMOS process. The response of the pass transistor array used in the fuzzy memory is shown in Fig. 6.6. There is some contrivance in the decoder, and while it is pMOS, it operates like CMOS; therefore, we can understand why there is almost no steady-state error. The results for speed of response were slow because the pattern was large due to the limitations of the production device. In SPICE II this was replaced by transistors with a gate length of 2 μm and a gate width of 5 μm, and the response speed was about 5 ns in a simulation of a device that included the decoder.

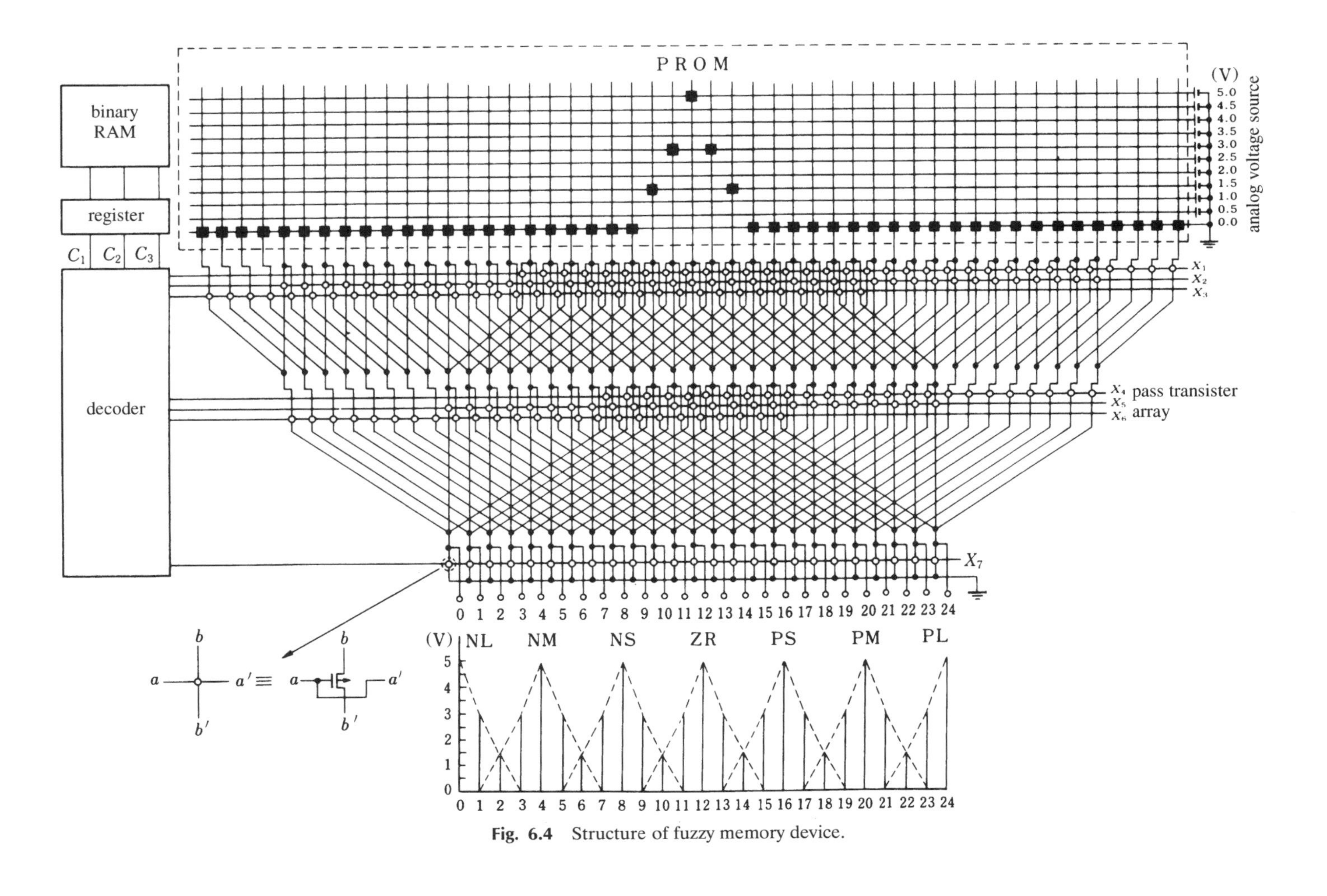

Fig. 6.4 Structure of fuzzy memory device.

Fig. 6.5 Photograph of fuzzy memory device.

(c) Fuzzy Inference Engine

A block diagram for a *fuzzy inference engine* for carrying out the forward fuzzy inference in (6.1) and the backward fuzzy inference in (6.2), using Eq. (6.11), comes out as shown in Fig. 6.7. Here we let the number of elements for fuzzy words A and A' and that for B and B' be m and n, respectively. C-MIN is m number of two-input MIN circuits in parallel and E-MAX is an m-input one-output MAX circuit. In addition, the TRUNCATION GATE is n two-input MIN circuits, in which one input for each is connected to a. A and B are input as "If x is A, then y is B," and then if we input "x is A'" for *real information* A', voltage for "y is B'" appears at n signal lines in the form of the sampled membership functions. The inference engine board for prototype *fuzzy computer YFC*-1 made in our lab in the fall of 1986 is shown in Fig. 6.8. This is a *basic fuzzy inference engine* that carries out fuzzy inference with antecedent propositions and one conclusion proposition. In this one inference engine board, 288 bipolar transistors are used, and the inference speed was 100 ns; in other

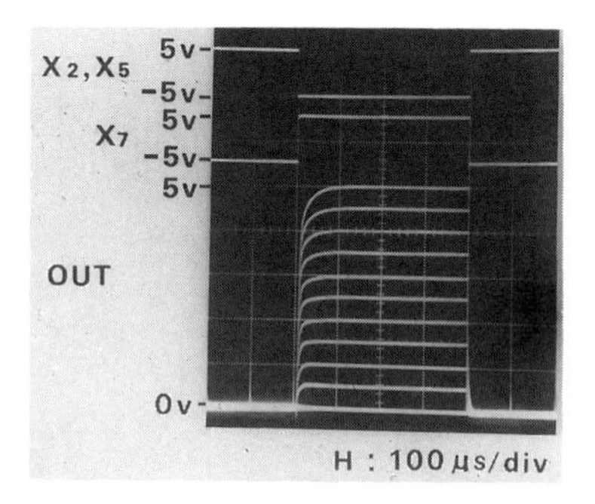

Fig. 6.6 Transition response of pass transition array used in fuzzy memory device. (Analog input voltages from 0 V to 5 V in 0.5 V steps.)

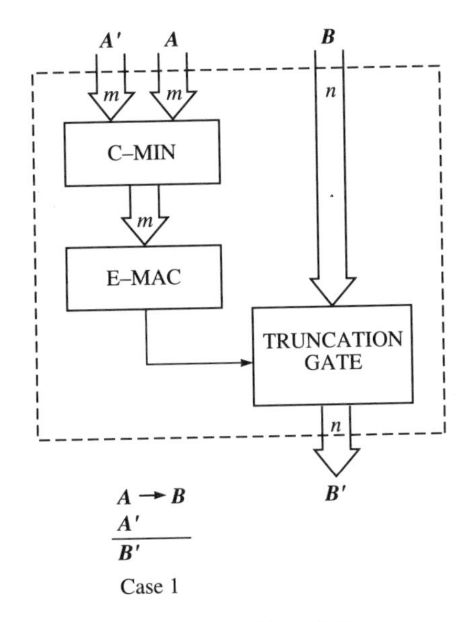

Fig. 6.7 Block diagram of fuzzy inference engine.

words, it is possible to do 10,000,000 fuzzy inferences per second (10 megaFIPS; FIPS: fuzzy inferences per second).

6.1.4 Structure of the Fuzzy Computer

In order to construct a high-speed fuzzy computer from the fuzzy building blocks discussed in Section 6.1.3, we must employ the fine-grain parallelism shown in Fig. 6.9. In other words, we must arrange r fuzzy inference engines in parallel and also input fuzzy words in parallel, in order to carry

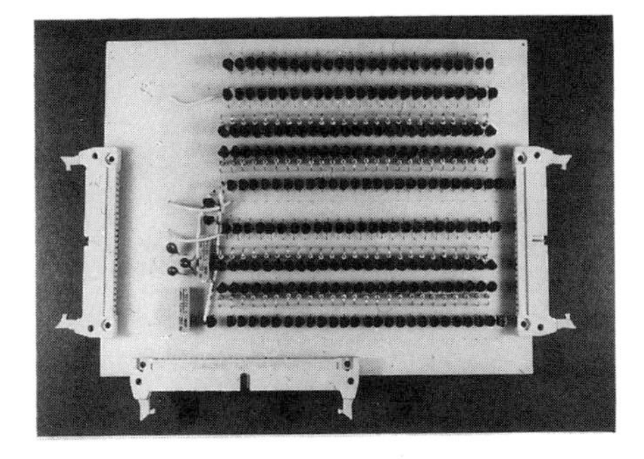

Fig. 6.8 Basic fuzzy inference engine host.

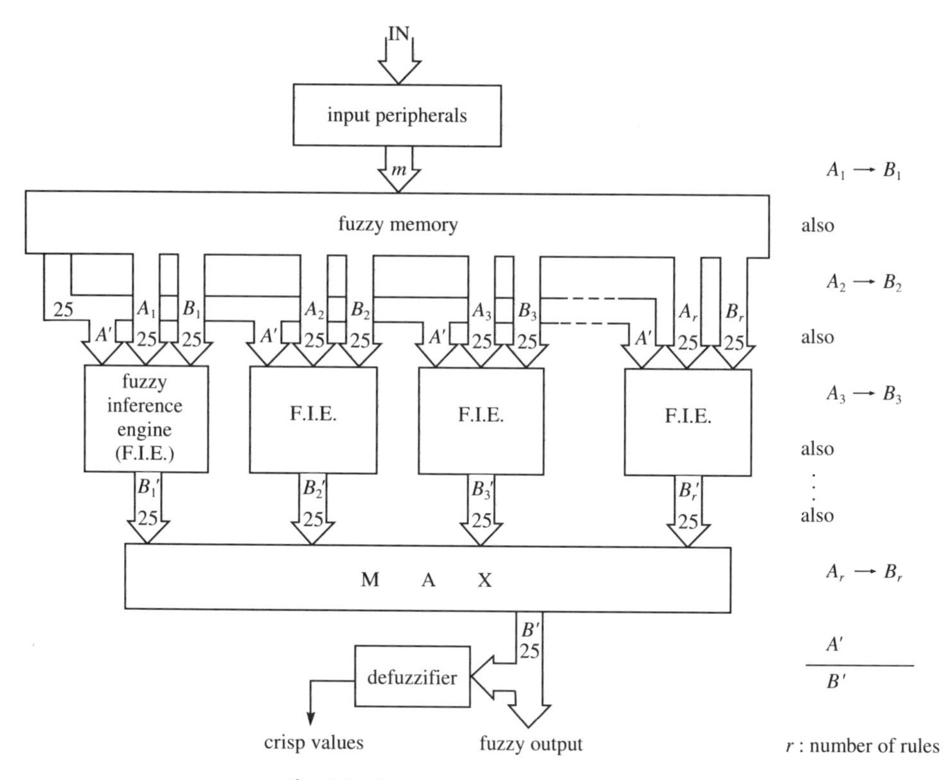

Fig. 6.9 Basic fuzzy computer architecture.

out r fuzzy inferences at the same time. In addition, the voltage for each fuzzy inference and each fuzzy word is distributed over m or n (here 25) signal lines, and the signals in these lines are processed in exact parallel. The MAX block in Fig. 6.9 is the block that carries out the ALSO inference rule, and it is the place where the logical sum of all of the inference results obtained from the inference engines is performed. Since the fuzzy output (inference result) B' cannot actually be used as it is in the form of a distributed voltage signal, a fixed value is output by means of a defuzzifier. The control unit and control bus that synchronize the operation of all lines in the fuzzy computer are left out of Fig. 6.9 for simplicity.

Figure 6.10 is the inference board for a demonstration model fuzzy computer that was displayed at the Second Annual Conference of the International Fuzzy Systems Society held in Tokyo on July 20–25, 1987. Since the input and output peripherals make use of a two-valued digital system, the speed of the entire system was determined by the speed of the digital system. It is set up for two antecedent propositions and one

Fig. 6.10 Extended fuzzy inference board used in FUZ-MMI fuzzy computer.

conclusion proposition with a choice of conjunctions AND or OR for connecting the two antecedent propositions.

6.1.5 The Future of Fuzzy Computers

In recent years, there has been a big shift from using computers for calculations to using them for inference. In the field of artificial intelligence, there has been a great blossoming of dedicated programs and hardware for inference. If we cannot transplant the experienced ambiguous know-how of human experts, there cannot be anything but simple artificial "hard intelligence." Flexible intelligence should be able to skillfully manipulate the fuzzy information included in natural language. In this way, fuzzy computers can be called the sixth generation. In 1965, L. A. Zadeh proposed the theory of fuzzy sets, and this is the 24th year since then. With the progress of semiconductors these days, it probably will not take that much time to make a real system. In 1995, the 30th anniversary of the advent of fuzzy set theory, there will probably be single-unit fuzzy computers or ones that work in combination with digital computers on the market.

6.1.6 Conclusion

In this section, we gave a simple explanation of fuzzy inference and hardware methods for carrying it out. The utility of fuzzy inference has been given in a number of instances so far. However, all of them are carried out with software, and the quality is not obvious. The quality of fuzzy processing is not processing based on deep logic, but rather it is short circuit type processing based on shallow logic; therefore, its life is speed. Because of this, if we do not carry it out using hardware rather than software, the authors are sure that the quality of fuzzy processing will not appear. However, we believe that the only ones in the whole world that are developing dedicated fuzzy chips are Togai Infralogic Co. and our laboratory. Our positions are different, since Togai Infralogic is working on it for efficient applications of digital systems and the authors are working on dedicated fuzzy hardware, but the goal of both is speed.

Even though it is still early, fuzzy computers are working. If they work, users will appear. If there are users, requirements will become stricter. Through this process fuzzy computers will come to be equipped with high-level functions. In a few years, single-board fuzzy computers will become popular among fans.

6.2 FUZZY SOFTWARE

6.2.1 Outline

(a) Handling Fuzziness by Means of Programs

Among uncertainties, randomness has been an object of calculation for a long time, using computer programming based on probability theory and statistics. Also, we can probably say that languages used for simulation are in some way programming languages for handling randomness. Fuzziness was probably handled individually by data bases, etc., in the past, but it was not openly and generally recognized. Conversely, uncertainty that is based on fuzziness was excluded in most cases, and examples of conscious efforts to deal with it were rare. With the proposal of fuzzy theory, the importance of the uncertainty known as fuzziness was confirmed, and with the working out of fuzzy theory and the corroboration of its usefulness, many programs for handling fuzziness have been created, beginning with fuzzy control, fuzzy mathematical programming, and fuzzy document retrieval.

However, these were all written with traditional programming languages. If there could be a capability for handling fuzzy sets themselves in a programming language, it should be helpful as a programming tool when fuzziness is the object. Furthermore, if fuzzy theory is used in programming itself, we can probably expect that programming languages with new capacities for dealing with fuzziness will be developed. In this section we will mainly call programming languages that handle fuzzy sets, and in addition programming languages that are based on fuzzy theory, *fuzzy software*. However, semispecial languages, such as fuzzy data base language, etc., have already been introduced [15], so here the fuzzy software that will be our objects will only be general-use programming languages.

(b) Fuzzy Software

As was mentioned before, fuzzy software can presently be divided into programming languages that can handle fuzzy sets and programming languages in which fuzzy theory is used in carrying out program control.

Since fuzzy sets are expressed using membership functions, it is possible to write fuzzy sets using traditional programming languages, based on two-valued logic. However, we must write cumbersome procedures for the operations among fuzzy sets because they are not designed for treating fuzzy sets. It would be ideal if the capacity for easily writing fuzzy sets and the various operations among fuzzy sets were built into the programming language. What has been developed with these kinds of goals are the former fuzzy programming languages, which can handle fuzzy sets. Of course these programming languages can normally express normal crisp constants and variables in the same way as they do fuzzy sets for these things. At present there are not very many actual applications [16], but in Section 6.2.3, we will introduce a language that handles fuzzy sets using LISP in detail.

It is natural that the latter programming languages, in which fuzzy sets are used in program control, can define fuzzy sets and use them, but interest is centered around those that control program branches by means of adding values between $[0, 1]$ or fuzzy sets on $[0, 1]$ to each statement (interpreted as truth values, belief certainty, and degree of ambiguity) and obtain results with degrees of these. Most of them are programming languages that are used for inference, and among these are fuzzy production systems [17] and fuzzy Prolog [18–20]. Fuzzy programming languages that handle fuzzy sets have as their objects the treating of fuzzy sets and have deterministic statements for implementation. On the other hand, in the latter programming languages, fuzzy truth values can be given to each statement itself. In Section 6.2.2, we will give a detailed introduction to fuzzy Prolog.

(c) Fuzzy Software and Hardware

The fuzzy software used up to now is actually operated on digital computers constructed on the basis of the two-valued logic. In other words, fuzzy software is written in the digital computer languages used up to now. This may seem somewhat strange, but if we consider that the way in which fuzzy theory handles uncertainty is based on strict theory, it is natural. However, there is a problem with efficiency, and in general fuzzy software is slow. This is because the value of the membership function for each element of a fuzzy set must be given by a binary expression, and strict calculations must be performed during operations between fuzzy sets with all of the binary expressions for the digits among the values of these membership functions.

The values for membership functions are subjectively determined numbers, and essentially not strict; it is not necessary to have such strict calculations. In this way, handling fuzzy sets with our current digital computers means excess expressions and excess calculations. Furthermore, another reason that the operation of fuzzy software is slow is that in operations between fuzzy sets, calculations must be done for each of the elements in the fuzzy sets, yet this is carried out in series. Essentially, these calculations have the characteristic of being carried out in parallel.

The speed problem can be solved by expressing the value of the membership function for each element physically as one analog quantity [21] or as a rough, several-bit expression [22], or making dedicated hardware for parallel calculations. This is the motivation of the fuzzy computer discussed in Section 6.1. However, there is at present no research into the tying together of fuzzy software and fuzzy computers.

(d) The Future of Fuzzy Software

Thre is at present little functioning, practical fuzzy software. What we are facing at present is making progress in the direction of adding the capability for handling fuzzy sets and fuzzy truth values to traditional programming languages just as they are. In terms of speed, which is one of the problems, there will probably be a solution in the direction of adding to our present computers dedicated hardware that is made up of either analog or digital circuits for handling fuzzy sets. If we consider the fact that fuzzy theory is a theory for clearly handling uncertain objects, the hardware for operating future fuzzy software would be fuzzy computers with a harmonious combination of analog and digital—analog for fuzzy sets and digital for exact control.

Zadeh proposed the concept of fuzzy algorithms [23] and the writing of natural-language propositions using fuzzy theory [24]. The ideal is probably the use of computers through the medium of natural language with its attendant ambiguity, but in order to do this, we can only wait for developments in research into the understanding of natural language along with research into fuzzy theory and fuzzy computers.

6.2.2 Fuzzy Prolog [20, 25]

(a) Introduction

Prolog is a programming language that is widely used in artificial intelligence systems and expert systems. Prolog is a nonprocedural language based on first-order predicate calculus, and the resolution principle is used for rules. First-order predicate calculus is a two-valued system, and it is usually assumed that values for the logical variables included in the various propositions and predications are either true (1) or false (0). Each of the statements for Prolog corresponds to a single logical equation, and these statements especially, in other words facts and rules, are assumed to be true (1).

On the other hand, in fuzzy logic, the truth values for the various logical variables are not just a two-valued logical true (1) or false (2); any value in the closed space [0, 1] is permissible. Accordingly, any value from [0, 1] is apportioned to each of the statements for Prolog, and if these numbers are interpreted as truth values or certainty, inference can be carried out with uncertain facts or uncertain rules. Since conclusions carrying values from [0, 1] are obtained, it seems that changes in program control could be made using these values. This was the motivation behind research on fuzzy Prolog.

A reinvestigation, from the standpoint of fuzzy theory, of the resolution principle that forms the basis of Prolog was necessary for the development of fuzzy Prolog [26, 27]. In this section we will give a simple explanation of the ideas behind fuzzy Prolog, which is based on a *fuzzy resolution principle*. The nature of the fuzzy Prolog discussed here is that of many-valued logic. However, instead of using any value from [0, 1] for truth values, we can consider using fuzzy sets on [0, 1] (known as *linguistic truth values*), but at present there has been no clarification of a fuzzy resolution principle for linguistic truth values.

(b) Resolution Certainty

With the resolution principle as a basis for inference, the law of complements $A \cdot \overline{A} = 0$ is tacitly assumed. The law of complements naturally holds in two-valued logic. For example, when we ask whether or not B arises given fact A with the rule $A \to B$, $A \to B$ is transformed into $\overline{A} \vee B$, $\overline{B}$ the negation of B added to the rule and fact, and a contradiction is derived from set $\{\overline{A} \cup B, A, \overline{B}\}$. In other words, B is derived from A, and $\overline{A} \vee B$ and a contradiction are derived from B and $\overline{B}$ by assuming $(B \cdot \overline{B} = 0)$ is derived. However, the law of complements does not always arise in cases of fuzzy sets, so this method cannot be used as is. In the fuzzy resolution principle, the A and B above, which are used for inference, are called *key words*. The resolution certainty of inference using key word A is defined as

$$\text{resolution certainty} = \max(C_A, C_{\overline{A}}).$$

Here C_A ($C_{\overline{A}}$) is called the *certainty* of $A(\overline{A})$, and it is defined by

$$C_A = (T(A) - 0.5) \times 2.$$

Since $T(A)$ is the truth value for A, it is usually 0 or 1 in two-valued logic, but in the case of fuzzy logic, we get any value from [0, 1]. Certainty C_A is 1 when $T(A)$ is 1, 0 when $T(A)$ is 0.5 and -1 when $T(A)$ is 0. If we consider the fact that $T(\overline{A}) = 1 - T(A)$, the preceding resolution certainty is equal to $|C_A|$ (see Fig. 6.11). In other words, when a key word in which the truth value of A is 0 (certainly false) or 1 (certainly true) is used for inference, the resolution certainty is 1, but when it is 0.5 (truth or falsity absolutely unknown), the resolution certainty is 0. With these key words, the total resolution certainty in inference that derives B from A and B is the value of the smaller of the resolution certainties for the two; in other words, it is defined as $\min(|C_A|, |C_B|)$. In this way we have an inference method in fuzzy logic for $A \cdot \overline{A}$ and $B \cdot \overline{B}$, information that expresses generally incomplete contradictions, that preserves the form of resolution certainty.

(c) An Example of Fuzzy Prolog

Let us explain the idea behind fuzzy Prolog with an extremely simple example. For simplicity we will consider only two parameters, youth and money, as conditions for lovers. Figure 6.12 shows the membership function for the fuzzy set that expresses "young." In actuality, this is defined as a membership function in fuzzy Prolog as follows:

```
/ *The membership function of young* /
young(Age){0/(0), 0/(13.5), 0.0155/(14.5), 0.2093/
(15.5),
0.3953 / (16.5), 0.5194 / (17.5), 0.09612 / (18.5),
0.0960/(19.5), 1/(20.5), 1/(24.5)}repeat#10.
young(Age){1/(24.5), 0.9922/(25.5), 0.7984/(26.5),
0.7829/(27.5),
0.7674/(28.5), 0.6202/(29.5), 0.5969/(30.5), 0.2093/
(31.5),
0.2093/(32.5), 0.2106/(33.5), 0.2106/(35.5), 0.0078/
(36.5), 0/(40),
0/(120)}repeat#14.
```

Here the numbers in parentheses are ages, and the numbers left of the slashes are membership values for those ages. Figure 6.12 is a continuous curve, but here it is defined dispersedly using representative points. Intermediate values are interpolated as values on a straight line. Next we let money be judged in terms of savings and the membership functions for three fuzzy sets, "rich," "general," and "poor," be given as shown in Figs. 6.13 through 6.15. These are written for fuzzy Prolog in the same way, as follows:

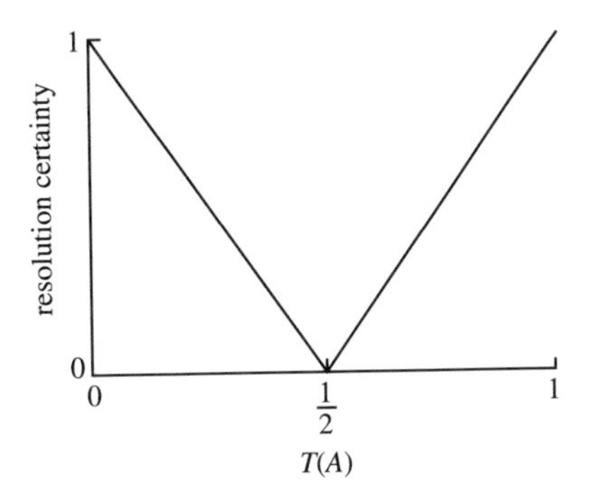

Fig. 6.11 Resolution certainty.

```
/ *The membership about money (¥10000)* /
rich(Money){0/ (0), 0 / (100), 0.1(200), 0.2 / (250),
0.35 / (300),
0.4(350), 0.55 / (400), 0.7 / (450), 0.085 / (500), 0.9 /
(600), 0.95 / (800),
1 / (1000), 1 / (10000000)}repeat#13
general(Money){0/ (0), 0 / (80), 0.1 / (90), 0.15 / (100),
0.25 / (120),
0.40 / (140), 0.65 / (160), 0.9(180), 1 / (200), 1 / (400),
0.8 / (450),
0.6 / (500), 0.35 / (600), 0.2 / (700), 0.1 / (800), 0.05 /
(900), 0 / (1000),
0 / (10000000)}repeat#18.
poor(Money){1/ (0), 1 / (50), 0.85 / (80), 0.8 / (100),
0.7 / (120),
0.55 / (140), 0.35 / (160), 0.1 / (180), 0 / (200), 0 /
(10000000)}repeat#10
```

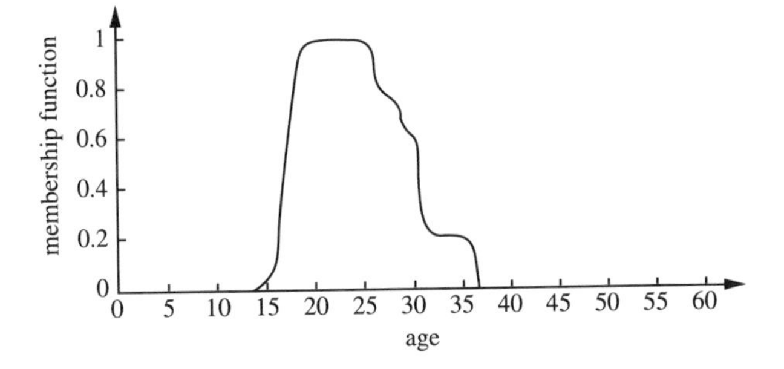

Fig. 6.12 Membership function of "young."

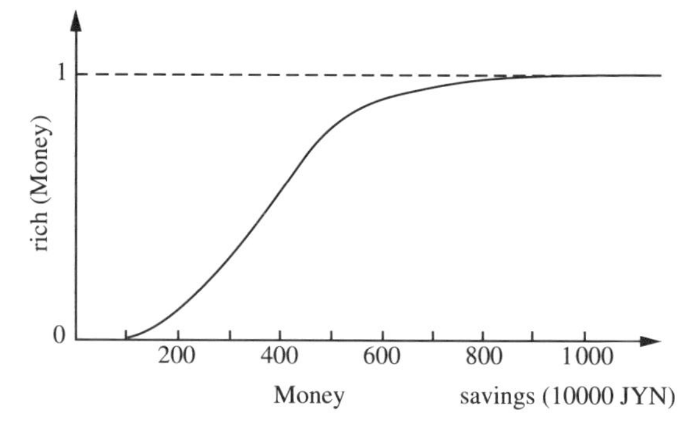

Fig. 6.13 Membership function of "rich."

Here, the numbers in parentheses are savings in units of ¥10,000, and the meaning of the other figures is the same as before.

Now let us say that we are given data for three women and two men. Ms. Yoshinaga is a woman who is 33 years old and has savings of ¥13,000,000. This is written as

```
/ *The names, the ages and the deposits (¥10000) of
persons* /
girl ( ' Yoshinaga ', 33, 1300).
```

In the same way, we let the other two women, Ms. Yamashita and Ms. Ishida, be

```
girl( ' Yamashita ', 25, 350)
girl( ' Ishida ', 20, 60)
```

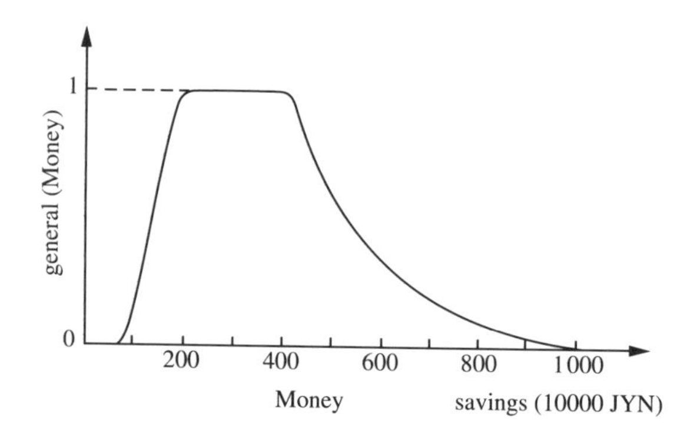

Fig. 6.14 Membership function of "general."

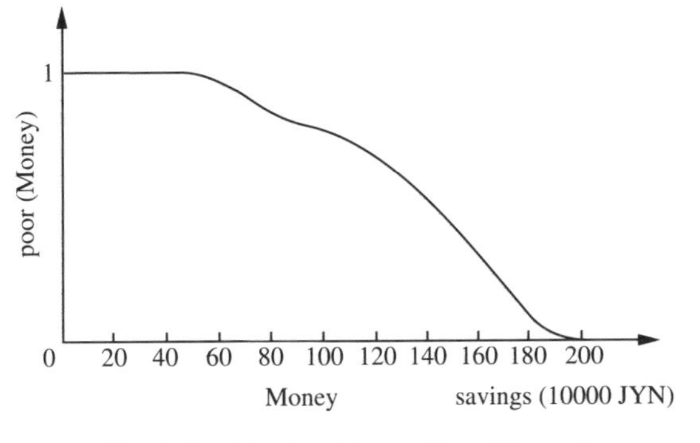

Fig. 6.15 Membership function of "poor."

Next we let the two men, Mr. Tanaka and Mr. Sato, be

```
boy( ' Tanaka ', 42, 4000)
boy( ' Sato ', 26, 250)
```

At this point we say that we have all of the actual data. However, since the facts concerning the "boys" and "girls" are clearly fixed facts, the degree of truth for all of them is 1 (not shown when 1).

Next, let us give the rules. Let us say that Yoshinaga thinks she wants a rich boyfriend, regardless of age. In the same way, Yamashita wants one who is young, as long as he is not poor, and Ishida will only accept a boyfriend who is both young and rich. These are written as follows in fuzzy Prolog.

```
/ *The girl who wants to find her boy- friend* /
boy_friend( ' Yoshinaga ', Boy):-boy(Boy,_, Money),
rich(Money).
boy_friend( ' Yamashita ', Boy):-boy(Boy, Age, Money),
not(poor(Money)),
young(Age).
boy_friend( ' Ishida ', Boy):-boy(Boy, Age, Money),
rich(Money),
young(Age).
```

We let the conditions for the men be as follows. That is,

```
/ *The boy who wants to find his girl- friend* /
girl_friend( ' Tanaka ', Girl):- girl(Girl, Age, _),
young (Age).
girl_friend( ' Sato ', Girl):-girl(Girl,_,_).
```

Tanaka wants a young girlfriend, and Sato wants a girlfriend with no other condition than that she is a woman.

Because this is a mutual relationship between girlfriends and boyfriends, we establish the following rule.

```
/ *The rule for introducing the boy and girl make
friend each other* /
introduce (Boy, Girl):- girl_friend(Boy, Girl),
boy_friend(Girl, Boy){0.6, 1}
```

The {0.6, 1} at the end is the *rule weight*, and this will be discussed later.

(d) Inference Method for Fuzzy Prolog

At this point we have defined the rules and facts for introducing boy- and girlfriends. Let us look at several problems here.

Question 1: Who are the rich men?

```
?- boy(Boy,_, Money), rich(Money).
(Boy = ' Tanaka ', Money = 4000) {MT = 1.0000, T = 1.0000
with CONF = 1.0000}
(Boy = ' Sato ', Money = 250) {MT = 0.3200, T = 0.2000 with
CONF = 0.6000}
```

In the answers, T is the truth value, CONF the resolution certainty of the inference that gave this result, and MT is the *overall truth value*. Truth value and certainty are values that should be thought of as strictly separate, but it is normally easier to understand them by bringing them together in one value; MT is defined as follows. In other words, since MT must be 0.5 when T is 0.5 and CONF is 0, MT is defined as

$$\begin{aligned} \mathrm{MT} &= (C_T \times \mathrm{CONF} \times 0.5 + 0.5) \\ &= (2T - 1) \times \mathrm{CONF} \times 0.5 + 0.5. \end{aligned}$$

In the preceding example, Tanaka is without doubt rich. Since Money = ¥2,500,000 is Sato's savings, and the value for the membership function of rich(250) is 0.2, the truth value in this instance is 0.2. In this case, the key word is rich(250), and by definition, the resolution certainty is CONF = $|C_{\mathrm{rich}(250)}| = |(0.2 = 0.5) \times 2| = 0.6$; therefore, the overall truth value is

$$\mathrm{MT} = (2 \times 0.2 - 1) \times 0.6 \times 0.5 + 0.5 = 0.32.$$

Here the fact that MT is less than 0.5 means that he is not rich, but we have to be careful of the uncertain conclusion. With Prolog, which uses two-valued logic, the derivation of this kind of result would be impossible. This is one of the characteristics of fuzzy Prolog.

Question 2: *Which are the young women*?

```
?-girl (Girl, Age,_), young(Age).
(Girl = ' Yoshinaga ', Age = 33) {MT = 0.3318, T = 0.2100
with CONF = 0.5801}
(Girl = ' Yamashita ', Age = 25) {MT = 0.9922, T = 0.9961
with CONF = 0.9922}
(Girl = ' Ishida ', Age = 20) {MT = 0.9695, T = 0.9845 with
CONF = 0.9690)
```

In this case, as before, we can say that it is almost completely true for Yamashita and Ishida, but we have an uncertain result for Yoshinaga.

Question 3: *Which people have the possibility of becoming boy- and girlfriend*?

```
?-Introduce(Boy,Girl).
(Boy = ' Tanaka ', Girl = ' Yoshinaga ') {MT = 0.2100, T =
0.0000 with CONF = 0.5801}
(Boy = ' Tanaka ', Girl = ' Yamashita ') {MT = 0.0039, T =
0.0000 with CONF = 0.9922}
Boy = ' Tanaka ', Girl = ' Ishida ') {MT = 0.0155, T = 0.0000
with CONF = 0.9690}
(Boy = ' Sato ', Girl = ' Yoshinaga ') {MT = 0.2000, T = 0.0000
with CONF = 0.6000}
(Boy = ' Sato ', Girl = ' Yamashita ') {MT = 0.8953, T = 1.0000
with CONF = 0.7906}
(Boy = ' Sato ', Girl = ' Ishida ') {MT = 0.2000, T = 0.0000
with CONF = 0.6000}
```

For all couples besides Sato and Yamashita, uncertainty is expressed, that is, little possibility of their becoming boy- and girlfriends. MT can be interpreted as the degree of possibility. For example, let us follow the Sato–Yoshinaga pairing. The premise of the rule for becoming boy- and girlfriend is girl_friend ('Sato', 'Yoshinaga') and boy_friend ('Yoshinaga', 'Sato'); the truth values are 1 and 0.2 for rich(250), and the smaller, 0.2, becomes the truth value of the premise. Here we will explain the weight of the rule. Rules can also be apportioned any numerical value from [0, 1].[1] This weight means "the degree of truth of the conclusion when the premise is true," and the following equation is used:

$$\text{rule weight} = \text{certainty of premise} \times \text{certainty of conclusion}.$$

By means of this, the certainty of the conclusion can be calculated from the certainty of the premise. However, we can specify the weight as an

[1]Actually, negative weights are recognized, and in this case, when the premise is positive (negative) it means that there is a negative (positive) degree for the conclusion.

interval with a certain width, and this means that the weight attached to the rule described earlier has an upper bound of 1 and a lower bound of 0.6. For $P \rightarrow Q$ {lower bound, upper bound}, when C_p, the certainty of P, is given, C_Q, the certainty of Q, is calculated using

$$|C_P| < \text{lower bound, inference impossible;} \tag{6.13}$$

$$\text{lower bound} \leqq |C_P| \leqq \text{upper bound}, \qquad C_Q = |C_P|/C_P; \tag{6.14}$$

$$|C_P| > \text{upper bound}, \qquad C_Q = \text{upper bound}/C_P. \tag{6.15}$$

When the weight is left out, it is assumed to be $\{0, 1\}$. In the preceding example, the certainty of the premise is $(0.2 - 0.5) \times 2 = -0.6$, and from Eq. (6.14) the certainty of the conclusion is -1, that is, a truth value of 0. In this case, the certainty of resolution from the key word rich(250) is $|-0.6| = 0.6$.

(e) Conclusion

With fuzzy Prolog, as introduced here, conclusions with truth values and resolution certainties can be obtained. In addition, statements for which the truth value is less than 0.5 can be handled, and these are things that were impossible with ordinary Prolog, which uses two-valued logic. However, there are a number of other methods that can be chosen for calculating truth values, but we can say that just how close these are to intuition must be justified through experiment. This is because there are many different uncertain conditions, and this corresponds to making clear just what kind of uncertainty is related to the object of the application of fuzzy Prolog.

6.2.3 Fuzzy Set Manipulation System in Lisp

If we build application systems [15] based on fuzzy sets [28], we first write a program representing and manipulating fuzzy sets on the computer, and then we go to work on building the application systems. The representation and manipulation of limited fuzzy sets are not so difficult [16], but if we want to use fuzzy relations, we have some difficulty. Furthermore, if we want to use type-2 fuzzy sets (fuzzy sets in which the membership grades are fuzzy sets) and level-2 fuzzy sets (fuzzy sets in which the elements are fuzzy sets), we have a great deal of difficulty.

Umano *et al.* have implemented a system that can manipulate very general fuzzy sets in Lisp [29–31]. There are several versions of this system, but here we will introduce a system in muLisp-86[2] (operating on

[2] muLisp is a trademark of Soft Warehouse Co.

MS-DOS)[3] for the NEC PC-9800 series. This system is distributed free of charge to interested parties [16]. (All manuals are presently in Japanese. No English ones are available.)

The facilities of this system are the following:

(1) It can represent and manipulate almost all of the fuzzy sets and fuzzy relations proposed up to now.
(2) Fuzzy sets and tuples can be written in a natural form, using { } and ⟨ ⟩, respectively.
(3) There are many built-in functions for fuzzy sets.
(4) One can easily define his or her own functions for fuzzy sets.
(5) Lisp functions can also be applied to fuzzy sets and tuples (by means of the extension principle [32]).
(6) Most Lisp functions can be used as is.

In the following we will demonstrate this system using several examples.

Example 1. Fuzzy Sets and Fuzzy Relations. In this system, fuzzy sets can be written using { } and fuzzy relations can be written as fuzzy sets of tuples using ⟨ ⟩ and { }. For example, we can write

$$\text{fuzzy set:} \quad \{0.1/a, 0.2/b, 0.3/c\},$$
$$\text{fuzzy relation:} \quad \{0.3/\langle a,1\rangle, 0.9/\langle b,2\rangle, 0.1/\langle c,3\rangle\}.$$

The Lisp function "setq" is used for assignment of these two variables. In order to assign fuzzy sets to symbols F1 and F2, we write

```
-> (setq F1 {0.1/a, 0.2/b, 0.3/c})
{0.1/A, 0.2/B, 0.3/C}
-> (setq F2 {0.7/b, 0.2/c, 0.4/d})
{0.7/B, 0.2/C, 0.4/D},
```

and to assign fuzzy relations to R1 and R2, we use

```
-> (setq R1 {0.3/⟨a, 1⟩, 0.9/⟨b, 2⟩, 0.1/⟨c, 3⟩})
{0.3/⟨A, 1⟩, 0.9/⟨B, 2⟩ 0.1/⟨C, 3⟩}
-> (setq R2 {0.5/⟨b, 2⟩, 0.8/⟨c, 3⟩, 1/⟨d, 4⟩})
{0.5/⟨B,2⟩, 0.8/⟨C, 3⟩, 1/⟨D, 4⟩}
```

where "−> " is a prompt from the system. When the underlined portion is input by the user, the system calculates (in Lisp we call it "evaluates") for the input and displays the result on the next line. The value for the setq function is the assigned one itself, so the assigned fuzzy sets and fuzzy

[3]MS-DOS is a trademark of Microsoft Corp.

relations are displayed. However, in muLisp-86 all letters become upper case, even if they are input in lower case.

Ordinary sets and relations can be expressed as special cases of fuzzy sets and fuzzy relations, type-*n* fuzzy sets and level-*m* fuzzy sets as nestings of { }, L-fuzzy sets as cases in which membership grades are tuples and general fuzzy sets as combinations of these [29].

Example 2. Operations on Fuzzy Sets. After example 1, we have fuzzy sets F1 and F2. We can find their union and intersection through

```
-> (union F1 F2)             ; grade operation is max
{0.1/A, 0.7/B, 0.3/C, 0.4/D}
-> (intersection F1 F2)      ; grade operation is min
{0.2/B, 0.2/C}
```

where the semicolon is followed by a note.

The same functions can be applied to fuzzy relations. After example 1 we have relations R1 and R2, and we can have

```
-> (union R1 R2)
{0.3/⟨A, 1⟩, 0.9/⟨B, 2⟩, 0.8/⟨C, 3⟩, 1/⟨D, 4⟩}
-> (intersection R1, R2)
{0.5/⟨B, 2⟩, 0.1/⟨C, 3⟩}
```

There are a lot of built-in functions including those in example 2. These can be divided into 10 groups, namely, (1) seven functions for fuzzy sets, (2) eleven functions for fuzzy relations, (3) five functions for membership grades, (4) four functions for elements, (5) five predicate functions, (6) ten comparative functions, (7) four transformation functions, (8) six input/output functions, (9) six functions to generate a new fuzzy set operation, and (10) five miscellaneous functions.

Example 3. Operations Using the Extension Principle. By adding "&" before Lisp function names, they can be used for fuzzy sets, and operations are executed through the extension principle [32].

In Lisp, addition and multiplication with ordinary numbers can be done with

```
-> (+ 3 4)    ; 3 + 4
7
-> (* 3 4)    ; 3 × 4
12
```

For fuzzy sets (fuzzy numbers), we first define the fuzzy numbers.

```
-> (setq about-3 {0.3/2, 1/3, 0.5/4}) ; about 3
{0.3/2, 1/3, 0.5/4}
-> (setq about-4 {0.4/3, 1/4, 0.6/5}) ; about 4
{0.4/3, 1/4, 0.6/5}
```

We can then use

```
-> (&+ about-3 about-4)          ; about 3 + about 4
{0.3/5, 0.4/6, 1/7, 0.6/8, 0.5/9}
-> (&* about-3 about-4)          ; about 3 × about 4
{0.3/6, 0.3/8, 0.4/9, 0.3/10, 1/12, 0.6/15, 0.5/
16, 0.5/20}
```

Example 4. Fuzzy Reasoning. Inference using a fuzzy rule and a fuzzy fact,

$$\begin{array}{l} \text{rule: if } x \text{ is large } y \text{ is small} \\ \text{fact: } x \text{ is very large} \\ \hline \text{conclusion: } y \text{ is?} \end{array} \tag{6.16}$$

was formulated in the following way by L. A. Zadeh [32].

Let us consider the general form of the above inference.

$$\begin{array}{l} \text{rule: if } x \text{ is } P \text{ then } y \text{ is } Q \\ \text{fact: } x \text{ is } A \\ \hline \text{conclusion: } y \text{ is } B \end{array} \tag{6.17}$$

where P and A are fuzzy sets in the universe of discourse U and Q and B are fuzzy sets in the universe of discourse V. Using the arithmetic rule, the rule can be translated into

$$R_a = (\sim P \times V) \oplus (U \times Q), \tag{6.18}$$

where $\sim$ means the complement, $\times$ the Cartesian product and $\oplus$ the bounded sum.

Now fuzzy set B for y can be found with

$$\begin{aligned} B &= A \circ R \\ &= \left\{ \max_{u \in U} (\mu_A(u) \wedge \mu_R(u,v))/v : v \in V \right\}, \end{aligned} \tag{6.19}$$

where $\wedge$ means the min. This operation is the image of fuzzy set A through fuzzy relation R, but this is called the compositional rule of inference, because it is a special case of composition of relations. The fuzzy reasoning just shown by Eq. (6.16) can be written as follows.

First, we define universes of discourse U and V. We let them be

```
-> (setq U {1, 2, 3, 4, 5})
{1, 2, 3, 4,5}
-> (setq V {1, 2, 3, 4, 5})
{1, 2, 3, 4, 5}
```

Second, we define the fuzzy sets "small" and "large" as

```
-> (setq small {1/1, 0.7/2, 0.1/3})
{1/1, 0.7/2, 0.1/3}
-> (setq large {1/5, 0.7/4, 0.1/3})
{1/5, 0.7/4, 0.1/3}
```

Third, we define the function "imply-arithmetic" that creates the fuzzy relation from the rule using Eq. (6.18). We use the Lisp function "defun" to define the function. Thus, we have

```
-> (defun imply-arithmetic (a b))        ; if a then b
   (bsum (cp (adif U a) V)(cp  U b)))
IMPLY-ARITHMETIC
```

where bsum, cp, and adif are, respectively, the bounded sum, Cartesian product, and absolute difference in this system. It is possible to define functions the values of which are fuzzy relations. Now we can create the fuzzy relation that corresponds to the rule "If x is large then y is small" from fuzzy sets small and large using this function. We get

```
-> (setq Ra (imply-arithmetic large small))
{1/⟨1, 1⟩, 1/⟨1, 2⟩, 1/⟨1, 3⟩, 1/⟨1, 4⟩, 1/⟨1, 5⟩,
1/⟨2, 1⟩, 1/⟨2, 2⟩, 1/⟨2, 3⟩, 1/⟨2, 4⟩, 1/⟨2, 5⟩,
1/⟨3, 1⟩, 1/⟨3, 2⟩, 1/⟨3, 3⟩, 0.9/⟨3, 4⟩, 0.9/⟨3, 5⟩,
1/⟨4, 1⟩, 1/⟨4, 2⟩, 0.4/⟨4, 3⟩, 0.3/⟨4, 4⟩, 0.3/⟨4, 5⟩,
1/⟨5, 1⟩, 0.7/⟨5, 2⟩, 0.1/⟨5, 3⟩}.
```

Fourth, we define the function that corresponds to the "very" in "very large." Since it is sufficient to square all of the grades, we get

```
-> (defun very F)
   (gfuncall 'expt F  2))
VERY
```

The function "gfuncall" is one of the functors (functions that take functions as parameters), and a parameter function is applied to all grades of fuzzy sets and the result is returned as the value of gfuncall (expt of F means the square of F, so all grades of fuzzy set F are squared). If we use this function for large, we get

```
-> (very large)
{0.01/3, 0.49/5, 1/5}.
```

Finally, we use the image function for fuzzy reasoning and we get

```
-> (image (very large)Ra)
{1/1, 0.7/2, 0.4/3, 0.3/5}
```

and we have obtained the inference result.

We have described a fuzzy set manipulation system in Lisp. This system can manipulate fuzzy sets, but the processing is deterministic, not fuzzy. For this reason, this system is not called "fuzzy Lisp." This system is a basic system (or tool) for building various "fuzzy... systems," including "fuzzy Lisp." Actually, this system has already been used to implement prototypes of a fuzzy production system [33] and fuzzy Prolog [34].

REFERENCES

1. Zadeh, L. A., "Outline of a New Approach to the Analysis of Complex Systems and Decision Process," *IEEE Trans. Systems, Man and Cybernetics* **SMC-3**(1), 28–44 (1973).
2. Zadeh, L. A., "Calculus of Fuzzy Restrictions," in *Fuzzy Sets and Their Applications to Cognitive and Decision Processes*, (L. A. Zadeh *et al.*, eds.), pp. 1–39, Academic Press (1975).
3. Zadeh, L. A., "The Concept of a Linguistic Variable and Its Application to Approximate Reasoning, Parts 1 & 2," *Information Sciences* **8**, 199–249; **8**, 301–351; **9**, 43–80 (1975).
4. Mamdani, E. H., "Application of Fuzzy Logic to Approximate Reasoning Using Linguistic Synthesis," *IEEE Trans. Computers* **C26**, 1182–1191 (December 1977).
5. Mizumoto, M. and Zimmerman, H. J., "Comparison of Fuzzy Reasoning Methods," *Fuzzy Sets and Systems* **8**(3), 253–283 (1982).
6. Fukami, S., Mizumoto, M., and Tanaka, K., "Some Considerations on Fuzzy Conditional Inference," *Fuzzy Sets and Systems* **4**(3), 243–273 (1980).
7. Yamakawa, T. and Sasaki, H., "Fuzzy Memory Device," *Proc. 2nd IFSA Congress*, Tokyo, Japan (July 1987).
8. Yamakawa, T., "An Approach to a Fuzzy Computer Hardware System," *Proc. 2nd International Conference on Artificial Intelligence*, Marseille, France, pp. 1–22 (December 1986).
9. Yamakawa, T., "High Speed Fuzzy Controller Hardware System," *Proc. 2nd Fuzzy System Symposium*, Tokyo, Japan, pp. 122–130 (June 1986).
10. Yamakawa, T., Japanese Patents No. 60-199225, No. 60-199229, No. 60-199230, No. 60-199231, No. 61-20428, No. 61-20429, No. 61-20430, No. 61-65525, No. 61-65526, No. 61-141085, No. 61-141214; European Patent Applications No. 016225A1, No. 0168004A2; United States Patents No. 4694418, No. 4716540.
11. Mizumoto, M., "Fuzzy Logic and Fuzzy Reasoning," *Mathematical Science* **284**, 10–18 (February 1987) (in Japanese).
12. Yamakawa, T., Miki, T., and Ueno, F., "Basic Fuzzy Logic Circuit Formed by Using p-MOS Current Mirror Circuits," *Trans. IECE of Japan* **J67-c**(12) 1022–1029 (December 1984).
13. Yamakawa, T. and Miki, T., "The Current Mode Fuzzy Logic Integrated Circuits Fabricated by the Standard CMOS Process," *IEEE Trans. Computers* **C-35**(2), 161–165 (February 1986).
14. Yamakawa, T., "A Simple Fuzzy Computer Hardware System Employing MIN & MAX Operations—A Challenge to the 6th Generation Computer," *Proc. 2nd IFSA Congress*, Tokyo, Japan (July 1987).
15. Terano, T., Asai, K., and Sugeno, M., eds. *Fuzzy Systems Theory and Its Applications* (trans. from the Japanese), Academic Press, Boston (1987).
16. Umano, M., "In Order to Use Fuzzy Sets on Computers," *Computer Today* **25**, 28–33 (1988) (in Japanese).
17. Umano, M., "A Fuzzy Production System," in *Fuzzy Logic in Knowledge Engineering* (H. Prade and C. V. Negoita, eds.), pp. 194–208, Verlag TUV Rheinland (1986).

18. Kanai and Ishizuka, "Prolog-EFL: Prolog with Built-in Fuzzy Logic," *Papers from the Society for Information Processing* **27**(4), 411–416 (1986).
19. Martin, T. P., Baldwin, J. F., and Pilsworth, B. W., "The Implementation of FPROLOG —A Fuzzy Prolog Interpreter," *Fuzzy Sets and Systems* **23**, 119–129 (1987).
20. Mukaidono, M., Shen, Z., and Ding, L., "Fuzzy Prolog," *Preprints of Second IFSA Congress*, pp. 452–455 (1987).
21. Yamakawa, Y., "An Approach to Fuzzy Computer Hardware Systems," *Proc. 2nd. Int. Conf. on AI* (1986).
22. Togai, M. and Watanabe, "A VLSI Implementation of Fuzzy Inference Engine toward an Expert System on a Chip," *2nd Int. Conf. on AI Applications* (1985).
23. Zadeh, L. A., "Outline of a New Approach to the Analysis of Complex Systems and Decision Process," *IEEE Trans. Systems, Man and Cybernetics* **3**(1) 28–44 (1973).
24. Zadeh, L. A., "PRUF—A Meaningful Representation Language for Natural Language," *Int. J. Man-Machine Studies* **10** (1978).
25. Mukaidono, M., "What Is Fuzzy PROLOG?," *Computer Today* **25**, 4–6 (1988) (in Japanese).
26. Mukaidono, M. and Mususawa, "About the Characteristics of Resolution Forms in Fuzzy Logic," *Papers from the Society for Electronic Communications* **J66-D**, 7 (1983) (in Japanese).
27. Shen, Z., Ding, L., and Mukaidono, M., "Fuzzy Resolution Principle," *Proceedings of 18th International Symposium on Multiple-Valued Logic, IEEE* (1988).
28. Zadeh, L. A., "Fuzzy Sets," *Information and Control* **8**, 338–353 (1965).
29. Umano, M., "Fuzzy-Set Manipulation System in Lisp," *Preprint of the Second IFSA Congress*, pp. 840–843 (1987).
30. Umano, M. and Kume, K., "Overview of Fuzzy Set Manipulation System in Lisp," *The 35th Annual Convention of the Information Processing Society of Japan*, No. 7Q-1, pp. 789–790 (1987) (in Japanese).
31. Kume, K. and Umano, M., "Implementation of Fuzzy Set Manipulation System in Lisp," *The 35th Annual Convention of the Information Processing Society of Japan*, No. 7Q-2, pp. 791–792 (1987) (in Japanese).
32. Zadeh, L. A., "The Concept of a Linguistic Variable and Its Application to Approximate Reasoning," *Information Sciences* **8**, 199–249; **8**, 301–357; **9**, 43–80 (1987).
33. Umano, M., "Implementation of Fuzzy Production System," *Proceedings of the Third IFSA Congress*, pp. 450–453 (1989).
34. Umano, M., "Fuzzy-Set Prolog," *Preprint of the Second IFSA Congress*, pp. 750–753 (1987).

Index